MICROWAVE DEVICES, CIRCUITS AND THEIR INTERACTION • *Charles A. Lee and G. Conrad Dalman*

ADVANCES IN MICROSTRIP AND PRINTED ANTENNAS • *Kai-Fong Lee and Wei Chen (eds.)*

SPHEROIDAL WAVE FUNCTIONS IN ELECTROMAGNETIC THEORY • *Le-Wei Li, Xiao-Kang Kang, and Mook-Seng Leong*

OPTICAL FILTER DESIGN AND ANALYSIS: A SIGNAL PROCESSING APPROACH • *Christi K. Madsen and Jian H. Zhao*

THEORY AND PRACTICE OF INFRARED TECHNOLOGY FOR NONDESTRUCTIVE TESTING • *Xavier P. V. Maldague*

OPTOELECTRONIC PACKAGING • *A. R. Mickelson, N. R. Basavanhally, and Y. C. Lee (eds.)*

OPTICAL CHARACTER RECOGNITION • *Shunji Mori, Hirobumi Nishida, and Hiromitsu Yamada*

ANTENNAS FOR RADAR AND COMMUNICATIONS: A POLARIMETRIC APPROACH • *Harold Mott*

INTEGRATED ACTIVE ANTENNAS AND SPATIAL POWER COMBINING • *Julio A. Navarro and Kai Chang*

ANALYSIS METHODS FOR RF, MICROWAVE, AND MILLIMETER-WAVE PLANAR TRANSMISSION LINE STRUCTURES • *Cam Nguyen*

FREQUENCY CONTROL OF SEMICONDUCTOR LASERS • *Motoichi Ohtsu (ed.)*

WAVELETS IN ELECTROMAGNETICS AND DEVICE MODELING • *George W. Pan*

SOLAR CELLS AND THEIR APPLICATIONS • *Larry D. Partain (ed.)*

ANALYSIS OF MULTICONDUCTOR TRANSMISSION LINES • *Clayton R. Paul*

INTRODUCTION TO ELECTROMAGNETIC COMPATIBILITY • *Clayton R. Paul*

ELECTROMAGNETIC OPTIMIZATION BY GENETIC ALGORITHMS • *Yahya Rahmat-Samii and Eric Michielssen (eds.)*

INTRODUCTION TO HIGH-SPEED ELECTRONICS AND OPTOELECTRONICS • *Leonard M. Riaziat*

NEW FRONTIERS IN MEDICAL DEVICE TECHNOLOGY • *Arye Rosen and Harel Rosen (eds.)*

ELECTROMAGNETIC PROPAGATION IN MULTI-MODE RANDOM MEDIA • *Harrison E. Rowe*

ELECTROMAGNETIC PROPAGATION IN ONE-DIMENSIONAL RANDOM MEDIA • *Harrison E. Rowe*

NONLINEAR OPTICS • *E. G. Sauter*

COPLANAR WAVEGUIDE CIRCUITS, COMPONENTS, AND SYSTEMS • *Rainee N. Simons*

ELECTROMAGNETIC FIELDS IN UNCONVENTIONAL MATERIALS AND STRUCTURES • *Onkar N. Singh and Akhlesh Lakhtakia (eds.)*

FUNDAMENTALS OF GLOBAL POSITIONING SYSTEM RECEIVERS: A SOFTWARE APPROACH • *James Bao-yen Tsui*

InP-BASED MATERIALS AND DEVICES: PHYSICS AND TECHNOLOGY • *Osamu Wada and Hideki Hasegawa (eds.)*

COMPACT AND BROADBAND MICROSTRIP ANTENNAS • *Kin-Lu Wong*

DESIGN OF NONPLANAR MICROSTRIP ANTENNAS AND TRANSMISSION LINES • *Kin-Lu Wong*

PLANAR ANTENNAS FOR WIRELESS COMMUNICATIONS • *Kin-Lu Wong*

FREQUENCY SELECTIVE SURFACE AND GRID ARRAY • *T. K. Wu (ed.)*

ACTIVE AND QUASI-OPTICAL ARRAYS FOR SOLID-STATE POWER COMBINING • *Robert A. York and Zoya B. Popović (eds.)*

OPTICAL SIGNAL PROCESSING, COMPUTING AND NEURAL NETWORKS • *Francis T. S. Yu and Suganda Jutamulia*

SiGe, GaAs, AND InP HETEROJUNCTION BIPOLAR TRANSISTORS • *Jiann Yuan*

ELECTRODYNAMICS OF SOLIDS AND MICROWAVE SUPERCONDUCTIVITY • *Shu-Ang Zhou*

SMART ANTENNAS • *Tapan K. Sarkar, Michael C. Wicks, Magdalena Salazar-Palma, and Robert J. Bonneau*

SMART ANTENNAS

SMART ANTENNAS

TAPAN K. SARKAR
MICHAEL C. WICKS
MAGDALENA SALAZAR-PALMA
ROBERT J. BONNEAU

With Contributions from:

Raviraj Adve, Paul Antonik, Russell D. Brown, Jeffrey Carlo,
Yongseek Chung, Todd B. Hale, Braham Himed, Zhong Ji,
Kyungjung Kim, Ralph E. Kohler, Eric Mokole,
Raul Fernandez-Recio, Richard A. Schneible,
Dipak Sengupta, and Hong Wang

IEEE PRESS

A JOHN WILEY & SONS, INC., PUBLICATION

Published by John Wiley & Sons, Inc., Hoboken, New Jersey
Published simultaneously in Canada.

For general information on our other products and services please contact our Customer Care Department within the U.S. at 877-762-2974, outside the U.S. at 317-572-3993 or fax 317-572-4002.

Wiley also publishes its books in a variety of electronic formats. Some content that appears in print, however, may not be available in electronic format.

Library of Congress Cataloging-in-Publication Data Is Available

ISBN 0-471-21010-2

Printed in the United States of America.

10 9 8 7 6 5 4 3 2 1

Contents

Preface

The term *smart antenna* is often used in mobile communications to describe an adaptive process designed to improve the capacity of a base station by focusing the radiated electromagnetic energy on transmit while improving the gain pattern on receive from a mobile system. This is called *space division multiple access*. Here, the transmitted signals from a base station are spatially directed to an intended mobile. In addition, the receive gain of the base station is also increased by spatially forming a beam along the direction of a mobile which is on a transmit mode. In this way the capacity of a base station can be increased, as it can now serve many mobile units simultaneously by directing a beam along each one of them. However, this promise of increased capacity through space division multiplexing can be further enhanced if one understands the true nature of an antenna (the source of radiating and/or the sensor of electromagnetic energy) which is the central point of this methodology. An antenna may be considered to be a device that maps spatial-temporal signals into the time domain, thus making them available for further analysis in a digital signal processor. In this philosophical framework, an ideal antenna is one that converts the spatial-temporal signals arriving at an antenna into a temporal signal without distortion. Hence, there is a tacit assumption that no information is destroyed by the antenna. This may be true when dealing with narrowband signals, but when considering the transmission of broadband signals, even a small radiator called a Hertzian dipole operating in free space behaves differently on transmit than it does on receive. It is important to note that in electromagnetics there does not exist any isotropic radiator, as even a *Hertzian dipole* has a directive pattern. However, along a certain plane the pattern can be omni-directional. On transmit the far field of an antenna (even that of a small Hertzian dipole operating in free space) is the time derivative of the input transient waveform fed to its input terminal. While on receive, the same antenna acts as a spatial integrator of the fields that are incident on it. Hence, the temporal and spatial properties of an antenna are intimately related and it is not advisable to separate them if one wants to realize the full potential of an antenna system. In this book the term *smart antenna* is used to imply that one is dealing appropriately with the dual spatial and temporal properties of an antenna on both transmit and receive.

An admirer of James Clerk Maxwell (the actual discoverer of electromagnetism) or Heinrich Hertz (the true father of radio, as he not only formulated the four equations of Maxwell that are available in electromagnetic textbooks today but also produced an experimental device to generate, transmit, propagate, and receive electromagnetic energy) will realize immediately that antennas act simultaneously as temporal and spatial filters. In addition, an antenna is a spatial sampler of the electric fields. One of the objectives of this book is to explain the basic difference between adaptive antennas and adaptive signal processing. Whereas for the former an antenna acts as a spatial filter, and therefore processing occurs in the angular domain, a signal-processing algorithm is applied in the temporal domain. To identify whether one is dealing with adaptive antennas or adaptive signal processing is to ask the following simple question: For a narrowband communication, can the adaptive system separate a desired signal from its coherent multipath components? In this case, there is not only a signal, but also multipath components that are correlated with the desired signal and interact (in either a constructive or destructive fashion) with the signal. Only an adaptive antenna can isolate the desired signal from its coherent multipath, as the information on how to separate them is contained in the angle of arrival (i.e., in the spatial domain). There is little information in the temporal domain for this case. In a conventional signal-processing algorithm, this type of coherent multipath separation is not trivial, and secondary processing that utilizes spatial concepts from electromagnetics is necessary. The critical point is that temporal processing cannot separate coherent signals spatially, since the differences between the signals manifest themselves in the spatial domain and not in the temporal domain. The signal-processing community sometimes views an antenna as a temporal channel, whereas practitioners of electromagnetics always consider an antenna to be a spatial filter. We want to distinguish between these disjoint temporal and spatial properties by adding the term *smart antennas* which we imply that we are merging these two distinct methodologies to provide better systems. In fact, in an adaptive system, one is shaping the spatial response of an antenna by processing the time domain signal. Hence, we do not treat these two spatial and temporal properties separately. An additional advantage to using this coupled spatial-temporal methodology is that we have a well-established mathematical tool, which treats this space-time continuum in an exact way. This mathematical framework for such a system is described by one of the oldest sets of equations in mathematical physics, equations that have withstood the test of erosion and corrosion of time. Even the advent of relativity has had little effect on them. This analytical framework is given by Maxwell's equations. A related problem that also needs to be addressed is what actually limits the speed of communication: is it based on the channel capacity defined by Shannon which does not include the speed of light or is it based on the dispersion introduced by the propagation medium as per Maxwell's equations? A moment of reflection on this critical question will reveal that we need to develop the problem along the space-time continuum as formulated by the Maxwell's equations.

Another objective of this book is to illustrate procedures for adaptive processing using directive elements in a conformal array. Under the current philosophy, it is uncommon to use directive elements in a phased array or antenna elements that are not uniformly spaced. The current thinking is that if one does not use omnidirectional antenna elements, it may not be possible to scan over wide angles. To increase the directive gain of the phased array, one increases the total number of elements by hundreds or even thousands. This increases the cost significantly, as one needs an analog-to-digital converter at each antenna element in addition to a complete receiver channel for downconversion of the radio-frequency signal to baseband. The complexity of a phased array can also be reduced if we employ directive antenna elements on a conformal surface. In addition, individual antenna elements may be nonuniformly spaced, or the conformal array can even be nonplanar. To treat such general array configurations in this book, we describe an electromagnetic preprocessing technique using an array transformation matrix which broadens the fundamental principles of adaptive antennas. Here we address phased array applications, including direction finding or angle-of-arrival estimation and adaptive processing utilizing directive elements that may be nonuniformly spaced and operating in the presence of near-field scatterers.

We also address problems in radar and mobile communications. To perform adaptive processing we need to have some *a priori* information about the signals that we are trying to detect. For dealing with phased array radars, we generally know or assume the direction of arrival of the signal of interest, as we know *a priori* along which direction the mainbeam of the array was pointing, or equivalently, along what spatial direction the energy was transmitted. Thus in radar, our goal is to estimate the strength of the reflected signal of interest, whose direction of arrival is known. What is unknown is the jammer interference and clutter scenario. Furthermore, we present a direct data domain approach that processes the data on a snapshot-by-snapshot basis to yield the desired information. Here, a snapshot is defined as the voltages available at the terminals of the antenna at a particular instance of time. Since we are processing the data in a batch mode, it is highly suitable for characterizing a dynamic environment where the nature of the interference and clutter may change over time. The direct data domain least squares approach presented in this book estimates the signal in the presence of jammer interference, clutter, and thermal noise. In this technique no statistical information about the clutter is necessary. Also, since no covariance matrix is formed in this procedure, the process can be implemented in real time on an inexpensive digital signal processing chip. We also present an extension of this technique to include traditional statistical processing when dealing with space-time adaptive processing.

Unlike radar, in mobile communications it is difficult to know *a priori* the direction of arrival of the signal. In this case, we exploit the temporal characteristics of the signal through introduction of the principles of cyclostationarity. Again a direct data domain method is presented to solve this problem on a snapshot-by-snapshot basis using the principles of cyclostaionarity. The advantage of exploiting the temporal characteristics of the signals is that the number of interferers can be greater than the number of antennas. However, the number of coherent interferers at the same frequency needs to be no more than half the number of antenna elements. Also shown is a method to incorporate the effects of mutual coupling between antenna elements and the effects of near-field scatterers, to improve the overall system performance.

One unique topic in this book is a multistage analysis procedure that combines electromagnetic analysis with signal processing. Initially, electromagnetic principles are applied to compensate for the effects of mutual coupling between antenna elements, including the effect of nonuniformity in the spacing between the elements and the presence of near-field scatterers. Then a direct data domain methodology is implemented to yield the signal of interest. A deterministic model for the signal of interest yields a lower value for the Cramer–

Rao bound than those using stochastic methods. In this approach, no statistical information about the interference environment is necessary. This makes it possible to perform real-time processing in a dynamic environment. These principles have been applied for space-time adaptive processing of experimental data obtained from an airborne multichannel radar system.

We also present a survey of various models for characterizing radio-wave propagation in urban and rural environments. We describe a method where it is possible to identify and eliminate multipath without spatial diversity and optimize the location of base stations in a complex environment.

Finally, it is demonstrated that in mobile communication where the transmit and receive ports can be clearly defined, it is possible to direct the signal from base stations to mobile units without having any *a priori* knowledge about their spatial coordinates or knowing the near-field electromagnetic environment in which they are radiating. This is possible through invocation of the principle of reciprocity. This approach will make space division multiplexing more than just an experimental concept but a commercial success.

Every attempt has been made to guarantee the accuracy of the material in the book. We would, however, appreciate readers bringing to our attention any errors that may have appeared in the final version. Errors and any comments may be e-mailed to either author.

Acknowledgments

We gratefully acknowledge Professors Carlos Hartmann (Syracuse University, Syracuse, New York), Felix Perez-Martinez (Polytechnic University of Madrid, Madrid, Spain), and Gerard J. Genello (Air Force Research Laboratory, Rome, New York) for their continued support in this endeavor.

Thanks are also due to Ms. Brenda Flowers, Ms. Maureen Marano, and Ms. Roni Balestra (Syracuse University) for their expert typing of the manuscript. We would also like to express sincere thanks to Santana Burintramart, Wonsuk Choi, Debalina Ghosh, Seongman Jang, Sheeyun Park, Cesar San Segundo, Harry Schwarzlander, Mengtao Yuan and Shengchun Zhao for their help with the book.

Tapan K. Sarkar (tksarkar@syr.edu)
Michael C. Wicks (Michael.Wicks@rl.af.mil)
Magdalena Salazar-Palma (salazar@gmr.ssr.upm.es)
Robert J. Bonneau (rbonneau@rl.af.mil)
New York
July 2002

1

INTRODUCTION

1.1   SOME REFLECTIONS ON CURRENT THOUGHTS

The fundamental bottleneck in mobile communication is that many users want to access the base station simultaneously and thereby establish the first link in the communication chain. The way the scarce resources of the base station are distributed to mobile users is through sharing. This is a technical definition of the term *multiple access*. Therefore, multiple accesses are implemented by sharing one or more of the four resources of the base station by the various mobile users randomly located in space and time. By *time* we imply that different users may start using the system at different times. This sharing can take place in any of the following four ways [1, 2]:

1.) *Bandwidth (Frequency Division Multiple Access* or in short, FDMA). Here, the frequency spectrum or the entire bandwidth is portioned off to different users and allocated for that communication duration. Hence each user communicates with the base station over an allocated narrow frequency band for the entire duration of the communication.

2.) *Time (Time Division Multiple Access* or in short, TDMA). Here, each mobile has the entire frequency resource of the base station for a short duration of the time (i.e., each user accesses the entire spectrum of the base station for a finite duration in an ordered sequence). With the advent of digital technology it is possible to have an intermittent connection for each mobile with the base station for a short period of time, and in this way the valuable frequency resource of the base station is shared.

3.) *Code (Code Division Multiple Access* or in short, CDMA). In this case, each user is assigned a unique code. In this way the user is allowed to access all the bandwidth, as in TDMA, and for the complete duration of the call, as in FDMA. All the users have access simultaneously to the entire spectrum for all the time. They are interfering with each other, and that is why this methodology was originally conceived as a covert mode

1

of communication. There are two main types of CDMA. One is called *direct sequence spread spectrum multiple access,* and the other is called *frequency hopped spread spectrum multiple access.* In the first case, two-way communication is accomplished through spread spectrum modulation where each user's digital waveform is spread over the entire frequency spectrum that is allocated to that base station. Typically, on transmit, the actual signal is coded and spread over the entire spectrum, where on receive, the intended user first detects the signal by convolving the received signal with his/her unique code and then demodulates the convolved signal. In the second case the transmit carrier frequency changes as a function of time in an ordered fashion so that the receiver can decode each narrowband transmission. At first glance it appears that CDMA is more complex than TDMA or FDMA, but with the advent of novel digital chip design, it is easy to implement CDMA in hardware.

4.) *Space (Space Division Multiple Access* or in short, SDMA). If a base station has to cover a large geographical area, the region is split into cells where the same carrier frequency can be reused in each cell. Therefore, for a large number of cells there is a high level of frequency reuse, which increases the capacity. In this primitive form the transmitted power of the base station limits the number of cells that may be associated with a base station since the level of interference at a base station is determined by the spatial separation between cells, as the mobiles are using the same frequency. This is one of the reasons that microcells and picocells have been proposed for personal communication systems. However, it was soon realized that the capacity of the base station could be increased further by spatially focusing the transmitted energy along the direction of the intended users. In this way, transmission can be achieved at the same carrier frequency simultaneously with different users. This can be accomplished by using an array of antennas at the base station and either a switched beam array or a tracking beam array can be used to direct the electromagnetic energy to the intended users.

In current times it appears that further enhancement in the capacity of a communication system can be achieved primarily in the implementation of SDMA. This is generally carried out using an adaptive process where we have a collection of antennas called *phased arrays.* One now dynamically combines the output from each antenna element using different weights. The weights modify the amplitude and phase of the voltages received at each antenna element. Through an appropriate combination of the voltages that are induced in them by the incident electromagnetic fields, one forms an antenna beam. This antenna beam can either be steered continuously or the beam can be switched along certain prefixed directions by selecting a set of *a priori* weights. This can be achieved in either of two ways.

The first way is to design an antenna with a narrow main beam. This is generally implemented by using a physically large antenna, as the width of the main lobe of the antenna is inversely proportional to the physical dimensions.

Hence, an electrically large antenna structure will have a very narrow beam and may also possess very low sidelobe levels. Creation of very low sidelobe levels may require extremely high tolerances in the variability of the actual physical dimensions of the radiating structures. This requires accurate design of the antenna elements in the phased array. Now one can mechanically steer this high-gain antenna to scan the entire geographical region of interest [3, 4]. This is actually done in developing the rotating antenna arrays in AWACS (Airborne Early Warning and Control System) radars [5]. Such a design makes the cost of the antennas very high. The other alternative is to use simple antenna elements such as dipoles, and then form the antenna beam by combining the received signals from a number of them by using a signal-processing methodology. This usually requires a receiver with an analog-to-digital converter (ADC) at the output of every antenna element, which also increases the cost. The signals from the antenna elements are now downconverted and sampled using an ADC, and then a digital beam-forming algorithm is used to form the main beam along the desired direction and to place nulls in the sidelobe regions along the direction of the interferers. The advantage of digital beam forming is that one can form any arbitrary low level of sidelobes with any width of the main lobe along the look direction [3, 4].

Historically, analog beam forming has been going on for a long time. Also, application of the Butler matrix to combine the outputs of the antenna elements is similar in principle to application of the fast Fourier transform (FFT) to the output voltages available at the antenna elements to form a beam [3, 4, 6]. This is because the far field is simply the Fourier transform of the induced current distribution on the radiating structure. Even though there is a one-to-one correspondence between the Butler matrix and the fast Fourier transform, there is an important fundamental difference. The Butler matrix processes the signals in the analog domain, whereas an FFT carries out similar processing in the digital domain. By processing signals in the analog domain, one is limited by the Rayleigh resolution criterion, which states that in order to resolve two closely spaced signals in space (i.e., their directions of arrival at the antenna array are very close to each other), one needs an antenna whose physical size is inversely proportional to the difference in the spatial angles of arrival at the array. Therefore, the closer two signals are located in space, the greater should be the physical size of the antenna in order to separate the two incoming signals. Therefore, the physical length of the array determines the angular resolution of a phased array performing analog processing. On the other hand, digital beam forming allows us to go beyond the curse of the Rayleigh limit if there is adequate signal strength and enough effective bits in the measured voltages (dynamic range) at each of the antenna elements to carry out beam forming [7].

Typically, adaptive beam forming is supposed to be synonymous with digital beam forming and smart antennas [1]. The term *smart antennas* implies that the antenna array can operate in any environment and has the capability to extract the signal of interest in the presence of interference and clutter and thus to adapt to the signal environment. However, a very important factor has been overlooked in the design process of adaptive systems. For example, if one observes a typical cellular phone, the chip and the signal processors that have been used in the system were probably developed within the last year, but the key ingredient (i.e., the antenna) currently used in many systems was developed about 100 years ago by Hertz, as it is a modification of a simple dipole. Nowadays, the dipole is being replaced by some form of helix (bifilar or quadrifilar), which had been used in AM radios for almost 75 years. The same disparity in technology can also be observed in television sets. Even though a modern television set may have advanced components both for video displays and for processing the video and audio signals, the very high frequency (VHF) antenna is still the "rabbit ear"—a dipole, and the ultra high frequency (UHF) antenna is a loop which was developed in the early nineteen hundreds. The principle behind such wide disparities in component technologies of modern communication systems lies primarily in the assumption that an antenna captures a spatial-temporal signal propagating through space and transforms it into a pure temporal signal without any distortion. This assumes that the antennas are essentially isotropic omni-directional point radiators. That is the reason why in contemporary literature the antenna is often referred to as a sensor of a temporal channel. In electromagnetics, the smallest source is an infinitesimally small dipole and it does not have an isotropic pattern, even though it is omnidirectional in certain planes.

An antenna to a spatial signal is equivalent to what an ADC is to a temporal signal. The purpose of an ADC is to produce high-fidelity temporal samples through the sample-and-hold mechanism of a temporal signal. For an ADC to be of good quality, it is essential that the sample time be much smaller than the hold period so that the sampled values provide a true representation of the analog signal. However, the quality of the ADC becomes questionable if the hold time is comparable to the sample time. In that case the temporal sample obtained from the ADC is not going to be representative of the true signal, as the ADC averages the output over the sample period, during which the signal of interest may have wide variations in amplitude. Under this scenario, where the hold time is comparable to the sample period, unless the effects of the ADC are removed through deconvolution, additional signal and data processing may not produce meaningful results.

This same problem arises in the practical application of antennas. An antenna is a spatial sampler of the electromagnetic fields propagating through space. A receiving antenna generally samples the electric field over its length and produces a voltage at the antenna terminals by integrating vectorially the electromagnetic fields incident upon it. When dealing with narrowband electromagnetic signals, a high-quality receiving antenna is often composed of an array of half-wavelength dipoles, typically spaced a half-wavelength apart [3–5]. So in an adaptive antenna environment, we are assuming the integrated value of the electrical field over a half wavelength to be equal to the actual value of the electric field at a point in space which corresponds to the feed point of the antenna. In other words, we are replacing the value of the incident electromagnetic field at the feed point of the antenna by a quantity that is the integral of the electromagnetic fields over a half wavelength in space. Thus by comparing the performance of a finite-sized antenna in spatial sampling of electromagnetic fields to that of temporal sampling of a signal by an ADC, it is quite clear that unless the effects of the antenna are removed from the measured data, signal and data processing may not result in the desired output. This is due to the basic premise that the spatial integral value of the electric field along the half-wavelength antenna is representative of the actual value of the electric field at the antenna feed point. This is not correct. Hence, one of the objectives of this book is to merge the electromagnetic analysis with the signal processing [7–11]. Now one can implement adaptive processing using realistic antenna elements operating in close proximity and incorporate mutual coupling effects. Moreover, there may be coupling between the antenna elements and the platform on which it is mounted. In addition, there may be near-field scatterers, including other antennas, buildings, trees, and so on, near the array that may again distort the beam. In this book we present and illustrate methods for adaptive processing incorporating near-field electromagnetic effects.

When dealing with broadband signals, we often assume that the omnidirectional isotropic point radiators have no effects on the signal. Such a simplistic assumption is seriously flawed. An antenna is not only a spatial sampler of the propagating electromagnetic field, it has a temporal response as well. It has a unique transfer function. For example, the far-field response from even an electrically small antenna is the result of a temporal differentiation of the driving time domain waveform. In addition, the radiated waveforms will have different signal shapes along different spatial directions. Moreover, an antenna of finite size will not mimic an omnidirectional point radiator in performance. On receive, an antenna vectorially integrates the spatial-temporal waveform that is incident on the structure. Therefore, unless the transfer function of the antenna is removed from the measured data, carrying out additional signal processing may not lead one to the correct solution to the problem at hand. This is not a simple problem, as the impulse response of both a transmitting and a receiving antenna of finite size is dependent simultaneously on both azimuth and elevation look directions. In a practical situation it is difficult to characterize the impulse response of either a transmitting or a receiving antenna, as it is difficult to know *a priori* at what azimuth and elevation angles the coupling is taking place. In broadband applications, the antenna responses must be accounted for. The easy way out in most practical systems in theory and in practice is to deal only with narrowband signals. One of the objectives of this book is to initiate a dialogue so that adaptive processing of the data collected through an antenna is performed in the correct fashion. Thus by combining electromagnetic analysis with signal processing, one can build toward a much more effective solution to the problem at hand.

A related problem in adaptive processing is that one often uses antenna elements that are very close to omnidirectional in nature. Even a dipole may have some directivity in elevation, but in azimuth it is still omnidirectional. This may not be an intelligent choice for cellular telephony, since a mobile will radiate most of the power in azimuth directions away from the intended user. The efficiency of a mobile communications system can be improved by using directive elements in a phased array. However, the problem with using directive elements is that it is not clear how to apply classical adaptive processing. Beamspace solutions offer one answer, but there may be others. One of the objectives of this book is to suggest adaptive systems with directive antenna elements and to illustrate how the measured outputs from directive elements can be combined when the directive element patterns of the antenna elements are properly oriented. Equivalently, the antenna elements in an array can be distributed nonuniformly to cover the physical structure and thereby further increase the radiation/receive efficiency. This is particularly useful in mobile systems, where the electromagnetic environment is not predictable nor may it be characterized in an accurate fashion. We address these issues and more in the following chapters.

1.2   ROADMAP OF THE BOOK

The book is organized as follows:

Chapter 2 provides a historical overview of Maxwell's equations and presents some simple formulas to calculate the impulse response of selected canonical antennas. Our purpose is to demonstrate that even an infinitesimally small point source radiating a broadband signal in free space has a nontrivial impulse response and that their effects must be included in channel characterization. Measured results are also provided to illustrate impulse responses of some typical antennas and physical platforms over which they may be mounted.

In Chapter 3 we describe the anatomy of an adaptive process and present classical historical developments (i.e., statistically based methodologies where one needs to have an aggregate of the voltages at the antenna elements over some spatial-temporal duration). Appendix A further delineates the differences between a deterministic approach and a stochastic approach and illustrates the strengths and weaknesses of each.

In Chapter 4 we describe a direct data domain least squares (D^3LS) approach, which operates, on a single snapshot of data. The advantages of using a D^3LS over conventional stochastic methodologies are explained in Appendix A. There are various compounding factors such as nonstationarity of the data and real-time signal processing issues that are aided by a deterministic model, as it is well suited to applications in a highly dynamic environment where processing data on a snapshot-by-snapshot basis is appropriate. In addition, there is no need to develop a stochastic model for clutter, which in a direct data domain approach is treated as an undesired signal just like interferers and thermal noise. For a conventional adaptive system a *snapshot* is defined as the set of voltages measured at the terminals of the antennas. Both the interference and clutter in this algorithm are treated as undesired electromagnetic signals impinging on the array. Since no covariance matrix is formed in this method, a least squares method operating on a single snapshot of the data can be implemented in real time using a modern digital signal-processing device. To this end a survey of the various forms of the adaptive conjugate gradient (CG) algorithm are presented in Appendix B. Their suitability when dealing with adaptive problems is also illustrated. In addition, several variants of this direct data domain least squares method can be implemented in parallel so that independent estimates of the same solution can be obtained. In reality, where the actual solution is unknown, different independent estimates can increase the level of confidence of the computed solution.

In Chapter 5 we illustrate how electromagnetic analysis can be utilized to correct for mutual coupling in an adaptive algorithm. Computed results with and without mutual coupling effects between the antenna elements are presented to illustrate the point that in a real system the finite size of the antenna must be accounted for. The efficiency of using a single snapshot of data for real-time processing is also discussed. It is shown that use of a direct data domain approach such as the Matrix Pencil along with an electromagnetic compensation technique leads to an accurate determination of the directions of arrival in a CDMA environment. The Matrix Pencil technique for both the one- and two-dimensional cases are described in Appendix C.

Chapter 6 demonstrates a two-step process implemented to extract the signal of interest in the presence of interferers and clutter when the antenna elements in the receiving array are nonuniformly spaced and operating in the presence of other near-field scatterers. The placement of the antenna elements in the array need not be coplanar. In addition, we illustrate how a conformal nonuniformly spaced microstrip patch array located on the side of an aircraft can be used for both direction finding and adaptive processing. Even though significant amounts of scattered energy are incident on the array from the wing and the fuselage, performance is not degraded and processing is carried out on a snapshot-by-snapshot basis. For these classes of radar-related problems, we generally assume that the direction of arrival of the signal of interest is known *a priori*. Additional examples are presented to illustrate how direction-of-arrival angle estimation can be carried out using conformal arrays on hemispherical and cylindrical surfaces having directive antenna elements with polarization diversity.

However, in a mobile communication environment we do not know *a priori* the direction of arrival of the signal of interest, and hence a different type of *a priori* knowledge about the signal is required. Chapter 7 describes the concept of cyclostationarity to illustrate how the D^3LS method can be applied to a set of received voltages at an antenna array which need not be coplanar and operates in the presence of mutual coupling and near-field scatterers. These techniques are called *blind methods*, as no training signals are necessary. We still carry out processing on a snapshot-by-snapshot basis, so this adaptive procedure is highly suitable in a dynamic environment.

Chapter 8 provides a survey of the various propagation models currently used in characterizing mobile communication channels. It includes both stochastic and numerical models.

Chapter 9 describes methods for optimizing the location of base stations for indoor wireless communications subject to a certain quality of service in a given environment. A survey of the various optimization techniques is presented to illustrate what class of methods is well suited for these types of problems.

In Chapter 10 we present a frequency diversity technique that can identify and eliminate various multipath components without spatial diversity.

In Chapter 11 we describe a methodology for directing the signal from a base station to a specified mobile user while simultaneously placing nulls along the direction of the other mobiles utilizing the principles of reciprocity. The advantage of this technique is that directing the electromagnetic signals to an intended user is possible without any knowledge of the physical location of the antennas or the electromagnetic multipath environment in which the system is operating. It is not even necessary to know the spatial coordinates of the transmitter or receiver.

Chapter 12 illustrates the extension of the D^3LS method to space-time adaptive processing (STAP). In this section the single snapshot-based direct data domain methodology is applied to the data collected by a side-looking airborne radar to detect a Saberliner aircraft in the presence of terrain and sea clutter. Several variations of the direct data domain methods are presented. Here also we use a single snapshot and model clutter in a deterministic fashion as unwanted electromagnetic signals. The voltages received at each antenna element in space and sampled in time corresponding to a single range cell characterize a single space-time snapshot corresponding to a specific range cell. Hence, in this approach we process the data on a range cell-by-range cell basis. Comparisons are also made with conventional statistical methods to illustrate the quality of the solution that can be obtained by applying this method to measured radar data. Another important factor is that direct data domain procedures require far fewer computational resources than a conventional stochastic covariance-based methodology. The direct method is further extended to carry out STAP using data from a circular array. Next, a hybrid STAP technique is described which utilizes the good points of both a direct method and a stochastic method. Finally, a knowledge-based STAP is described which is capable of automatically selecting the most appropriate method for a given data set.

The unique features of this book are:
1.) Electromagnetic analysis and signal-processing techniques are combined to analyze and design adaptive systems. Thus the presence of mutual coupling between antenna elements and the presence of near-field scatterers can be incorporated in the analysis.
2.) A direct data domain least squares algorithm is developed which processes the data on a snapshot-by-snapshot basis. Thus it is quite suitable for real-time implementation the use of *a priori* information either through the direction of arrival or through use of the concept of cyclostationarity and processing a single snapshot of data.
3.) The principle of reciprocity is exploited to direct a signal to a mobile user while simultaneously placing nulls along other directions without any spatial information about the base station or the mobile user or exact characterization of the electromagnetic environment in which they are operating.

4.) The direct data domain approach is extended to include space-time adaptive processing for dealing with side-looking radars to carry out filtering in space and time to detect weak signals in the presence of terrain and sea clutter using either linear or circular arrays. A knowledge-based STAP approach is described which is quite suitable for making use of a variety of algorithms, depending on the given data set. One advantage of this methodology is that it is transparent to a user.

REFERENCES

[1]   J. Litva and T. K. Lo, *Digital Beam Forming in Wireless Communications*, Artech House, Norwood, MA, 1996.

[2]   S. R. Saunders, *Antennas and Propagation for Wireless Communication Systems*, Wiley, New York, 1999.

[3]   R. J. Mailloux, *Phased Array Antenna Handbook*, Artech House, Norwood, MA, 1994.

[4]   R. C. Hansen, *Phased Array Antennas*, Wiley, New York, 1998.

[5]   G. W. Stimson, *Introduction to Airborne Radar*, 2nd ed., SciTech Publishing, Mendham, NJ, 1998.

[6]   H. L. Van Trees, *Optimum Array Processing*, Wiley, New York, 2002.

[7]   S. Haykin, *Adaptive Filter Theory*, 4th ed., Prentice Hall, Upper Saddle River, NJ, 2002.

[8]   J. C. Liberti and T. S., Rappaport, *Smart Antennas for Wireless Communications*, Prentice Hall, Upper Saddle River, NJ, 1999.

[9]   G. V. Tsoulos, *Adaptive Antennas for Wireless Communications*, IEEE Press, Piscataway, NJ, 2001.

[10] V. Solo and X. Kong, *Adaptive Signal Processing Algorithms*, Prentice Hall, Upper Saddle River, NJ, 1995.

[11] W. Webb, *The Complete Wireless Communications Professional*, Artech House, Norwood, MA, 1999.

2

WHAT IS AN ANTENNA AND HOW DOES IT WORK

SUMMARY

An antenna consists of any structure made of material bodies that can be composed of either conducting or dielectric materials or may be a combination of both. However, the structure should be matched to the source of the electromagnetic energy so that it can radiate or receive the electromagnetic fields in an efficient manner. The interesting phenomenon is that an antenna displays selectivity properties not only in frequency but also in space. In the frequency domain an antenna is capable of displaying a resonance phenomenon where at a particular frequency the current density induced on it can be sufficiently significant to cause radiation of electromagnetic fields from that structure. An antenna thus possesses an impulse response which is a function of both the azimuth and elevation angles. An antenna also displays spatial selectivity as it generates a radiation pattern which can selectively transmit or receive electromagnetic energy along certain spatial directions. As a receiver of electromagnetic fields, an antenna also acts as a spatial sampler of the electromagnetic fields propagating through space. The voltage induced in the antenna is related to the polarization and the strength of the incident electromagnetic fields. The objective of this chapter is to illustrate how the impulse response of an antenna can be determined. Another goal is to demonstrate that the impulse response of an antenna when it is transmitting is different from its response when the same structure operates in the receive mode. An antenna provides the matching necessary between the various electrical components associated with the transmitter and receiver and the free space where the electromagnetic wave is propagating. From a functional perspective an antenna is thus related to a loudspeaker, which matches the acoustic generation/ receiving devices to the open space. An antenna is like our lips, whose instantaneous change of shapes provides the necessary match between the vocal cord and the outside environment as the frequency of the voice changes. So by proper shaping of the antenna structure one can focus the radiated energy along certain specific directions in space. This spatial directivity occurs only at certain specific frequencies, providing selectivity in frequency. The interesting point is that it is difficult to separate these two spatial and temporal properties of the antenna, even though in the literature they are treated separately. The tools that deal with the dual coupled space-time analysis are *Maxwell's equations*. We first present the background of Maxwell's equations and illustrate how to solve for them analytically. Then we utilize them to illustrate how to obtain the impulse responses of antennas both as transmitting and receiving elements and illustrate their relevance in the saga of smart antennas.

2.1 HISTORICAL OVERVIEW OF MAXWELL'S EQUATIONS

In the year 1864, James Clerk Maxwell (1831–1879) read his "Dynamical Theory of the Electromagnetic Field" [1] at the Royal Society (London). He observed theoretically that electromagnetic disturbance travels in free space with the velocity of light. He then conjectured that light is a transverse electromagnetic wave. In his original theory Maxwell introduced 20 equations involving 20 variables [1]. These equations together expressed mathematically virtually all that was known about electricity and magnetism. Through these equations Maxwell essentially summarized the work of Hans C. Oersted (1777–1851), Karl F. Gauss (1777–1855), André M. Ampère (1775–1836), Michael Faraday (1791–1867), and others, and added his own radical concept of *displacement current* to complete the theory.

Maxwell assigned strong physical significance to the magnetic vector and electric scalar potentials *A* and *ψ*, respectively (bold variables denote vectors; italic denotes that they are function of both time and space, whereas roman variables are a function of space only), both of which played dominant roles in his formulation. He did not put any emphasis on the sources of these electromagnetic potentials, namely the currents and the charges. He also assumed a hypothetical mechanical medium called *ether* to justify the existence of displacement currents in free space. This assumption produced a strong opposition to Maxwell's theory from many scientists of his time. It is well known now that Maxwell's equations, as we know them now, do not contain any potential variables; neither does his electromagnetic theory require any assumption of an artificial medium to sustain his displacement current in free space. The original interpretation given to the displacement current by Maxwell is no longer used; however, we retain the term in honor of Maxwell. Although modern Maxwell's equations appear in modified form, the equations introduced by Maxwell in 1864 formed the foundation of electromagnetic theory, which together is popularly referred to as *Maxwell's electromagnetic theory*.

Maxwell's original equations were modified and later expressed in the form we now know as Maxwell's equations independently by Heinrich Hertz (1857–1894) and Oliver Heaviside (1850–1925). Their work discarded the requirement of a medium for the existence of displacement current in free space, and they also eliminated the vector and scalar potentials from the fundamental equations. Their derivations were based on the impressed sources, namely the current and the charge. Thus, Hertz and Heaviside, independently, expressed Maxwell's equations involving only the four field vectors E, H, B, and D: the electric field intensity, the magnetic field intensity, the magnetic flux density and the electric flux density or displacement, respectively. Although priority is given to Heaviside for the vector form of Maxwell's equations, it is important to note that Hertz's 1884 paper [2] provided the Cartesian form of Maxwell's equations, which also appeared in his later paper of 1890 [3]. It is important to note that the coordinate forms of the four equations that we use nowadays were first obtained by Hertz [2] in scalar form and then by Heaviside in 1888 in vector form [4].

It is appropriate to mention here that the importance of Hertz's theoretical work [2] and its significance appear not to have been fully recognized [5]. In this paper [2] Hertz started from the older action-at-a-distance theories of electromagnetism and proceeded to obtain Maxwell's equations in an alternative way that avoided the mechanical models that Maxwell used originally and formed the basis for all his future contributions to electromagnetism, both theoretical and experimental. In contrast to the 1884 paper, in his 1890 paper [3] Hertz postulated Maxwell's equations rather than deriving them alternatively. The equations, written in component forms rather than in vector form as done by Heaviside [4], brought unparalleled clarity to Maxwell's theory. The four equations in vector notation containing the four electromagnetic field vectors are now commonly known as Maxwell's equations. However, Einstein referred to them as *Maxwell–Heaviside–Hertz equations* [6].

Although the idea of electromagnetic waves was hidden in the set of 20 equations proposed by Maxwell, he had in fact said virtually nothing about electromagnetic waves other than light, nor did he propose any idea to generate such waves electromagnetically. It has been stated [6, Ch. 2, p. 24]: *"There is even some reason to think that he [Maxwell] regarded the electrical production of such waves as an impossibility."* There is no indication left behind by him that he believed such was even possible. Maxwell did not live to see his prediction confirmed experimentally and his electromagnetic theory fully accepted. The former was confirmed by Hertz's brilliant experiments, his theory received universal acceptance, and his original equations in a modified form became the language of electromagnetic waves and electromagnetics, due mainly to the efforts of Hertz and Heaviside [7].

Hertz discovered electromagnetic waves around the year 1888 [8]; the results of his epoch-making experiments and his related theoretical work (based on the sources of the electromagnetic waves rather than on the potentials) confirmed Maxwell's prediction and helped the general acceptance of Maxwell's electromagnetic theory. However, it is not commonly appreciated that *"Maxwell's theory that Hertz's brilliant experiments confirmed was not quite the same as the one Maxwell left at his death in the year 1879"* [6]. It is interesting to note how the relevance of electromagnetic waves to Maxwell and his theory prior to Hertz's experiments and findings are described in [6]: *"Thus Maxwell missed what is now regarded as the most exciting implication of his theory, and one with enormous practical consequences. That relatively long electromagnetic waves or perhaps light itself, could be generated in the laboratory with ordinary electrical apparatus was unsuspected through most of the 1870's."*

Maxwell's predictions and theory were thus confirmed by a set of brilliant experiments conceived and performed by Hertz, who generated, radiated (transmitted), and received (detected) electromagnetic waves of frequencies lower than light. His initial experiment started in 1887, and the decisive paper on the finite velocity of electromagnetic waves in air was published in 1888 [3]. After the 1888 results, Hertz continued his work at higher frequencies, and his later papers proved conclusively the optical properties (reflection, polarization, etc.) of electromagnetic waves and thereby provided unimpeachable confirmation of Maxwell's theory and predictions. English translation of Hertz's original publications [9] on experimental and theoretical investigation of electric waves is still a decisive source of the history of electromagnetic waves and Maxwell's theory. Hertz's experimental setup and his epoch-making findings are described in [10].

Maxwell's ideas and equations were expanded, modified, and made understandable after his death mainly by the efforts of Heinrich Hertz, George Francis Fitzgerald (1851–1901), Oliver Lodge (1851–1940), and Oliver Heaviside. These three have been christened as "the Maxwellians" by Heaviside [2, 11].

Next we review the four equations that we use today due to Hertz and Heaviside, which resulted from the reformulation of Maxwell's original theory. Here in all the expressions we use SI units.

2.2   REVIEW OF MAXWELL–HEAVISIDE–HERTZ EQUATIONS

The four Maxwell's equations are one of the oldest sets of equations in mathematical physics, having withstood the erosion and corrosion of time. Even with the advent of relativity, there was no change in their form. We briefly review the derivation of the four equations and illustrate how to solve them analytically [12]. The four equations consist of Faraday's law, generalized Ampère's law, generalized Gauss's law of electrostatics, and Gauss's law of magnetostatics, respectively.

2.2.1   Faraday's Law

Michael Faraday (1791–1867) observed that when a bar magnet was moved near a loop composed of a metallic wire, there appeared to be a voltage induced between the terminals of the wire loop. In this way, Faraday showed that a magnetic field produced by the bar magnet under some special circumstances can indeed generate an electric field to cause the induced voltage in the loop of wire and there is a connection between the electric and magnetic fields. This physical principle was then put in the following mathematical form:

$$V = - \oint_L \mathbf{E} \cdot \mathbf{d}\ell = - \frac{\partial \Phi_m}{\partial t} = - \frac{\partial}{\partial t} \iint_S \mathbf{B} \cdot \mathbf{d}s \qquad (2.1)$$

where V =   voltage induced in the wire loop of length L
$\mathbf{d}\ell$ =   differential length vector along the axis of the wire loop
$\mathbf{E}$ =   electric field along the wire loop
Φ_m =   magnetic flux linkage with the loop of surface area S
$\mathbf{B}$ =   magnetic flux density
S =   surface over which the magnetic flux is integrated (this surface is bounded by the contour of the wire loop)
L =   total length of the loop of wire
$\bullet$ =   scalar dot product between two vectors
$\mathbf{d}s$ =   differential surface vector normal to the surface

This is the integral form of Faraday's law, which implies that this relationship is valid over a region.   It states that the line integral of the electric field is equivalent to the rate of change of the magnetic flux passing through an open surface S, the contour of which is the path of the line integral. The variables in italic, for example $\mathbf{B}$, indicate that they are a function of four variables, x, y, z, t. This consists of three space variables (x, y, z) and a time variable, t. When the vector variable is written as $\mathbf{B}$, it is a function of the three spatial variables (x, y, z) only. This nomenclature between the variables denoted by italic as opposed to roman is used to distinguish their functional dependence on spatial-temporal variables or spatial variables, respectively.
    To extend this relationship to a point, we now establish the differential form of Faraday's law by invoking Stokes' theorem for the electric field. Stokes' theorem relates the line integral of a vector over a closed contour to a surface integral of the curl of the vector, which is defined as the rate of spatial change of the vector along a direction perpendicular to its orientation (which provides a rotary motion, and hence the term *curl* is used), so that

$$\oint_L \mathbf{E} \cdot \mathbf{d}\ell = \iint_S (\nabla \times \mathbf{E}) \cdot \mathbf{d}s \qquad (2.2)$$

where the curl of a vector in the Cartesian coordinates is defined by

$$\nabla \times E \ (x,y,z,t) = determinant \ of \ \begin{vmatrix} \hat{x} & \hat{y} & \hat{z} \\ \dfrac{\partial}{\partial x} & \dfrac{\partial}{\partial y} & \dfrac{\partial}{\partial z} \\ E_x & E_y & E_z \end{vmatrix} \qquad (2.3)$$

$$= \hat{x} \left[\frac{\partial E_z}{\partial y} - \frac{\partial E_y}{\partial z} \right] + \hat{y} \left[\frac{\partial E_x}{\partial z} - \frac{\partial E_z}{\partial x} \right] + \hat{z} \left[\frac{\partial E_y}{\partial x} - \frac{\partial E_x}{\partial y} \right]$$

Here $\hat{x}$, $\hat{y}$, and $\hat{z}$ represent the unit vectors along the respective coordinate axes, and E_x, E_y, and E_z represent the x, y, and z components of the electric field intensity along the respective coordinate directions. The surface S is limited by the contour L. ∇ stands for the operator $[\hat{x}(\partial/\partial x) + \hat{y}(\partial/\partial y) + \hat{z}(\partial/\partial z)]$. Using (2.2), (2.1) can be expressed as

$$\oint_L E \cdot d\ell = \iint_S (\nabla \times E) \cdot ds = -\frac{\partial}{\partial t} \iint_S B \cdot ds \qquad (2.4)$$

If we assume that the surface S does not change with time and in the limit making it shrink to a point, we get Faraday's law at a point in space and time as

$$\nabla \times E (x, y, z, t) = \frac{1}{\varepsilon} \nabla \times D(x, y, z, t)$$

$$= -\frac{\partial B(x, y, z, t)}{\partial t} = -\mu \frac{\partial H(x, y, z, t)}{\partial t} \qquad (2.5)$$

where the constitutive relationships between the flux densities and the field intensities are given by

$$B = \mu H = \mu_0 \mu_r H \qquad (2.6a)$$

$$D = \varepsilon E = \varepsilon_0 \varepsilon_r E \qquad (2.6b)$$

D is the electric flux density and H is the magnetic field intensity. Here, ε_0 and μ_0 are the permittivity and permeability of vacuum, respectively, and ε_r and μ_r are the relative permittivity and permeability of the medium through which the wave is propagating.

Equation (2.5) is the point form of Faraday's law or the first of Maxwell's equations. It states that at a point the negative rate of the temporal variation of the magnetic flux density is related to the spatial change of the electric field along a direction perpendicular to the orientation of the electric field (termed the curl of a vector) at that same point.

2.2.2   Generalized Ampère's Law

André M. Ampère observed that when a wire carrying current is brought near a magnetic needle, the magnetic needle is deflected in a very specific way determined by the direction of the flow of the current with respect to the magnetic needle. In this way Ampère established the complementary connection between the magnetic field generated by an electric current created by an electric field which is the result of applying a voltage difference between the two ends of the wire. Ampère first illustrated how to generate a magnetic field using the electric field or current. Ampère's law can be stated mathematically as

$$I = \oint_L \boldsymbol{H} \cdot \mathbf{d}\ell \qquad (2.7)$$

where I is the total current encircled by the contour. We call this the *generalized Ampère's law* because we use the total current, which includes the displacement current due to Maxwell and the conduction current. In principle, Ampère's law is connected strictly with the conduction current. Since we use the term *total current*, we use the prefix *generalized*. Therefore, the line integral of $\boldsymbol{H}$, the magnetic field intensity along any closed contour L, is equal to the total current flowing through that contour.

To obtain a point form of Ampère's law, we employ Stokes' theorem to the magnetic field intensity and integrate the current density $\boldsymbol{J}$ over a surface to obtain

$$I = \iint_S \boldsymbol{J} \cdot \mathbf{d}s = \oint_L \boldsymbol{H} \cdot \mathbf{d}\ell = \iint_S (\nabla \times \boldsymbol{H}) \cdot \mathbf{d}s$$

$$= \frac{1}{\mu} \iint_S (\nabla \times \boldsymbol{B}) \cdot \mathbf{d}s \qquad (2.8)$$

This is the integral form of Ampère's law, and by shrinking S to a point, one obtains a relationship between the electric current density and the magnetic field intensity at the same point, resulting in

$$\boldsymbol{J}(x,y,z,t) = \nabla \times \boldsymbol{H}(x,y,z,t) \qquad (2.9)$$

Physically, it states that the spatial derivative of the magnetic field intensity along a direction perpendicular to the orientation of the magnetic field intensity is related to the electric current density at that point. Now the electric current density $\boldsymbol{J}$ may consist of different components. This may include the conduction current (current flowing through a conductor) density $\boldsymbol{J}_c$ and displacement current density (current flowing through air, as from a transmitter to a receiver without any physical connection, or current flowing through the dielectric between the plates of a capacitor) J_d, in addition to an externally applied impressed current density J_i. So in this case we have

$$J = J_i + J_c + J_d = J_i + \sigma E + \frac{\partial D}{\partial t} = \nabla \times H \qquad (2.10)$$

where D is the electric flux density or electric displacement and σ is the conductivity of the medium. The conduction current density is given by *Ohm's law*, which states that at a point the conduction current density is related to the electric field intensity by

$$J_c = \sigma E \qquad (2.11)$$

The displacement current density introduced by Maxwell is defined by

$$J = \frac{\partial D}{\partial t} \qquad (2.12)$$

We are neglecting the convection current density, which is due to the diffusion of the charge density at that point. We consider the impressed current density as the source of all the electromagnetic fields.

2.2.3 Generalized Gauss's Law of Electrostatics

Karl Friedrich Gauss established the following relation between the total charge enclosed by a surface and the electric flux density or displacement D passing through that surface through the following relationship:

$$\oiint_S D \cdot ds = Q \qquad (2.13)$$

where integration of the electric displacement is carried over a closed surface and is equal to the total charge Q enclosed by that surface S.

We now employ the divergence theorem. This is a relation between the flux of a vector function through a closed surface S and the integral of the divergence of the same vector over the volume V enclosed by S. The divergence of a vector is the rate of change of the vector along its orientation. It is given by

$$\oiint_S D \cdot ds = \iiint_V \nabla \cdot D \, dv \qquad (2.14)$$

Here dv represents the differential volume. In Cartesian coordinates the divergence of a vector, which represents the rate of spatial variation of the vector along its orientation, is given by

$$\nabla \cdot \boldsymbol{D} = \left[\hat{x}\frac{\partial}{\partial x} + \hat{y}\frac{\partial}{\partial y} + \hat{z}\frac{\partial}{\partial z}\right] \cdot \left[\hat{x}D_x + \hat{y}D_y + \hat{z}D_z\right]$$

$$= \frac{\partial D_x(x,y,z,t)}{\partial x} + \frac{\partial D_y(x,y,z,t)}{\partial y} + \frac{\partial D_z(x,y,z,t)}{\partial z}$$

(2.15)

So the divergence ($\nabla \cdot$) of a vector represents the spatial rate of change of the vector along its direction, and hence it is a scalar quantity, whereas the curl ($\nabla \times$) of a vector is related to the rate of spatial change of the vector perpendicular to its orientation, which is a vector quantity and so possesses both a magnitude and a direction.

By applying the divergence theorem to the vector $\boldsymbol{D}$, we get

$$\oiint_S \boldsymbol{D} \cdot \mathbf{ds} = \iiint_V \nabla \cdot \boldsymbol{D} \, dv = Q = \iiint_V q_v \, dv \qquad (2.16)$$

Here q_v is the volume charge density and V is the volume enclosed by the surface S. Therefore, if we shrink the volume in (2.16) to a point, we obtain

$$\nabla \cdot \boldsymbol{D} = \frac{\partial D_x(x,y,z,t)}{\partial x} + \frac{\partial D_y(x,y,z,t)}{\partial y} + \frac{\partial D_z(x,y,z,t)}{\partial z}$$

$$= q_v(x,y,z,t) \qquad (2.17)$$

This implies that the rate change of the electric flux density along its orientation is influenced only by the presence of a free charge density at that point.

2.2.4 Generalized Gauss's Law of Magnetostatics

Gauss's law of magnetostatics is similar to the law of electrostatics defined in Section 2.2.3. If one uses the closed surface integral for the magnetic flux density $\boldsymbol{B}$, its integral over a closed surface is equal to zero, as no free magnetic charges occur in nature. Typically, magnetic charges appear as pole pairs. Therefore, we have

$$\oiint \boldsymbol{B} \cdot \mathbf{ds} = 0 \qquad (2.18)$$

From the application of the divergence theorem to (2.18), one obtains

$$\iiint_V \nabla \cdot \boldsymbol{B} \, dv = 0 \qquad (2.19)$$

which results in

$$\nabla \cdot \boldsymbol{B} = 0 \tag{2.20}$$

Equivalently in Cartesian coordinates, this becomes

$$\frac{\partial B_x(x,y,z,t)}{\partial x} + \frac{\partial B_y(x,y,z,t)}{\partial y} + \frac{\partial B_z(x,y,z,t)}{\partial z} = 0 \tag{2.21}$$

This completes the presentation of the four equations, which are popularly referred to as Maxwell's equations, which really were developed by Hertz in scalar form and cast by Heaviside into the vector form that we use today. These four equations relate all the spatial-temporal relationships between the electric and magnetic fields.

2.2.5   Equation of Continuity

Often, the equation of continuity is used in addition to equations (2.18)–(2.21) to relate the impressed current density $\boldsymbol{J}_i$ to the free charge density q_v at that point. The equation of continuity states that the total current is related to the negative of the time derivative of the total charge by

$$I = \frac{-\partial Q}{\partial t} \tag{2.22}$$

By applying the divergence theorem to the current density, we obtain

$$I = \oiint_S \boldsymbol{J} \cdot \mathbf{d}s = \iiint_V (\nabla \cdot \boldsymbol{J}) \, dv = \frac{-\partial}{\partial t} \iiint_V q_v \, dv \tag{2.23}$$

Now shrinking the volume V to a point results in

$$\nabla \cdot \boldsymbol{J} = \frac{-\partial q_v}{\partial t} \tag{2.24}$$

In Cartesian coordinates this becomes

$$\frac{\partial J_x(x,y,z,t)}{\partial x} + \frac{\partial J_y(x,y,z,t)}{\partial y} + \frac{\partial J_z(x,y,z,t)}{\partial z} = -\frac{\partial q_v(x,y,z,t)}{\partial t} \tag{2.25}$$

This states that there will be a spatial change of the current density along the direction of its flow if there is a temporal change in the charge density at that point.

Next we obtain the solution of Maxwell's equations.

2.3 SOLUTION OF MAXWELL'S EQUATIONS

Instead of solving the four coupled differential Maxwell's equations directly dealing with the electric and magnetic fields, we introduce two additional variables A and ψ. Here A is the magnetic vector potential and ψ is the scalar electric potential. The introduction of these two auxiliary variables facilitates solution of the four equations.

We start with the generalized Gauss's law of magnetostatics, which states that

$$\nabla \cdot B(x, y, z, t) = 0 \qquad (2.26)$$

Since the divergence of the curl of any vector A is always zero, that is,

$$\nabla \cdot \nabla \times A(x, y, z, t) = 0 \qquad (2.27)$$

one can always write

$$B(x, y, z, t) = \nabla \times A\ (x, y, z, t) \qquad (2.28)$$

which states that the magnetic flux density can be obtained from the curl of the magnetic vector potential A. So if we can solve for A, we obtain B by a simple differentiation. It is important to note that at this point A is still an unknown quantity. In Cartesian coordinates this relationship becomes

$$\nabla \times B(x, y, z, t) = \hat{x} B_x(x, y, z, t) + \hat{y} B_y(x, y, z, t) + \hat{z} B_z(x, y, z, t)$$

$$= determinant\ of \begin{bmatrix} \hat{x} & \hat{y} & \hat{z} \\ \dfrac{\partial}{\partial x} & \dfrac{\partial}{\partial y} & \dfrac{\partial}{\partial z} \\ A_x & A_y & A_z \end{bmatrix}$$

$$= \hat{x} \left[\frac{\partial A_z(x, y, z, t)}{\partial y} - \frac{\partial A_y(x, y, z, t)}{\partial z} \right] \qquad (2.29)$$

$$+ \hat{y} \left[\frac{\partial A_x(x, y, z, t)}{\partial z} - \frac{\partial A_z(x, y, z, t)}{\partial x} \right]$$

$$+ \hat{z} \left[\frac{\partial A_y(x, y, z, t)}{\partial x} - \frac{\partial A_x(x, y, z, t)}{\partial y} \right]$$

Note that if we substitute **B** from (2.28) into Faraday's law given by (2.5), we obtain

$$\nabla \times E = -\frac{\partial B}{\partial t} = -\frac{\partial}{\partial t}\left[\nabla \times A\right] = -\nabla \times \frac{\partial A}{\partial t} \qquad (2.30)$$

or equivalently,

$$\nabla \times \left[E + \frac{\partial A}{\partial t}\right] = 0 \qquad (2.31)$$

If the curl of a vector is zero, that vector can always be written in terms of the gradient of a scalar function ψ, since it is always true that the curl of the gradient of a scalar function ψ is always zero, that is,

$$\nabla \times \nabla \psi(x, y, z, t) = 0 \qquad (2.32)$$

where the gradient of a vector is defined through

$$\nabla \psi = \left(\hat{x}\frac{\partial}{\partial x} + \hat{y}\frac{\partial}{\partial y} + \hat{z}\frac{\partial}{\partial z}\right)\psi(x, y, z, t) \qquad (2.33)$$

We call ψ the electric scalar potential. Therefore, we can write the following (we choose a negative sign in front of the term on the right-hand side of the equation for convenience):

$$E + \frac{\partial A}{\partial t} = -\nabla \psi \qquad (2.34)$$

or

$$E(x, y, z, t) = -\frac{\partial A(x, y, z, t)}{\partial t} - \nabla \psi(x, y, z, t) \qquad (2.35)$$

This states that the electric field at any point can be given by the time derivative of the magnetic vector potential and the gradient of the scalar electric potential. So we have the solution for both **B** from (2.28) and **E** from (2.35) in terms of **A** and ψ. The problem now is how we solve for **A** and ψ. Once **A** and ψ are known, **E** and **B** can be obtained through simple differentiation, as in (2.35) and (2.28), respectively.

Next we substitute the solution for both **E** [using (2.35)] and **B** [using (2.28)] into Ampère's law, which is given by (2.10), to obtain

$$J_i(x,y,z,t) + \sigma E(x,y,z,t) + \frac{\partial D(x,y,z,t)}{\partial t} = \nabla \times H(x,y,z,t) \qquad (2.36)$$

Since the constitutive relationships are given by (2.6) (i.e., $D = \varepsilon E$ and $B = \mu H$), then

$$J_i + \sigma E + \varepsilon \frac{\partial E}{\partial t} = \frac{1}{\mu} \nabla \times \nabla \times A \qquad (2.37)$$

Here we will set $\sigma = 0$, so that the medium in which the wave is propagating is assumed to be free space, and therefore it has no conductivity. So we are looking for the solution for an electromagnetic wave propagating in free space. In addition, we use the following vector identity:

$$\nabla \times \nabla \times A = \nabla (\nabla \cdot A) - (\nabla \cdot \nabla) A \qquad (2.38)$$

By using (2.38) in (2.37), one obtains

$$\nabla \times \nabla \times A = \nabla(\nabla \cdot A) - (\nabla \cdot \nabla) A$$
$$= \mu J_i + \mu \varepsilon \frac{\partial}{\partial t} \left[-\frac{\partial A}{\partial t} - \nabla \psi \right] = \mu J_i - \mu \varepsilon \frac{\partial^2 A}{\partial t^2} - \mu \varepsilon \nabla \frac{\partial \psi}{\partial t}$$

$$(2.39)$$

or equivalently,

$$(\nabla \cdot \nabla) A - \mu \varepsilon \frac{\partial^2 A}{\partial t^2} + \mu J_i = \nabla \left[\nabla \cdot A + \mu \varepsilon \frac{\partial \psi}{\partial t} \right] \qquad (2.40)$$

Since we have introduced two additional new variables, A and ψ, we can without any problem impose a constraint between these two variables or these two potentials. This can be achieved by setting the right-hand side of the expression in (2.40) equal to zero. This results in

$$\nabla \cdot A + \mu \varepsilon \frac{\partial \psi}{\partial t} = 0 \qquad (2.41)$$

which is known as the *Lorenz gauge condition* [13]. It is important to note that this is not the only constraint that is possible between the two newly introduced variables A and ψ. This is only a particular assumption, and other choices will yield different forms of the solution of the Maxwell–Heaviside–Hertz equations.

Next, we observe that by using (2.41) in (2.40), one obtains

$$(\nabla \cdot \nabla) \, A \, - \, \mu\varepsilon \, \frac{\partial^2 A}{\partial t^2} = - \, \mu \, J_i \tag{2.42}$$

In summary, the solution of Maxwell's equations starts with the solution of equation (2.42) first, for A given the impressed current J_i. Then the scalar potential ψ is solved for by using (2.41). Once A and ψ are obtained, the electric and magnetic field intensities are derived from

$$H = \frac{1}{\mu} \, B = \frac{1}{\mu} \, \nabla \times A \tag{2.43}$$

$$E = -\frac{\partial A}{\partial t} - \nabla \psi \tag{2.44}$$

This completes the solution in the time domain, even though we have not yet provided an explicit form of the solution. We now derive the explicit form of the solution in the frequency domain and from that obtain the time domain representation. We assume the temporal variation of all the fields to be time harmonic in nature, so that

$$E(x,y,z,t) = \mathbf{E}(x,y,z) \, e^{j\omega t} \tag{2.45}$$

$$B(x,y,z,t) = \mathbf{B}(x,y,z) \, e^{j\omega t} \tag{2.46}$$

where $\omega = 2\pi f$ and f is the frequency (hertz) of the electromagnetic fields. By assuming a time variation of the form $e^{j\omega t}$, we now have an explicit form for the time differentiations, resulting in

$$\begin{aligned} \frac{\partial}{\partial t} \left[A \, (x,y,z,t) \right] &= \frac{\partial}{\partial t} \left[\mathbf{A}(x,y,z) \, e^{j\omega t} \right] \\ &= j\omega \, \mathbf{A}(x,y,z) \, e^{j\omega t} \end{aligned} \tag{2.47}$$

Therefore, (2.43) and (2.44) are simplified in the frequency domain after eliminating the common time variations of $e^{j\omega t}$ from both sides to form

$$\mathbf{H}(x,y,z) = \frac{1}{\mu} \, \mathbf{B}(x,y,z) = \frac{1}{\mu} \, \nabla \times \mathbf{A}(x,y,z) \tag{2.48}$$

$$\mathbf{E}(x, y, z) = -j\omega \mathbf{A}(x, y, z) - \nabla\psi(x, y, z) \tag{2.49}$$

Furthermore, in the frequency domain (2.41) transforms into

$$\nabla \cdot \mathbf{A} + j\omega\mu\varepsilon\psi = 0$$

or equivalently,

$$\psi = -\frac{\nabla \cdot \mathbf{A}}{j\omega\mu\varepsilon} \tag{2.50}$$

In the frequency domain, (2.42) transforms into

$$\nabla^2 \mathbf{A} + \omega^2\mu\varepsilon\mathbf{A} = -\mu\mathbf{J}_i \tag{2.51}$$

The solution for $\mathbf{A}$ in (2.51) can now be written explicitly in an analytical form through [12]

$$\mathbf{A}(x, y, z) = \frac{\mu}{4\pi} \int_V \frac{\mathbf{J}_i(x', y', z') \, e^{-jkR}}{R} \, dv \tag{2.52}$$

where

$$\mathbf{r} = \hat{x}x + \hat{y}y + \hat{z}z \tag{2.53}$$

$$\mathbf{r}' = \hat{x}x' + \hat{y}y' + \hat{z}z' \tag{2.54}$$

$$R = \left|\mathbf{r} - \mathbf{r}'\right| = \sqrt{(x - x')^2 + (y - y')^2 + (z - z')^2} \tag{2.55}$$

$$k = \frac{2\pi}{\lambda} = \frac{2\pi f}{c} = \sqrt{\omega^2\mu\varepsilon} = \sqrt{(2\pi)^2 f^2\mu\varepsilon} \tag{2.56}$$

$$c = \text{velocity of light in the medium} = \frac{1}{\sqrt{\mu\varepsilon}} \tag{2.57}$$

$$\lambda = \text{wavelength in the medium} \tag{2.58}$$

In summary, first the magnetic vector potential $\mathbf{A}$ is solved for in the frequency domain given the impressed currents $\mathbf{J}_i(\mathbf{r})$ through

$$\mathbf{A}(\mathbf{r}) = \mathbf{A}(x, y, z) = \frac{\mu}{4\pi} \iiint_V \frac{\mathbf{J}_i(\mathbf{r}') \, e^{-jk|\mathbf{r} - \mathbf{r}'|}}{\left|\mathbf{r} - \mathbf{r}'\right|} d\mathbf{r}' \tag{2.59}$$

then the scalar electric potential ψ is obtained from (2.50). Next, the electric field intensity $\mathbf{E}$ is computed from (2.49) and the magnetic field intensity $\mathbf{H}$ from (2.48).

In the time domain the equivalent solution for the magnetic vector potential $\mathbf{A}$ is then given by the time-retarded potentials:

$$\mathbf{A}(\mathbf{r},t) = \mathbf{A}(x,y,z,t) = \frac{\mu}{4\pi} \iiint\limits_{V} \frac{\mathbf{J}_i\left(\mathbf{r}', t - \dfrac{|\mathbf{r} - \mathbf{r}'|}{c}\right)}{|\mathbf{r} - \mathbf{r}'|} d\,\mathbf{r}' \qquad (2.60)$$

It is interesting to note that the time and space variables are now coupled and they are not separable. That is why in the time domain the spatial and temporal responses of an antenna are intimately connected and one needs to look at the complete solution. From the magnetic vector potential we obtain the scalar potential ψ from (2.41). From the two vector and scalar potentials the electric field intensity $\mathbf{E}$ is obtained from (2.44) and the magnetic field intensity $\mathbf{H}$ from (2.43).

We now use these expressions to calculate the impulse response of some typical antennas in both the transmit and receive modes of operations. The reason that impulse response of an antenna is different in the transmit mode than in the receive mode is because the reciprocity principle in the time domain contains an integral over time. The reciprocity theorem in the time domain is quite different from its counterpart in the frequency domain. For the former a time integral is involved, whereas for the latter no such relationship is involved. Because of the frequency domain reciprocity theorem, the antenna radiation pattern when in the transmit mode is equal to the antenna pattern in the receive mode. However, this is not true in the time domain, as we shall now see through examples.

2.4 RADIATION AND RECEPTION PROPERTIES OF POINT SOURCE ANTENNAS

2.4.1 Radiation of Fields from Point Sources

In this section we first define what is meant by the term *radiation* and then observe the nature of the fields radiated by point sources and the temporal nature of the voltages induced when electromagnetic fields are incident on them. In contrast to the acoustic case (where an isotropic source exist), in the electromagnetic case there are no isotropic point sources. Even for a point source, which in the electromagnetic case is called a *Hertzian dipole*, the radiation pattern is not isotropic, but it can be omnidirectional in certain planes.

We describe the solution in both the frequency and time domains for such classes of problems.

Any element of current or charge located in a medium will produce electric and magnetic fields. However, by the term *radiation* we imply the amount of finite energy transmitted to infinity from these currents. Hence, radiation is related to the far fields or the fields at infinity. A static charge may generate near fields, but it does not produce radiation, as the field at infinity due to this charge is zero. Therefore, radiated fields or far fields are synonymous. We will also explore the sources of a radiating field.

2.4.1.1 Far Field in Frequency Domain of a Point Radiator. If we consider a delta element of current or a Hertzian dipole located at the origin represented by $\mathbf{J}\delta(0, 0, 0)$, the magnetic vector potential from that current element is given by

$$\mathbf{A}(x,y,z) = \frac{\mu}{4\pi} \frac{e^{-jkR}}{R} \mathbf{J}_i \tag{2.61}$$

where

$$R = \sqrt{x^2 + y^2 + z^2} \tag{2.62}$$

Here limit our attention to the electric field. The electric field at any point in space is then given by

$$\mathbf{E}(x,y,z) = -j\omega\mathbf{A} - \nabla\psi = -j\omega\mathbf{A} + \frac{\nabla(\nabla \cdot \mathbf{A})}{j\omega\mu\varepsilon}$$
$$= \frac{1}{j\omega\mu\varepsilon}\left[k^2\mathbf{A} + \nabla(\nabla \cdot \mathbf{A})\right] \tag{2.63}$$

In rectangular coordinates, the fields at any point located in space will be

$$\mathbf{E}(x,y,z) = \frac{1}{j\omega\mu\varepsilon}\left[\begin{array}{c} k^2\mathbf{A}(x,y,z) + \left\{\hat{x}\frac{\partial}{\partial x} + \hat{y}\frac{\partial}{\partial y} + \hat{z}\frac{\partial}{\partial z}\right\} \\ \times \left\{\frac{\partial A_x(x,y,z)}{\partial x} + \frac{\partial A_y(x,y,z)}{\partial y} + \frac{\partial A_z(x,y,z)}{\partial z}\right\} \end{array}\right] \tag{2.64}$$

However, some simplifications are possible for the far field (i.e., if we are observing the fields radiated by a point source at a distance of $2D^2/\lambda$ from the source, where D is the largest physical dimension of the source and λ is the wavelength). For a point source, everything is in the far field. Therefore, for all practical purposes, observing the fields at a distance $2D^2/\lambda$ from a source is equivalent to observing the fields from the same source at infinity. In that case, the far fields can be obtained from the first term only in (2.63) or (2.64). This first term due to the magnetic vector potential is responsible for the far field and there is no contribution from the scalar electric potential ψ. Hence,

$$\mathbf{E}_{\text{far}}(x,y,z) = -j\,\omega\mathbf{A} = -j\,\frac{\omega\mu}{4\pi}\,\mathbf{J}_i\,\frac{e^{-jkR}}{R} \tag{2.65}$$

and one obtains a spherical wavefront in the far field for a point source. However, the power density radiated is proportional to E_θ and that is clearly zero along $\theta = 0°$ and is maximum in the azimuth plane where $\theta = 90°$. The characteristic feature is that the far field is polarized and the orientation of the field is along the direction of the current element. It is also clear that one obtains a spherical wavefront in the far field radiated by a point source.

The situation is quite different in the time domain, as the presence of the term ω in the front of the expression of the magnetic vector potential will illustrate.

2.4.1.2 Far Field in Time Domain of a Point Radiator. We consider a delta current source at the origin of the form

$$\mathbf{J}_i\,\delta(0,0,0,t) = \hat{z}\,\delta(0,0,0)\,f(t) \tag{2.66}$$

where $\hat{z}$ is the direction of the orientation of the elemental current element and $f(t)$ is the temporal variation for the current fed to the point source located at the origin. The magnetic vector potential in this case is given by

$$A(\mathbf{r},t) = \frac{\mu}{4\pi}\,\frac{\hat{z}\,f(t - |R|/c)}{R} \tag{2.67}$$

There will be a time retardation factor due to the space-time connection of the electromagnetic wave that is propagating, where R is given by (2.62).

Now the transient far field due to this impulsive current will be given by

$$E(\mathbf{r},t) = -\frac{\partial A(\mathbf{r},t)}{\partial t} = -\frac{\mu\hat{z}}{4\pi R}\,\frac{\partial f(t - |R|/c)}{\partial t} \tag{2.68}$$

Hence, the time domain field radiated by a point source is given by the time derivative of the transient variation of the elemental current element. Therefore, a time-varying current element will always produce a far field and hence will cause radiation. However, if the current element is not changing with time, there will be no radiation from it. Equivalently, the current density $\mathbf{J}_i$ can be expressed in terms of the flow of charges; thus it is equivalent to ρv, where ρ is the charge density and v is its velocity. Therefore, radiation from a time-varying current element in (2.68) can occur if any of the following three scenarios occur:

1.) The charge density ρ may change as a function of time.
2.) The direction of the velocity vector v may change as a function of time.
3.) The velocity v may change as a function of time, or equivalently, the charge is accelerated or decelerated.

Therefore, in theory any one of these three scenarios can cause radiation. For example, in a dipole the current goes to zero at the ends of the structure and hence the charges decelerate when they come to the end of a wire. That is why radiation seems to emanate from the ends of the wire and also from the feed point of a dipole where a current is injected or a voltage is applied and where the charges are induced and hence accelerated. Current flowing in a loop of wire can also radiate as the direction of the velocity is changing as a function of time even though its magnitude is constant. So a current flowing in a loop of wire may have a constant angular velocity, but the temporal change in the orientation of the velocity vector may cause radiation. To maintain the same current along a cross section of the wire loop, the charges located along the inner circumference of the loop have to decelerate, whereas the charges on the outer boundary have to accelerate. This will cause radiation. In a klystron, by modulating the velocity of the electrons, one can have bunching or change of the electron density with time. This also causes radiation. In summary, if any one of the three conditions described above occurs, there will be radiation.

By observing (2.68), we see that a transmitting antenna acts as a differentiator of the transient waveform fed to its input. The important point to note is that an antenna acts as a differentiator on transmit, and therefore in all broadband simulations the differential nature of the point source must be taken into account. This implies that if the input to a point radiator is a pulse, it will radiate two impulses of opposite polarities—a derivative of the pulse. Therefore, when a modulated digital signal is fed to an antenna, what comes out is the derivative of that pulse. It is rather unfortunate that very few simulations in mobile communications really take this property of an antenna into account in designing multiple input/multiple output (MIMO) systems.

2.4.2   Reception Properties of a Point Receiver

On receive, an antenna behaves in a completely different way than on transmit. We observed that an antenna acts as a differentiator on transmit. On receive, the voltage received at the terminals of the antenna is given by

$$V = \int E \cdot d\ell \qquad (2.69)$$

where the path of the integral is along the length of the antenna. Equivalently, this voltage, which is called the *open-circuit voltage* V_{oc} , is equivalent to the dot product of the incident field vector and the effective height of the antenna and is given by [14, 15]

$$V_{oc} = \mathbf{E} \cdot \mathbf{H}_{eff}$$

The effective height of an antenna is defined by

$$H_{eff} = \int_0^H I(z) \, dz = H \, I_{av} \qquad (2.70)$$

where H is the length of the antenna and it is assumed that the maximum value of the current along the length of the antenna $I(z)$ is unity. I_{av} then is the average value of the current on the antenna. This equation is valid at only a single frequency. Therefore, when an electric field E^{inc} is incident on a small dipole of total length L from a broadside direction, it induces approximately a triangular current on the structure [15]. Therefore, the effective height in this case is $L/2$ and the open-circuit voltage induced on the structure in the frequency domain is given by

$$V_{oc}(\omega) = -\frac{L E^{inc}(\omega)}{2} \qquad (2.71)$$

and in the time domain as the effective height now becomes an impulse-like function, we have

$$V_{oc}(t) = -\frac{L E^{inc}(t)}{2} \qquad (2.72)$$

Therefore, in an electrically small receiving antenna called a *voltage probe* the induced waveform will be a replica of the incident field provided that the frequency spectrum of the incident electric field lies mainly in the low-frequency region, so that the concept of an electrically small antenna is still applicable.

2.5   RADIATION AND RECEPTION PROPERTIES OF ELECTRICALLY SMALL DIPOLE-LIKE STRUCTURES

Now we are going to provide simplified far-field expressions for the electromagnetic fields from simple dipole-like structures. The objective is to provide physical insight into the impulse response of these structures and to illustrate the point that any broadband application must take their electromagnetic properties into account. Otherwise, the system is not going to perform at all.

2.5.1   Radiation Fields from Electrically Small Dipoles in the Frequency Domain

In this section we look into the properties of a radiation field from simple dipole-like structures. For example, consider a wire antenna of length L that is z-directed so that the antenna stretches from $z = -L/2$ to $z = +L/2$. We assume a spatial distribution of the current as

$$I(z) = \cos\frac{\pi z}{L} \tag{2.73}$$

on the wire. We want to point out that this expression is drastically different from the typical expressions for currents that are available in electromagnetic text-books [12], which are generally written in the form

$$I(z) = \sin\left[k\left(\frac{L}{2} - |z|\right)\right] \tag{2.74}$$

where $k = 2\pi/\lambda$ and λ is the wavelength for the current at the frequency of operation. It is important to note that if the length of the antenna does not exceed half a wavelength at the highest frequency of interest, only (2.73) is valid. The rationale for not choosing the latter even though it provides a more compact form and also has an explicit frequency dependence for the far field is that we are going to use the same mathematical expression for the spatial variation when we address the transient problems. The far field for a z-directed current element with a current distribution of the form (2.73) is given by

$$E_\theta = -E_z \sin\theta$$

$$= \frac{-j\omega\mu e^{-jkR}\sin\theta}{R} \int_{-L/2}^{L/2} \cos\frac{\pi z'}{L} e^{j(2\pi z'/\lambda)} \cos\theta\, dz' \tag{2.75}$$

where θ is the angle defined from the z-axis, and the far-field approximation has been used in extracting the term $\exp(-jkR)/R$ out of the integral. R is the distance from the origin to the point where the far field has been calculated. This yields

$$E_\theta \cong -j\omega\mu\,\sin\theta\,\frac{e^{-jkR}}{R}\frac{\cos\left[\dfrac{\pi L}{\lambda}\cos\theta\right]}{\dfrac{2\pi}{L}\left[\dfrac{1}{4} - \left(\dfrac{L\cos\theta}{\lambda}\right)^2\right]} \tag{2.76}$$

Hence the radiation pattern from a dipole at a wavelength λ is primarily a donut-shaped pattern. For the specific case of $L \approx \lambda/2$, it is zero at $\theta = 0°$ and it is maximum at $\pi/2$. For dipoles of very small length L, it is seen that the pattern is dominated by the term $\sin \theta$.

2.5.2 Radiation Fields from Electrically Small Wire-like Structures in the Time Domain

To calculate the transient radiated fields from a dipole structure with a pulsed excitation, we assume that the spatial and temporal variation of the current for the dipole is given by [16, 17]

$$\boldsymbol{J}(z, t) = \hat{z} \cos\frac{\pi z}{L} P\{z\} f(t) P\{t\} \tag{2.77}$$

where the function P denotes a pulse function so that

$$P\{z\} = 1 \quad \text{for} \quad -L/2 \leq z \leq L/2$$
$$= 0 \quad \text{otherwise} \tag{2.78}$$

and

$$P\{t\} = 1 \quad \text{for} \quad -T/2 \leq t \leq T/2$$
$$= 0 \quad \text{otherwise} \tag{2.79}$$

where $f(t)$ represents the temporal variation for the current. $L/2$ is the half-length of the dipole. Here it has been assumed that the length of the antenna is much smaller than the wavelength at the highest frequency of interest. This derivation is valid for an electrically small antenna. The duration of the excitation pulse is for a time T and it is centered at the origin. So the temporal function is not causal in a strict mathematical sense. However, we choose it in this way for computational convenience. The function $f(t)$ is next assumed to be a sinusoidal carrier, that is,

$$f(t) = \sin \omega_0 t \tag{2.80}$$

So for a digital signal, this carrier at frequency f_0 is on for a time duration T. Our goal is to find the radiated far-field waveshape from a dipole when such a pulsed current distribution exists on the dipole.

We define

$$\zeta = \frac{L \cos \theta}{2c} \tag{2.81}$$

Then the transient far field at a distance R is given by

$$E_z(\theta,t) \approx -\frac{\partial A_z(\theta,t)}{\partial t}$$

$$\approx -\frac{\mu}{4\pi R}\frac{\partial}{\partial t}\int_{-\infty}^{\infty}\frac{J\left(z',t-\frac{R-z'\cos\theta}{c}\right)}{R-z'\cos\theta}\,dz' \qquad (2.82)$$

$$= -\frac{\mu}{4\pi R}\frac{\partial}{\partial t}\int_{-\infty}^{\infty}J\left(z',\tau+\frac{R-\zeta\,z'}{L}\right)dz'$$

where

$$\tau = t - \frac{R}{c} \qquad (2.83)$$

Equivalently,

$$E_z(\theta,t) = -\frac{\mu}{4\pi R}\int_{-\infty}^{\infty}\cos\frac{\pi z'}{L}\ P\{z'\}\frac{\partial}{\partial t}\left\{\sin\left[\omega_0\left(\tau+\frac{2\zeta z'}{L}\right)\right]\right\}d z' \qquad (2.84)$$

Since

$$\frac{\partial}{\partial t}\left[f\left(\tau+\frac{2\zeta z'}{L}\right)\right] = \frac{L}{2\zeta}\frac{\partial}{\partial z'}\left[f\left(\tau+\frac{2\zeta z'}{L}\right)\right]$$

and after integrating by parts, one obtains

$$E_z(\theta,t) \approx -\frac{\mu}{8R\zeta}\int_{z_1}^{z_2}\sin\frac{\pi z'}{L}\ P(z')\sin\left[\omega_0\left(\tau+\frac{2\zeta z'}{L}\right)\right]dz' \qquad (2.85)$$

where

$$z_1 = -\left(\tau+\frac{T}{2}\right)\frac{L}{2\zeta}$$

$$z_2 = -\left(\tau-\frac{T}{2}\right)\frac{L}{2\zeta} \qquad (2.86)$$

The integral needs to be calculated with extreme caution. We consider two different distinct situations:

A. $\zeta > T/2$
B. $\zeta < T/2$

For each of these two cases there are five different situations, which need to be addressed separately [14].

Case A: $\zeta > 0;\ \zeta \le T/2$

i. $z_1 = \dfrac{L}{2}$. In this case we have $\tau \le -\left[\dfrac{T}{2} + \zeta\right]$.

The transient fields in this region is going to be zero, so that $E_z(\theta, t) = 0$.

ii. $-\dfrac{L}{2} \le z_1 \le \dfrac{L}{2} \le z_2$. In this case we have $-\dfrac{T}{2} - \zeta \le \tau \le \left[-\dfrac{T}{2} + \zeta\right]$.

So the transient far field for this case is given by the expression

$$E_z(\theta, t) = -\frac{\mu}{8R\zeta} \int_{z_1}^{L/2} \sin\frac{\pi z'}{L} \sin\left[\omega_0\left(\tau + \frac{2\zeta z'}{L}\right)\right] dz'$$

$$= -\frac{\mu L \omega_0}{4R} \frac{\cos\left[\omega_0(\tau + \zeta)\right]}{\pi^2 - 4\zeta^2\omega_0^2}$$

$$+ \frac{\mu L}{16R\zeta} \left[\frac{\sin\left\{\dfrac{\omega_0 T}{2} - \dfrac{\pi}{2\zeta}\left(\tau + \dfrac{T}{2}\right)\right\}}{\pi - 2\zeta\omega_0} + \frac{\sin\left\{\dfrac{\omega_0 T}{2} + \dfrac{\pi}{2\zeta}\left(\tau + \dfrac{T}{2}\right)\right\}}{\pi + 2\zeta\omega_0} \right]$$

(2.87)

iii. $z_1 \le -\dfrac{L}{2} \le \dfrac{L}{2} \le z_2$. In this case we have $-\dfrac{T}{2} + \zeta \le \tau \le \dfrac{T}{2} - \zeta$.

The transient far field is then given by

$$E_z(\theta, t) = -\frac{\mu}{8R\zeta} \int_{-L/2}^{L/2} \sin\frac{\pi z'}{L} \sin\left[\omega_0\left(\tau + \frac{2\zeta z'}{L}\right)\right] dz'$$

$$= -\frac{\mu L \omega_0}{2R} \frac{\cos(\omega_0\zeta)}{\pi^2 - 4\zeta^2\omega_0^2} \cos(\omega_0\tau)$$

(2.88)

This is the steady-state response.

iv. $z_1 \le -\dfrac{L}{2} \le z_2 \le \dfrac{L}{2}$. In this case $\dfrac{T}{2} - \zeta \le \tau \le \dfrac{T}{2} + \zeta$.

So the transient field can be expressed as

$$E_z(\theta, t) = -\frac{\mu}{8R\zeta} \int_{-L/2}^{z_2} \sin\frac{\pi z'}{L} \sin\left[\omega_0\left(\tau + \frac{2\zeta z'}{L}\right)\right] dz'$$

$$= -\frac{\mu L \omega_0}{4R} \frac{\cos\left[\omega_0(\tau - \zeta)\right]}{\pi^2 - 4\zeta^2 \omega_0^2}$$

(2.89)

$$+ \frac{\mu L}{16R\zeta} \left[\frac{\sin\left\{\frac{\omega_0 T}{2} + \frac{\pi}{2\zeta}\left(t - \frac{T}{2}\right)\right\}}{\pi - 2\zeta\omega_0} + \frac{\sin\left\{\frac{\omega_0 T}{2} - \frac{\pi}{2\zeta}\left(t - \frac{T}{2}\right)\right\}}{\pi + 2\zeta\omega_0} \right]$$

and finally,

v. $z_2 \leq -\dfrac{L}{2}$ or $\tau \geq \zeta + \dfrac{T}{2}$.

In this case

$$E_z(\theta, t) = 0$$

Thus, the region defined by case iii, $-\zeta \leq \tau \leq +\zeta$, is the steady-state region and the other two regions where the fields are finite are the transient regions.

Next we consider case B.

Case B: $\zeta > T/2$

In this case, all the situations are exactly the same as in case A, except for situation iii. For this situation, we now have

iii. $-\dfrac{L}{2} \leq z_1 \leq z_2 \leq \dfrac{L}{2}$, or equivalently, $\dfrac{T}{2} - \zeta \leq \tau \leq \zeta - \dfrac{T}{2}$.

The transient field for this case is given by

$$E_z(\theta, t) = -\frac{\mu}{8R\zeta}\int_{z_1}^{z_2}\sin\frac{\pi z'}{L}\sin\left[\omega_0\left(\tau + \frac{2\zeta z'}{L}\right)\right]dz'$$

$$= -\frac{\mu_0 L}{8R\zeta}\left[\frac{\sin\left\{\frac{T}{4\zeta}(\pi - 2\omega_0 u)\right\}}{\pi - 2\omega_0\zeta} - \frac{\sin\left\{\frac{T}{4u}(\pi + 2\omega_0\zeta)\right\}}{\pi + 2\omega_0\zeta}\right]\cos\left(\frac{\pi}{2}\frac{\tau}{\zeta}\right)$$

$$(2.90)$$

It is interesting to observe that in this case, the transient radiation is caused by an aperture of a reduced width of dimension $z_2 - z_1 = TL/2\zeta$, and the transient response lasts for a duration equal to $(2\zeta - T)$, during which time the effective part of the radiation shifts from one end of the physical aperture to the other.

As an example, consider several cycles of a sine wave of the form shown in Figure 2.1. We consider a dipole of length $L = 1$ m. We consider f_0 to be 2 GHz. If we consider the far field along the angle $\theta = 30°$, we see that $\zeta = 8.33 \times 10^{-10}$. We now choose T to be 7.0 ns. The z-component of the electric field will then be given by case A. For $R = 1$ m it is shown in Figure 2.2. It is seen that initially there is a transient region and after that the steady-state region of the fields is reached. There is also a derivative operation going on, as the waveshape is more of the form of a cosine than a sine. If we now shorten the pulse by making $T = 1$ ns so that the pulse is much narrower in time, as shown in Figure 2.3, we will not have any steady-state regions. In this case Figure 2.4 gives the far field.

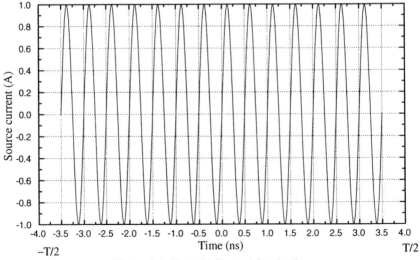

Figure 2.1. Sinusoidally modulated pulse.

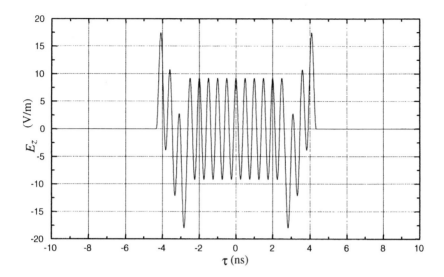

Figure 2.2. Far field along the elevation angle $\theta = 30°$ for case A.

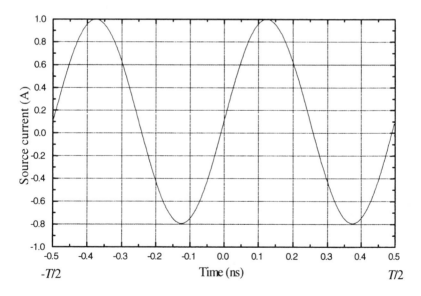

Figure 2.3. Sinusoidally modulated short pulse.

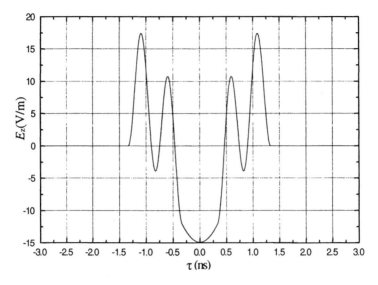

Figure 2.4. Far field along the elevation angle $\theta = 30°$ for case B.

As a second example we consider the response of the dipole to a pulse excitation. If the transient excitation is a simple rectangular pulse rather than part of a sine wave, then in that case, we have

$$E_z(\theta, t) = -\frac{\mu}{8R\zeta} \int_{z_1}^{z_2} \sin\frac{\pi z'}{L} dz' \qquad (2.91)$$

where

$$
\begin{aligned}
z_1 &= -\left[\tau + \frac{T}{2}\right]\frac{L}{2\zeta} \\
z_2 &= -\left[\tau - \frac{T}{2}\right]\frac{L}{2\zeta}
\end{aligned}
\qquad (2.92)
$$

In this case we have five different situations, as before.

Case A: $\zeta > 0$; $\zeta \leq T/2$

i. $z_1 \geq L/2$, $\tau \leq -\left[\dfrac{T}{2} + \zeta\right]$, and $E_z(\theta, t) = 0$.

ii. $-\dfrac{L}{2} \leq z_1 \leq \dfrac{L}{2} \leq z_2$.

In this case, $-T/2 - \zeta \leq \tau \leq -T/2 + \zeta$ and the transient fields are given by

$$E_z\left(\theta, t\right) = \frac{\mu}{8R\zeta} \int_{z_1}^{L/2} \sin\frac{\pi z'}{L} \, dz' = \frac{\mu L}{8\pi R\zeta} \cos\left[\frac{\pi}{2\zeta}\left(\tau + \frac{T}{2}\right)\right] \qquad (2.93)$$

iii. $z_1 \le -\dfrac{L}{2} \le \dfrac{L}{2} \le z_2$ and $-\dfrac{T}{2} + \zeta \le \tau \le \dfrac{T}{2} - \zeta$.

The transient field in this case is $E_z\left(\theta, t\right) = 0$.

iv. $z_1 \le -\dfrac{L}{2} \le z_2 \le \dfrac{L}{2}$ and $\dfrac{T}{2} - \zeta \le \tau \le \dfrac{T}{2} + \zeta$.

We have

$$E_z\left(\theta, t\right) = -\frac{\mu}{8R\zeta} \int_{-L/2}^{z_2} \sin\frac{\pi z'}{L} \, dz' = \frac{\mu L}{8\pi R\zeta} \cos\frac{\pi}{2\zeta}\left(\tau - \frac{T}{2}\right) \qquad (2.94)$$

v. $z_2 \le -\dfrac{L}{2}$ or $\tau \ge \zeta + \dfrac{T}{2}$, this results in $E_z\left(6, t\right) = 0$.

So in this case, the steady-state response is zero. Therefore, when using digital modulation techniques for channel characterization, unless the transient response of the antennas is included in the model it is difficult to see how one can practically characterize the propagation channel without them.

For case B, we have the same expressions as in Case A, except for situation iii.

Case B: $\zeta > T/2$

iii. $-\dfrac{L}{2} \le z_1 \le z_2 \le \dfrac{L}{2}$ or equivalently, $\dfrac{T}{2} - \zeta \le \tau \le \zeta - \dfrac{T}{2}$.

In this case

$$\begin{aligned}
E_z(\theta, t) &= -\frac{\mu L}{8\pi R\zeta}\left[\cos\left\{\frac{\pi}{2\zeta}\left(\tau + \frac{T}{2}\right)\right\} - \cos\left\{\frac{\pi}{2\zeta}\left(\tau - \frac{T}{2}\right)\right\}\right] \\
&= \frac{\mu L}{4\pi R\zeta} \sin\frac{\pi\tau}{2\zeta} \sin\frac{\pi T}{4\zeta}
\end{aligned} \qquad (2.95)$$

As an example we consider the pulse shown in Figure 2.5, having a width T of 0.67 ns. In this case, when $T \approx 4\zeta$ the radiated far field along the elevation angle $\theta = 30°$ for an input pulse is as shown in Figure 2.6. It is interesting to observe that the steady-state response is zero, as the derivative of a constant is zero. However, there is a transient response. When $T = 1.6\zeta$, the far-field waveshape is given by Figure 2.7, as there is no steady-state region. The waveshape is given by Figure 2.7, as there is no steady-state region. The important point to note is that the impulse response of the antennas needs to be taken into account in the design of systems, particularly when they are designed to transmit pulse-like waveforms.

In short, the transient responses from dipoles are quite complex, and except for a few simple cases, it is not possible in general to know *a priori* the waveshapes that come out of these structures. Hence, one needs to use numerical techniques for the solution of a general class of problems.

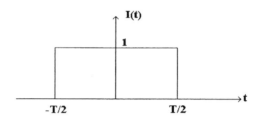

Figure 2 .5. Rectangular pulse.

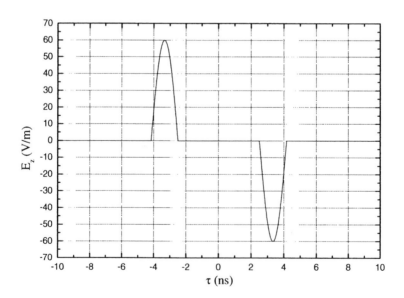

Figure 2.6. Far field along the elevation angle $\theta = 30°$ for case A.

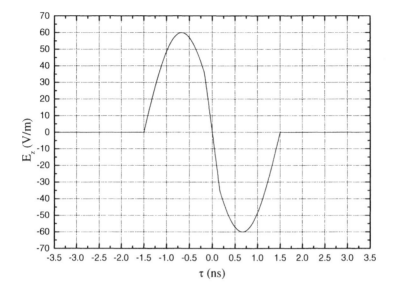

Figure 2.7. Far field along the elevation angle $\theta = 30°$ for case B.

The other important point is to note that the dipole on transmit actually differentiates the waveshape that is fed to it if the pulse width of the input waveform is larger than the transient time across the dipole.

2.6   RADIATION AND RECEPTION PROPERTIES OF FINITE-SIZED DIPOLE-LIKE STRUCTURES

In this section we describe the impulse responses of transmitting and receiving dipole-like structures whose dimensions are comparable to a wavelength. Therefore, these structures are not electrically small. The reason for choosing finite-sized structures is that the impulse responses of these wire-like structures are quite different from the cases described in the preceding section. For a finite-sized antenna structure, which is comparable to the wavelength at the frequency of operation, the current distribution on the structure can no longer be taken to be independent of frequency.   Hence the frequency term must explicitly be incorporated in the expression of the current.

2.6.1   Radiation Fields from Wire-like Structures in the Frequency Domain

For a finite-sized dipole, the current distribution that is induced on it can be represented mathematically to be of the form (2.74). We here assume that the current distribution is known. However, in a general situation we have to use a numerical technique to solve for the current distribution on the structure before we can solve for the far fields. This is particularly important when mutual coupling effects are present or there are other near-field scatterers. For a current distribution given by (2.74), the far fields can be obtained [15] as

$$E_\theta = \frac{I_0 \eta \, e^{j\omega[t-r/c]}}{2\pi r} \left[\frac{\cos\left(\dfrac{kL}{2}\cos\theta\right) - \cos\dfrac{kL}{2}}{\sin\theta} \right] \qquad (2.96)$$

where η is the characteristic impendence of free space and I_0 represents the maximum value of the current. Here L is the length of the antenna. k is the free-space wavenumber and is equal to $2\pi/\lambda = \omega/c$, where c is the velocity of light in that medium. It is important to note that only along the broadside direction and in the azimuth plane of $\theta = \pi/2 = 90°$ is the radiated electric field omni-directional in nature.

2.6.2   Radiation Fields from Wire-like Structures in the Time Domain

When the current induced on the dipole is a function of frequency, the far-zone time-dependent electric field at a spatial location r is given by [14]

$$E_\theta(t) = \frac{\eta}{2\pi r \sin\theta} \left\{ \begin{array}{l} I\left(t - \dfrac{r}{c}\right) + I\left[t - \dfrac{r}{c} - \dfrac{L}{c}\right] \\[2mm] \quad - I\left[t - \dfrac{r}{c} - \dfrac{L}{2c}(1+\cos\theta)\right] \\[2mm] \quad - I\left[t - \dfrac{r}{c} - \dfrac{L}{2c}(1-\cos\theta)\right] \end{array} \right\} \qquad (2.97)$$

where $I(t)$ is the transient current distribution on the structure. It is interesting to note that for L/c small compared to the pulse duration of the transient current distribution on the structure, then from [14] the far field can be written as

$$E_\theta \simeq \frac{\eta}{2\pi r} \left(\frac{L}{2c} \right)^2 \frac{\partial^2 I \left(t - \frac{r}{c} \right)}{\partial t^2} \sin \theta \qquad (2.98)$$

that is, the far-field now is proportional to the second temporal derivative of the transient current on the structure. It is important to note that this expression is quite different from (2.68), which is applicable for electrically small structures.

2.6.3 Induced Voltage on a Finite-Sized Receive Wire-like Structure due to a Transient Incident Field

For a finite-sized antenna of total length L, the effective height will be a function of frequency and it is given by

$$H_{\text{eff}}(\omega) = \int_{-L/2}^{L/2} \sin \left[k \left(\frac{L}{2} - |z| \right) \right] dz = \frac{2c}{\omega} \left[1 - \cos \left(\frac{kL}{2} \right) \right] \qquad (2.99)$$

Hence the induced voltage for a broadside incidence will be given approximately by

$$V_{\text{oc}}(\omega) = -H_{\text{eff}}(\omega) E^{\text{inc}}(\omega) \qquad (2.100)$$

In the time domain, the effective height will be given by

$$H_{\text{eff}}(t) = jc \begin{cases} +1 & 0 < t < \dfrac{L}{2c} \\[2ex] -1 & \dfrac{-L}{2c} < t < 0 \end{cases} \qquad (2.101)$$

Hence the transient received voltage in the antenna due to an incident field will result in the following convolution (defined by the symbol ✪) between the incident electric field and the effective height, resulting in

$$V_{\text{oc}}(t) = -E^{\text{inc}}(t) \ \text{✪} \ H_{\text{eff}}(t) \qquad (2.102)$$

This illustrates that when (2.101) is used in (2.102), the received open-circuit voltage will be approximately the derivative of the incident field when L/c is small compared to the duration of the initial duration of the incident pulse.

2.7 TRANSIENT RESPONSES FROM DIFFERENT ANTENNA SHAPES

In this section we observe the transient response from realistic structures. Because the impulse responses from antennas are very complex, it is not possible to obtain an analytic solution as before, but it is necessary to carry out a numerical simulation. Such numerical analysis is quite possible nowadays through the use of computer programs defined in [18, 19], which can carry out analysis of electromagnetic radiation from any arbitrary-shaped composite metallic and dielectric structures. The general-purpose code described in [18] can perform the analysis in the time domain, whereas for frequency domain analysis the computer code described in [19] can handle any arbitrary-shaped composite geometries. Here we use these two codes to present typical results for far-field radiation from multiple antennas.

In the preceding section we assumed the current distribution on the structure and then developed the expression for the far fields from those currents. However, in a real situation, one needs to solve for the actual current distribution on the antenna, and what we performed previously is only approximate. It is difficult to obtain the physics of the problem from a purely numerical solution, and the simplified expressions of the preceding section may provide more intuitive feeling as to the actual nature of the radiation from the antenna structures.

As an example, consider a Gaussian pulse of the form given in Figure 2.8, which is used to excite an antenna. The spectrum of this pulse is shown in Figure 2.9 and contains a wide band of frequencies. This pulse is applied to the center feed point of a strip dipole oriented along the z-direction, as shown in Figure 2.10. The strip is 1 m long and 0.04 m wide. It is subdivided into small triangular patches. The center point corresponds to edge 41, where this Gaussian waveform is applied. We now observe the far field along various elevation directions. Figure 2.11 illustrates that different waveshapes are radiated in the far field along different elevation directions and the fields oscillate for a long time even after the excitation has died down. As expected, the fields are stronger along the broadside direction.

Next, we consider this dipole not as a transmitting antenna but as a receiving antenna. In this case we have an incident electric field coming from different elevation angles and we want to observe the form of the current that is induced at the center point of the dipole due to different incident fields. The form of the incident pulse is exactly the same as in Figure 2.8. The plot of the current at the feed point due to different angles of incidence of the Gaussian pulse is shown in Figure 2.12. In summary, the impulse response of a dipole is different on transmit than on receive.

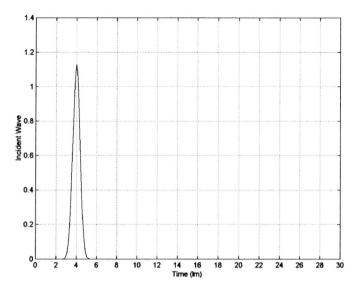

Figure 2.8. Gaussian-shaped pulse used either as a voltage source or as an incident wave.

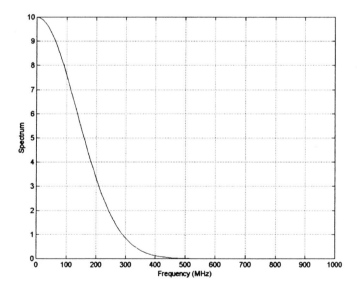

Figure 2.9. Spectrum of the Gaussian pulse which is used as an input excitation.

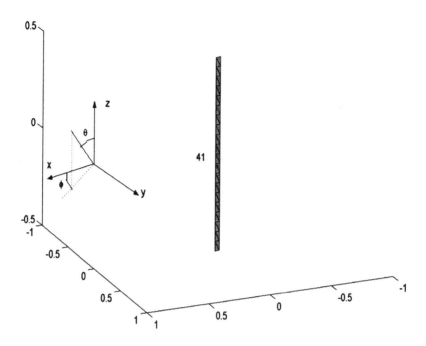

Figure 2.10. Single dipole of length 1m and width 0.04 m.

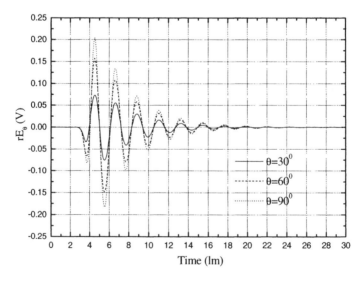

Figure 2.11. Radiated electric far-field along $\varphi = 0°$ for different elevation angles for the incident Gaussian pulse.

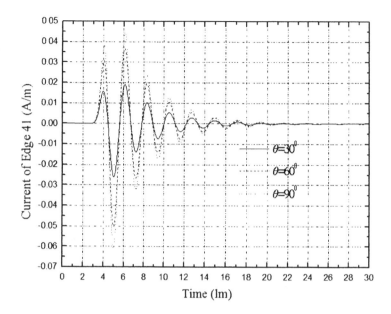

Figure 2.12. Induced current at edge 41 due to a θ-polarized incident Gaussian pulse arriving from $\varphi = 0°$ and for different elevation angles.

For the next example we consider two strip dipoles of length 1 m and separated from each other by the same distance, as shown in Figure 2.13. The widths of the dipoles are 0.04 m. We apply the same Gaussian pulse at edge 41, as shown in Figure 2.8. The other dipole acts as a parasitic element. We observe the various polarizations of the far field and study its temporal shapes along different elevation and azimuth angles. In Figure 2.14 the transient waveshapes for the E_θ component are observed for a fixed azimuth angle of 45° and for different elevation angles. Figure 2.15 provides the other component of the electric field E_ϕ for the same set of angles. Figures 2.16 and 2.17 provide the electric far-field pattern for the azimuth angle of 90° for different elevation angles for the E_θ and E_ϕ components, respectively. Finally, Figures 2.18 and 2.19 provide electric far-field pattern for the two components of the electric field for the azimuth angle of 135°. The interesting point here is that even for such simple structures without a numerical electromagnetic code as described in [18, 19], it will be almost impossible to predict the nature of the radiated electromagnetic fields accurately.

For the receiving case, we observe the induced current at edge 41 on one of these strip dipoles when they are illuminated by the Gaussian pulse of Figure 2.8 from different angles of elevation. In this case the broadside of the strips, as shown in Figure 2.20, are oriented along the polarization of the incident field; otherwise, the induced current on the strip will be zero. The induced current along edge 41 for different elevation angles is presented in Figure 2.21.

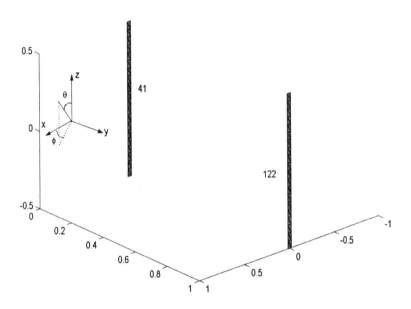

Figure 2.13. Two dipoles separated by 1 m along the y-direction.

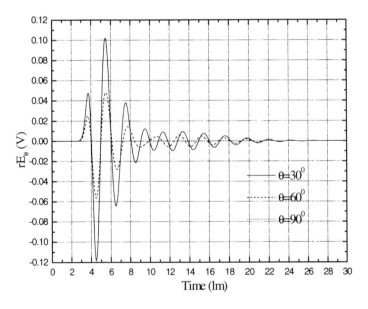

Figure 2.14. Radiated far field for E_θ along $\varphi = 45°$ for different elevation angles from the two dipoles when one of them is center fed with a Gaussian pulse of Figure 2.8.

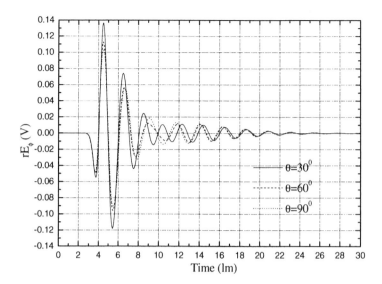

Figure 2.15. Radiated far field for E_φ along $\varphi = 45°$ for different elevation angles from the two dipoles when one of them is center fed with a Gaussian pulse of Figure 2.8.

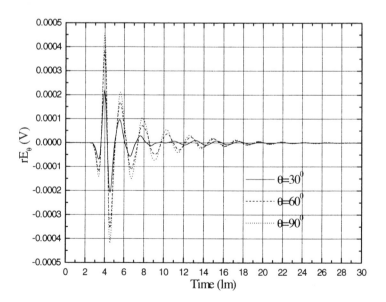

Figure 2.16. Radiated far field for E_θ along $\varphi = 90°$ for different elevation angles from the two dipoles when one of them is center fed with a Gaussian pulse of Figure 2.8.

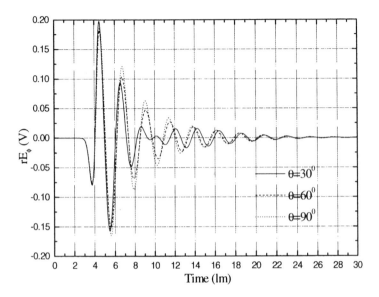

Figure 2.17. Radiated far field for E_φ along $\varphi = 90°$ for different elevation angles from the two dipoles when one of them is center fed with a Gaussian pulse of Figure 2.8.

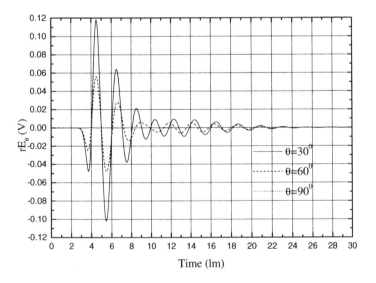

Figure 2.18. Radiated far field for E_θ along $\varphi = 135°$ for different elevation angles from the two dipoles when one of them is center fed with a Gaussian pulse of Figure 2.8.

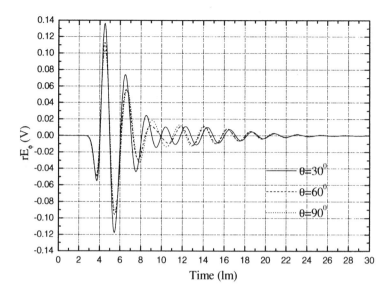

Figure 2.19. Radiated far field for E_φ along $\varphi = 135°$ for different elevation angles from the two dipoles when one of them is center fed with a Gaussian pulse of Figure 2.8.

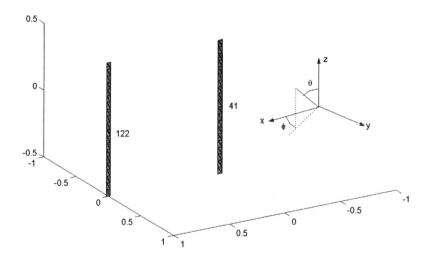

Figure 2.20. Two dipoles separated by 1 m along the x-direction.

Finally, we look at the fields radiated by an eight-turn helix located over a ground plane. The pitch of the helix s is 0.23 m. The diameter D is 0.32 m, as shown in Figure 2.22. The circumference of the helix is 1m. The wavelength is also 1 m. The pitch angle is approximately given by $12° \leq \alpha \leq 14°$ with $\tan \alpha =$

$s/\pi D$. The real and imaginary parts of the far-field pattern along $\theta = 30°$ are shown in Figure 2.23 and the magnitude and phase in Figure 2.24. Finally, Figure 2.25 presents the transient far field along the elevation angle $\theta = 30°$. It is seen that the impulse response of even very simple antennas is quite complex, and the numerical tools presented in [18, 19] are absolutely essential to obtain accurate responses from various composite electromagnetic structures.

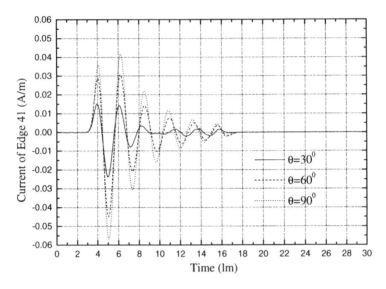

Figure 2.21. Induced current at edge 41 for one of the dipoles due to a θ-polarized incident Gaussian pulse arriving from $\varphi = 0°$ for different elevation angles.

$$\lambda = 1 \text{ m}, f = 300 \text{ MHz}, C_\lambda = \frac{\pi D}{\lambda} = 1, 12° \le \alpha \le 14°, \tan \alpha = \frac{s}{\pi D},$$

$$D = \frac{\lambda}{\pi} \approx 0.32 \text{ m}, s = 0.23 \text{ m}$$

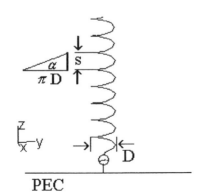

Figure 2.22. Helical antenna.

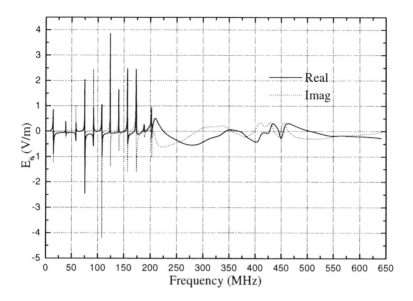

Figure 2.23. Real and imaginary parts of the far field along $\theta = 30°$.

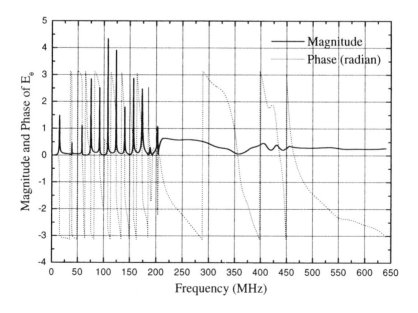

Figure 2.24. Magnitude and phase of the far field along $\theta = 30°$.

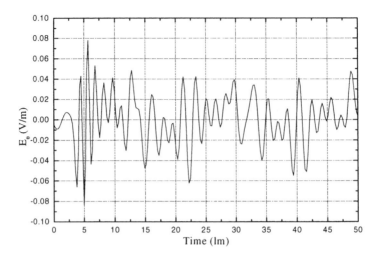

Figure 2.25. Transient far-field response from the helix along $\theta = 30°$ due to a Gaussian shaped voltage excitation.

The responses of the simple antennas typically used in mobile communications illustrate that they all have narrowband responses and for wideband applications proper attention must be paid to their magnitude and phase responses

2.8 MEASURED IMPULSE RESPONSES OF SOME
REPRESENTATIVE STRUCTURES

In this section we observe measured impulse responses from some representative electromagnetic systems.

Figure 2.26 shows the impulse response of a pair of coplanar log periodic antennas due to an input of a 100-ps impulse. The log periodic antennas in this case are two AEL Model APN 995Bs, as shown in the figure. They operate from 50 MHz to 1.1 GHz. The interesting point to note is that the response of this antenna to a 100-ps input pulse rings for several nanoseconds.

As a second example we observe the radiated pulse from a log periodic model APN 502A antenna due to the same 100-ps pulse excitation. It is clear from Figure 2.27 that the log periodic antennas are highly dispersive devices, and one has to be extremely careful when dealing with a broadband operation using them.

Figure 2.28 displays the transfer function between two straight 20-ft planar horn antennas. In Figure 2.28 we have the measured impulse response between the two horn antennas facing each other when their separation is 97 ft in the first case and 56 ft for the second case. The scales for the time and voltages are also shown in the figures.

Impulse Response of a Pair of Coplanar Log Periodic Antennas
AEL Model # APN 995B (50 MHz to 1.1 GHz)

 Output-Sweep Speed: 1ns/div

Input: 100ps impulse

Figure 2.26. Impulse response of a pair of coplanar log periodic antennas.

Temporal - Spatial Compression

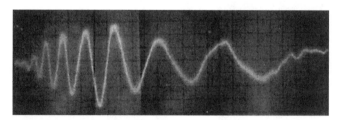

400 ps/div

Transmitting Transfer function with 100ps Pulse
Excitation

Log -Periodic Antenna - Model APN-502A

Figure 2.27. Transmitted waveform from a log periodic antenna due to a 100-ps pulse excitation.

Transfer Function of 20-foot TEM Horns

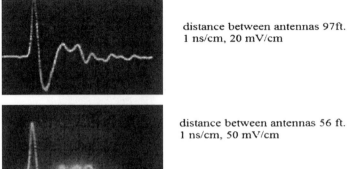

distance between antennas 97ft.
1 ns/cm, 20 mV/cm

distance between antennas 56 ft.
1 ns/cm, 50 mV/cm

Figure 2.28. Transfer function between two planar 20-ft horn antennas.

In Figure 2.29 the return pulses due to an impulse excitation from a 2-inch-thick ice slab are shown. The right figure shows the return from the front face and the delayed one from the other face of the ice slab.

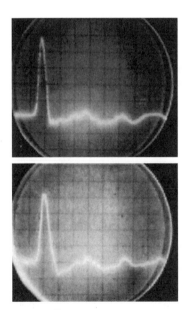

Radar Transparency of 2-inch Thick Ice

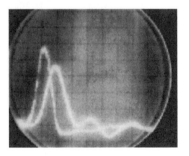

200 ps/div

400 ps/div

Figure 2.29. Time domain reflections of an impulse from a 2-inch-thick ice.

Finally Figure 2.30 shows the electromagnetic return of a 100-ps impulse from a B-57 aircraft from the front and wing sides. It is seen that the impulse responses from various objects indeed contain structural information of the object.

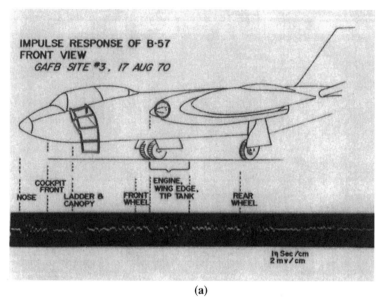

(a)

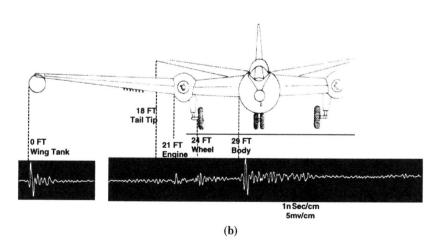

(b)

Figure 2.30. (a) Side and (b) front impulse responses of a B-57 aircraft.

From these sample measurements it is seen that high bandwidth excitations can indeed provide high-fidelity displays if the dispersion in the systems can be accounted for appropriately.

2.9   CONCLUSION

The objective of this chapter is to present the necessary mathematical formulations, popularly known as Maxwell's equations, which dictate the space-time behavior of antennas. Additionally, some examples are presented to note that the impulse response of antennas is quite complicated and the waveshapes depend on both the observation and the incident angles in azimuth and elevation of the electric fields. Any broadband processing must deal with a deconvolution process that factors out the impulse response of both the transmitting and receiving antennas. The examples presented in this chapter do reveal that the waveshape of the impulse response is indeed different for both transmit and receive modes, which are again dependent on both the azimuth and elevation angles. For an electrically small antenna, the radiated fields produced by it in the broadside direction are simply the differentiation of the pulse that is fed to it. While on receive it samples the field incident on it. However, for a finite-sized antenna, the radiated field is proportional to the temporal double derivative of the current induced on it, and on receives, the same antenna differentiates the transient electric field that is incident on it. Hence all broadband applications should deal with the complex problem of determining the impulse responses of the transmitting and receiving antennas. For the narrowband case, determination of the impulse response is not necessary. The goal of this chapter is to outline the methodology that will be necessary to determine the impulse response of the transmit/receive antennas. By thus combining the electromagnetic analysis with the signal-processing algorithms, it will be possible to design better systems.

REFERENCES

[1]   J. C. Maxwell, "A Dynamical Theory of the Electromagnetic Field," Philosophical Transactions, Vol. 166, pp. 459–512, 1865 (reprinted in the *Scientific Papers of James Clerk Maxwell*, Vol. 1, pp. 528–597, Dover, New York, 1952).

[2]   H. Hertz, "On the Relations between Maxwell's Fundamental Equations of the Opposing Electromagnetics," (in German), *Wiedemann's Annalen*, Vol. 23, pp. 84–103, 1884. (English translation in [9, pp. 127–145]).

[3]   H. Hertz, "On the Fundamental Equations of Electromagnetics for Bodies at Rest," in [9, pp. 195–240].

[4]   P. J. Nahin, *Oliver Heaviside: Sage in Solitude*, IEEE Press, New York, 1988.

[5]   C-T Tai and J. H. Bryant, "New Insights into Hertz's Theory of Electromagnetism," *Radio Science*, Vol. 29, No. 4, pp. 685–690, July–Aug. 1994.

[6]   B. J. Hunt, *The Maxwellians*, Chap. 2, p. 24, Cornell University Press, Ithaca, NY, 1991.

[7]   D. L. Sengupta and T. K. Sarkar "Maxwell, Hertz, "The Maxwellians and the Early History of Electromagnetic Waves," *IEEE Antennas and Propagation Magazine*, (to be published).

[8]   H. Hertz, "On the Finite Velocity of Propagation of Electromagnetic Action," *Sitzungsber ichte der Berliner Academic der Wissenschaften*, Feb. 2, 1888; *Wiedemann's Annalen*, Vol. 24, pp. 551, reprinted in H. Hertz translated by D. E. Jones, *Electric Waves*, Chap. 7, pp. 107–123, Dover, New York, 1962.

[9]   H. Hertz, *Electric Waves* (authorized English translation by D. E. Jones), Dover, New York, 1962.

[10] J. H. Bryant, *Heinrich Hertz: The Beginning of Microwaves*, IEEE Service Center, Piscataway, NJ, 1988.

[11] J. G. O'Hara and W. Pritcha, *Hertz and Maxwellians*, Peter Peregrinus, London, 1987.

[12] K. Kraus, *Electromagnetics*, McGraw-Hill, New York, 1980.

[13] R. Nevels and C. Shin, "Lorenz, Lorentz, and the Gauge," *IEEE Antennas and Propagat. Magazine*, Vol. 43, No. 3, pp. 70–72, June 2001.

[14] D. L. Sengupta and C. T. Tai, "Radiation and Reception of Transients by Linear Antennas," Chap. 4, pp. 182–234 in L. B. Felsen (ed.), *Transient Electromagnetic Fields*, Springer-Verlag, New York, pp. 182–234, 1976.

[15] J. D. Kraus, *Antennas*, McGraw-Hill, New York, 1988.

[16] D. K. Cheng and F. I. Tseng, "Transient and Steady State Antenna Pattern Characteristics for Arbitrary Time Signals," *IEEE Transactions on Antennas and Propagation*, Vol. 12, pp. 492–493, 1964.

[17] F. I. Tseng and D. K. Cheng, "Antenna Pattern Response to Arbitrary Time Signals," *Canadian Journal of Physics*, Vol. 42, pp. 1358–1368, 1964.

[18] T. K. Sarkar, W. Lee, and S. M. Rao, "Analysis of Transient Scattering from Composite Arbitrarily Shaped Complex Structures," *IEEE Transactions on Antennas and Propagation*, Vol. 48, No. 10, pp. 1625–1634, Oct. 2000.

[19] B. M. Kolundzija, J. S. Ognjanovic, and T. K. Sarkar, *WIPL-D: Electromagnetic Modeling of Composite Metallic and Dielectric Structures*, Artech House, Norwood, MA, 2000 (*http://wipl-d.com*).

3

ANATOMY OF AN ADAPTIVE ALGORITHM

SUMMARY

This chapter provides the historical background of an adaptive algorithm and presents a description of the statistical methodology developed for analog systems. Then a prelude to direct data domain approaches is provided.

3.1 INTRODUCTION

Basically, in an adaptive methodology, the goal is to estimate the desired response in an adaptive fashion, using a model transfer function. Historically, the first method to be developed was the Wiener filter. Below, it is shown how this methodology has progressed over the years, and its relationship to new spatially based adaptive techniques, as opposed to time variable-based methodology.

The anatomy of an adaptive technique is shown in Figure 3.1. The input signal, $x(t)$, is used to track/approximate a desired signal, $d(t)$, through a linear filter, whose impulse response is $h(t)$. The characteristics of the linear filter are changed by a controller. The controller, in turn, is affected by the error signal that is generated by taking the instantaneous difference between the desired signal, $d(t)$, and the output, $y(t)$, from the linear filter. Historically, this problem of finding the linear filter $h(t)$ was solved by Kolmogorov [1, 2], in the analysis of stationary time series, and simultaneously, by Wiener [3], in the control of antiaircraft guns.

The approaches by Kolmogorov and Wiener are very similar. The methodology starts by defining the error signal, $e(t)$, as

$$e(t) = d(t) - y(t) = d(t) - x(t) \otimes h(t) \tag{3.1}$$

where $\otimes$ represents a convolution, and therefore,

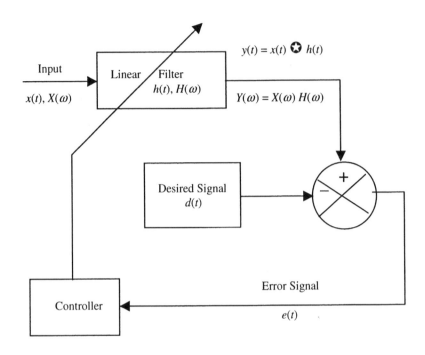

Figure 3.1. Anatomy of an adaptive algorithm.

$$e(t) = d(t) - \int_{-\infty}^{\infty} x(t - \tau) h(\tau)\, d\tau \tag{3.2}$$

Next, the expected value of the squared error is taken, to yield $\mathcal{E}$, that is,

$$\mathcal{E} = \mathcal{E}\left[e(t)\, e*(t)\right] = \mathcal{E}\left[\left|e(t)\right|^2\right] \tag{3.3}$$

where $\mathcal{E}[\bullet]$ is the expectation operator, which has been described in Appendix A, and * represents the complex conjugate.

To find the optimum filter $h(t)$, the error, $\mathcal{E}$, needs to be differentiated with respect to h, and this leads to the orthogonality of the error with the input $x(t)$. In other words,

$$\mathcal{E}\left[x(t)\, e*(t)\right] = 0 \tag{3.4}$$

If we define

$$R_{xd}(t) = \mathcal{E}[x(t)\, d*(t)] \tag{3.5}$$

$$R_{xx}(t) = \mathcal{E}[x(t)\, x*(t)] \tag{3.6}$$

We obtain

$$R_{xd}(t) = \int_{-\infty}^{\infty} R_{xx}(t-\tau)\, h(\tau)\, d\tau \tag{3.7}$$

This integral equation provides the solution for the filter $h(t)$ that is going to match $d(t)$ for a given input $x(t)$ in optimum fashion [4].

The basic principles of an adaptive technique thus illustrate that something must be known about the desired signal in order to define the adaptive process. This knowledge could be about the constant amplitude that we are trying to estimate (as in a digital signal) or some hidden structure in the spectral characteristics (e.g., cyclostationarity, and so on). An adaptive procedure cannot be completely defined without some knowledge of the desired signal. What that information is may change from problem to problem, but it has to exist.

In adaptive-antenna problems, it is not enough to assume that there exists a signal $d(t)$, or to match the output to a desired signal. We need to know more about the signal. Such information may be related to the angle of arrival of the desired signal. Or, it may be associated with the modulation technique used. In many digital-communications applications, the binary signal may be of constant magnitude, and what needs to be solved for is the sign of the signal. Alternatively, for other types of signals used in mobile communications, the spectrum of the signal may have conjugate-transpose symmetry, namely, the spectrum may be cyclostationary. This may be equivalent, in some cases, to saying that the autocorrelation function of the desired signal may have periodic properties.

To further explain the situation, for most adaptive-antenna problems encountered in phased array radars we know the Doppler and direction of arrival of the signal, and the goal is to estimate its strength in the presence of jammers, clutter, and noise. An important class of jammers is multipath signals, which may be coherent with the signal of interest. Sometimes, we will also deal with jammers that may be in the main lobe of the antenna and could be intermittent (i.e., blinking jammers). In mobile communications, the direction of arrival of the signal is not known *a priori*. There, we know that for digitally transmitted signals, in many cases, the signal has a constant magnitude only, and its sign needs to be estimated. This leads to the class of constant-modulus algorithms.

In some other applications, such as, binary phase shift keying (BPSK) or quadrature phase shift keying (QPSK) transmission, we know that the spectrum has some conjugate symmetric property, or equivalently, that the autocorrelation function of the signal of interest may be periodic or cyclostationary. We exploit this information to extract the signal in the presence of jammers, clutter, and noise. In the literature, these algorithms are called *blind-equalization techniques*.

In summary, all adaptive techniques require some knowledge about the signal of interest, in order to estimate it in the presence of interference and thermal noise. Without such information, an adaptive procedure is not defined.

Thus in an adaptive problem dealing particularly with phased array radars, the function $d(t)$ is generally associated with the look direction constraint, that is, we know from what direction the signal of interest is coming. It may also be associated with the spectral properties or amplitudes of the signal of interest that we are trying to estimate. The autocorrelation matrix $[R_{xx}]$ deals with the covariance matrix of the voltages that are measured at each antenna element at different times.

3.2   HISTORICAL BACKGROUND

We now illustrate how the Wiener filter was modified to deal with digital data. An adaptive beamformer is a device that is able to separate signals which have the same frequency content but are separated in the spatial domain. This provides a means for separating a desired signal from interfering signals. An adaptive beamformer is able to optimize the array pattern automatically by adjusting the weights associated with each antenna element until a prescribed objective function is satisfied.

Traditionally, adaptive beamforming has been employed in sonar and radar. It started with the invention of the intermediate frequency sidelobe canceller (SLC) in 1959 by Howells [5] at Syracuse University Research Corporation. Applebaum [6] developed the concept of a fully adaptive array in 1966. This algorithm is based on the general problem of maximization of signal-to-noise ratio (SNR) at the output of the antenna. The SLC was included as a special case in Applebaum's work. Another independent approach to adaptivity used the least mean squares (LMS) error algorithm, which was invented by Widrow et. al. [7]. Despite its simplicity, the LMS algorithm is capable of achieving satisfactory performance under the right set of conditions. Constraints are used to ensure that the desired signals are not filtered out along with the unwanted signals. Although Applebaum's maximum SNR algorithm and Widrow's LMS algorithm were discovered independently and were developed using two different approaches, they are basically similar. For stationary signals, both algorithms converge to the optimum Wiener solution.

A different technique for solving the adaptive beamforming problem was proposed in 1969 by Capon. His algorithm leads to an adaptive beamformer with a minimum variance-distortion-less response (MVDR). This has also been referred to by some as the maximum likelihood method (MLM) because the algorithm maximizes the likelihood function of the input signal vector. It is also one of the earliest adaptive beamforming techniques that offer the ability to resolve signals that are separated by a fraction of an antenna beamwidth. In 1974, Reed and his co-workers showed that fast adaptivity is achieved by using the sample matrix inversion (SMI) technique. Using this technique, the adaptive weights can be computed directly, unlike the maximum SNR algorithm and the LMS algorithm, which may suffer from slow convergence if the eigenvalue spread of the sample covariance matrix $[R_{xx}]$ is relatively large. By inverting the covariance matrix directly, the performance of the SMI scheme becomes independent of the eigenvalue spread of $[R_{xx}]$.

It is interesting to note that all these algorithms were developed prior to the digital revolution and they really do not exploit adequately the digital nature of the data. It is for this reason that a direct data domain algorithm has been developed. It is conceptually simple and is computationally very fast through implementation of the FFT and the conjugate gradient method to solve the appropriate equations for the weights. Most of the techniques were based on statistical methodologies because in those years it was not easy to quantify the analog signals of interest.

With the advent of digital technology, these techniques were reemployed, this time dealing with digitally sampled data. However, with the design of faster processors, the Wiener filter theory, developed in the preceding section, also became available for the enhancement of signals in a noisy environment. With the availability of high-speed signal processors and analog-to-digital converters, these techniques were reemployed in the digital domain. It can be seen [8, 9] that the speed of the adaptive processes were greatly enhanced by replacing the LMS algorithm by a conjugate gradient method, saving several orders of magnitude of central processing unit (CPU) time. Also, the method can be used to improve the reliability of the estimate [10] while performing adaptive processing. However, these methods were basically applications of the methodology described earlier for calculating the instantaneous error signal and then applying a "forgetting factor" to decimate the old data as new data arrived. There is a plethora of books and published articles on the adaptive methodologies that depend on the statistical methodologies. A good summary of such techniques is available in [11].

In addition, Haykin [11] points out that a stochastic methodology leads to the design of an adaptive filter that will operate in a probabilistic sense on average for all the operational environments, assumed to be wide-sense stationary. On the other hand, a deterministic approach provides the solution for the given data at hand, and without invoking any of the stochastic methodology and without assuming the nature of the probability density functions. For example, if one takes a normal coin, when tossed up, on the average it will fall with the "head" facing up 50% of the time and with the "tail" up the other 50%. However, it is not known a priori what is going to happen on a single toss. A deterministic approach provides the solution for that single realization, which operates on the given data for one snapshot only. This philosophy has been further amplified by Hofstetter and Gardner [12–14] and is explained in Appendix A.

We now deal with the class of algorithms based on the method of least squares. This class of solution techniques is easy to implement for real-time processing and is computationally very robust if we want to develop an estimate for the solution on a snapshot-by-snapshot basis. In contrast, conventional statistical methodologies require data measured over the antennas using multiple snapshots. Here a snapshot is defined as the set of voltages measured across all the antenna elements at a particular instant of time. Hence, for the single snapshot-based methods, one can efficiently use digital signal processors to solve adaptive problems. In these procedures, a model-dependent methodology using the method of least squares (without invoking any assumptions about the statistics of the signals that are to be tracked) is utilized. This gives rise to the minimum-variance distortionless response (MVDR), based on a statistical methodology, but using the data in the test snapshot only. As pointed out by Gardner [13], the stochastic approach has become prevalent because one is dealing with analog signals and second, because communications engineers want to design systems that will perform well, on average, over the ensemble. However, since it is not feasible to make measurements over many realizations (systems), the communications engineers have settled on characterizing system performance, in practice, by averaging over time for a single system. In order to replace the ensemble averages by time averages, one needs to assume wide-sense stationarity. Furthermore, since the measurement is limited to one system, one has to invoke the concept of ergodicity. This is equivalent to using stationary stochastic-process models that are ergodic, so that the mathematically calculated expected values (ensemble averages) will equal the measured time averages. The mathematical descriptions of these statements are available in Appendix A. Hofstetter [12] states that

...Unfortunately, however the logic seems to have stopped at this point. It apparently was not recognized (except by too few to make a difference) that once consideration was restricted to ergodic stationary models, the stochastic process and its associated ensemble could be dispensed with because a completely equivalent theory of statistical interference and decision that is based entirely on time averages over a single record of data could be used.

Gardner [13] further points out that

Any calculations made using a model based on the time average theory could be applied to any one member of an ensemble if one so desires because the arguments that justify the ergodic stochastic model also guarantee that the time-average for one ensemble member will be the same (with probability one) as the time average for any other ensemble member. Whenever transient behavior is of interest ergodic models are ruled out, because all transient behavior is lost in an infinitely long time-average. Thus to counter the conceptual simplicity and realism offered by the time-average approach the stochastic-process approach offers the advantage of more general applicability.

These considerations lead us to apply the direct data domain approach to adaptive processing, which we discuss in the next chapter. The philosophical differences and the mathematical subtleties between a statistical covariance-based methodology and the direct data domain approach are described in Appendix A. Furthermore, in this method we do away with the time averages and focus on spatial averaging. In addition, we solve an estimation problem rather than a detection problem. It can also be shown that a direct data domain methodology provides a lower value for the Cramer–Rao bound than does a stochastic-based methodology [20].

3.3 MINIMUM VARIANCE DISTORTIONLESS RESPONSE TECHNIQUE

The minimum variance distortionless technique (MVDR) is based on the statistical methodology of Capon [11, 15], but without invoking the statistics of the underlying signal. It is assumed that the direction of the arrival of the signal is known. The goal is to estimate its strength in the presence of jammers, clutter, and thermal noise. We deal with discrete signals, and the linear filter of Figure 3.1 is now replaced by the weight vectors, $w(q)$, and the received signal is replaced by $x(q)$. The goal is to estimate the weights and to use them to find the signal of interest, $s(q)$, embedded in the received signals, $x(q)$. The desired signal, $d(q)$, is now a function of the input signals. As pointed out by Haykin [11],

This methodology may be viewed as an alternate to Wiener filter theory. Basically, Wiener filters are derived from ensemble averages (which is achieved by taking the expected value) with the result that one filter (in a probabilistic sense) is obtained for all realizations of the operational environment, assumed to be wide-sense stationary. On the other hand, the method of least squares is deterministic in approach. Specifically, it involves the use of time averages, with the result that the filter depends on the number of samples used in the computation.

This has been implemented by Owsley [16, 17].

Consider a linear array of $N + 1$ uniformly spaced isotropic omnidirectional receiving elements, separated by a distance Δ, as shown in Figure 3.2. We have $N + 1$ sensors in the array. We further assume that narrowband signals, consisting of a desired signal and interference with center frequency f_0, are impinging on the array from various azimuth angles φ, measured from the end-fire direction of the array, with the constraint $0° \leq \varphi \leq 180°$. The signal we want to estimate is arriving from an *a priori* known angle, φ_s, and the various jamming signals are arriving from various angles φ_j, including coherent/noncoherent multipaths. The jamming signals may be located in the main beam of the array, that is, $\varphi_s - \varphi_j < 50.8°/(L/\lambda)$, where L is the length of the array, λ is the wavelength corresponding to the frequency f_0 (i.e., $\lambda = 2\pi c/f_0$, where c is the velocity of light), and, typically, Δ (the spacing between the elements) is chosen to be $\lambda/2$.

The signals received in the $N + 1$ sensors are $x(m)$, $x(m-1)$, ..., $x(m-N)$ at the mth time instance. We consider the case where a set of N weights is attached to the last N sensors. For the first sensor, the weight is equal to unity. We then define

$$y(m) = \sum_{q=1}^{N} w^*(q) x(m-q) = 0 \qquad (3.8)$$

It is further assumed that we can predict the first sample from the next N samples. Hence,

$$d(m) = x(m) \qquad (3.9)$$

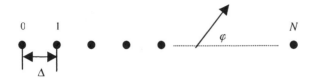

Figure 3.2. Uniform linear array of omnidirectional isotropic point radiators.

Therefore, in this system we are using the last N samples $x(m-1), \ldots, x(m-N)$ to predict the mth sample $x(m)$. Then, the error is defined [with $w(0) = 1$] to be

$$e(m) = d(m) - y(m) = \sum_{q=0}^{N} w^*(q)x(m-q) = [w]^H [x] \tag{3.10}$$

where H denotes the conjugate transpose and $[\cdot]$ denotes a matrix. Without any loss of generality, we no longer assume that $w(0) = 1$, but that it has some value that could be different than 1. The goal is to extract the signal of interest, $s(q)$, which is embedded in the received signal, $x(q)$, in the presence of interferers, jammers, and the like. Next, the objective is to find the weights $w(q)$ by minimizing the sum of the errors $e(m)$:

$$\Xi = \sum_{m=i_1}^{i_2} |e(m)|^2 \tag{3.11}$$

where the summation runs over indices from i_1, to i_2. The error Ξ is different from $\mathcal{E}$, where the former forms a sum of a series and the latter is an integral. The exact values of i_1 and i_2 are determined from the given data.

However, this information by itself is not sufficient to solve the adaptive problem. Additionally, one needs to specify the direction of arrival of the signal of interest, $s(q)$. This is achieved by maintaining the main-beam gain of the array along the direction of arrival φ_s:

$$\sum_{q=0}^{N} w^*(q) e^{-j\frac{q 2\pi \Delta}{\lambda} \cos \varphi_s} = 1 \tag{3.12}$$

where Δ/λ is the interelement spacing of the sensors in terms of the wavelength of the operating frequency, and φ_s is the direction of arrival of the signal from the end-fire direction of the array (here, $\varphi = 90°$ is the broadside direction; also note that a linear array cannot resolve the ambiguity of $+\varphi$ or $-\varphi$, that is, from which side of the array the signal is coming). That is why we restrict the scan of the azimuth angle from $0° \leq \varphi \leq 180°$. The objective of the weights is to minimize the cost function, Ξ, formed through the Lagrange multiplier, χ (a complex constant):

$$\Xi = \sum_{m=N+1}^{I} |e(m)|^2 + \chi \left(\sum_{q=0}^{N} w^*(q) e^{-j\frac{2\pi\Delta q}{\lambda} \cos \varphi_s} - 1 \right) \quad (3.13)$$

where I is the total number of time samples available. Here, the index m runs from $N + 1$ to I because of equation (3.10). Let us define

$$\xi_0 = \frac{2\pi\Delta}{\lambda} \cos \varphi_s \quad (3.14)$$

Then the error is defined by

$$\Xi = \sum_{m=N+1}^{I} \left| \sum_{q=0}^{N} w^*(q) x(m-q) \right|^2 + \chi \left(\sum_{q=0}^{N} w^*(q) e^{-j\xi_0 q} - 1 \right) \quad (3.15)$$

By minimizing the cost function Ξ in terms of the weights $w(q)$, one obtains

$$\frac{\partial \Xi}{\partial w^*(q)} = 2 \sum_{m=N+1}^{I} x(m-q) \left\{ \sum_{p=0}^{N} w(p) x^*(m-p) \right\} + \chi e^{-j q \xi_0}$$

$$= 2 \sum_{p=0}^{N} w(p) \sum_{m=N+1}^{I} x(m-q) x^*(m-p) + \chi e^{-j q \xi_0} \quad (3.16)$$

Since at the minimum, the first derivative is zero, this yields

$$\sum_{p=0}^{N} w(p) R(p, q) = -\frac{1}{2} \chi e^{-j q \xi_0} \quad \text{for} \quad q = 0, 1, ..., N \quad (3.17)$$

where

$$R(p, q) = \sum_{m=N+1}^{I} x(m-q) x^*(m-p) \quad (3.18)$$

Or, utilizing a matrix notation,

$$[R]_{(N+1)\times(N+1)} [w]_{(N+1)\times 1} = -\frac{1}{2} \chi [s(\xi_0)]_{(N+1)\times 1} \quad (3.19)$$

where

$$[s(\xi_0)]_{(N+1)\times 1} = \left[1, e^{-j\xi_0}, ..., e^{-j N\xi_0} \right]^T \quad (3.20)$$

and T denotes the transpose of a matrix. The optimum weight vectors are given by

$$[w]_{opt} = -\frac{1}{2} \chi [R]^{-1} [s(\xi_0)] \tag{3.21}$$

If H represents the conjugate transpose of a matrix, then from equation (3.12),

$$[w]^H [s(\xi_0)] = 1 \tag{3.22}$$

Utilization of equation (3.21) in equation (3.22) results in

$$\chi = \frac{-2}{[s(\xi_0)]^H [R]^{-1} [s(\xi_0)]} \tag{3.23}$$

Therefore, the final result for the optimum weights is given by

$$[w]_{opt} = \frac{[R]^{-1} [s(\xi_0)]}{[s(\xi_0)]^H [R]^{-1} [s(\xi_0)]} \tag{3.24}$$

and the estimated value (denoted by a subscript *est*) of the desired signal denoted by s is given by

$$[s(\xi_0)]_{est} = [w]^H [R][w] = \frac{1}{[s(\xi_0)]^H [R]^{-1} [s(\xi_0)]} \tag{3.25}$$

and the minimum value of the error is given by

$$e_{min}(m) = \sum_{q=0}^{N} w^{*}_{opt}(q) x(m-q) = [w_{opt}]^H [x] \tag{3.26}$$

It is interesting to note that by varying the parameter ξ_0 or by changing the angle φ it is possible to get the adapted antenna pattern. Because we are plotting the inverse of a function that goes to zero, it provides large value for the peaks, as we are essentially plotting the numerical error in the function. Hence, in these plots the peaks have no physical meaning, as the plots represent the power density. However, the integral under the curve, which represents power, has physical significance.

This optimum solution has several interesting properties, as originally outlined by Capon [15], implemented by Owsley [16, 17], and summarized by Haykin [11]. Namely, the optimum weights are unbiased if the sequence $x(k)$ contains noise that is zero mean. In addition, this least squares estimate, w_{opt}, is the best linear unbiased estimate. Finally, when the additive noise in $x(k)$ is white and Gaussian with zero mean, the least squares estimate achieves the Cramer–Rao lower bound (the Cramer–Rao bound provides the smallest value of the variance of an unbiased estimator in Gaussian noise) for unbiased estimates. In addition to these advantages, there are some drawbacks, which are as follows:

1.) The computation of the matrix [R], used in the evaluation of the optimum weights in equation (3.21), is a $(N + 1)^2$ $(I - N + 1)$ process, which is difficult to carry out in real time.

2.) The computation of $[R]^{-1}$ can also be expensive and computationally unstable. For example, evaluation of the inverse requires an $\sim \Theta (N + 1)^3$ process [here $\Theta (\bullet)$ denotes "on the order of"], as the dimension of [R] is $N + 1$. In addition, in the absence of noise, [R] is singular. The presence of additive noise may make it nonsingular, but this could be numerically unstable.

3.) In the evaluation of the elements of matrix [R], time averaging is carried out, as shown in equation (3.18). Hence, if there are intermittent (blinking) jammers or a coherent multipath, this method cannot eliminate them. Coherent multipath depicts a signal $s(t)$ in terms of the multipath, $gs(t)$. If g is -1, then (complete) fading occurs and the signal is canceled and, hence, the adaptive technique cannot reconstitute the signal, as the multipath can only be detected in the spatial domain of the arrays.

4.) Inherent in this development is the assumption that the signal of interest is arriving from an angle φ_s. However, due to misadjustment or for other reasons, the signal may be arriving from an angle $\varphi_s + \Delta\varphi$, and not exactly at φ_s. In this case, the adaptive processor considers the actual signal at $\varphi_s + \Delta\varphi$ as a jammer, and cancels it. This issue results in a problem of signal cancellation due to mismatch. A possible solution is to have a number of constraints, instead of a single constraint as given by equation (3.12). This is equivalent to defining the number of constraints required to characterize the 3-dB beamwidth of the adaptive array. This would require modifying equation (3.13) with a number of Lagrange multipliers for a number of points as constraints defining the 3-dB beamwidth.

Other applications of using transient samples to carry out spatial processing may be found in [18, 19]. In the next chapter, an alternative methodology is presented where many of these problems can be mitigated. This new method is based on spatial analysis of the data rather than dealing with the time variable. Therefore, we are processing the data on a snapshot-by-snapshot basis. A snapshot is defined as consisting of the voltages induced in the $N + 1$ elements of the array at a particular time instance $t = T_0$ (for example).

3.4   CONCLUSION

A brief survey of the conventional methodology is presented to illustrate the problems in real-time implementation of the adaptive technique. Since time averaging is carried out implicitly to estimate the signals of interest, it may be difficult to handle coherent or blinking interferers.

REFERENCES

[1]   A. N. Kolmogorov, "Sur l'Interpolation et Extrapolation des Suites Stationaries," *Comptes Rendus de l'Academica des Sciences*, Vol. 208, pp. 2043–2045, 1939 (English translation in [2]).

[2]   T. Kailath (ed.), "Linear Least-Squares Estimation," in *Benchmark Papers in Electrical Engineering and Computer Science*, Dowden, Hutchinson & Ross, Stroudsburg, PA, 1977.

[3]   N. Wiener, *Extrapolation, Interpolation and Smoothing of Stationary Time Series with Engineering Applications*, MIT Press, Cambridge, MA, 1949 (originally published as a classified report in 1942).

[4]   T. K. Sarkar, J. Nebat, D. D. Weiner, and V. K. Jain, "A Discussion of Various Approaches to the Identification/Approximation Problem," *IEEE Transactions on Antennas and Propagation*, Vol. 30, No. 2, pp. 267–272, March 1982.

[5]   P. Howells, "Intermediate Frequency Sidelobe Canceller", U.S. patent 3,202,990, filed May 4, 1959 and granted Aug. 24, 1965. Also in *IEEE Transactions on Antennas and Propagation*, Vol. 24, No. 5, pp. 575–584, Sept. 1976.

[6]   S. P. Applebaum *Adaptive Arrays*, Syracuse University Research Corporation Report SPL, TR 66-1, 1966. Also in *IEEE Transactions on Antennas and Propagation*, Vol. 24, No. 5, pp. 585–598, Sept. 1976.

[7]   B. Widrow, P. Mantay, L. Griffiths and B. Goode, "Adaptive Antenna Systems," *Proceedings of the IEEE*, Vol. 55, pp. 2143–2159, Dec. 1967.

[8]   H. Chen, T. K. Sarkar, S. A. Dianat, and J. D. Brule, "Adaptive Spectral Estimation by the Conjugate Gradient Method," *IEEE Transactions on Acoustics, Speech and Signal Processing*, Vol. 34, No. 2, pp. 272–284, Apr. 1986.

[9]   M. G. Bellanger, *Adaptive Digital Filters and Signal Analysis*, Marcel Dekker, New York, 1987.

[10]   J. Guangqing, T. K. Sarkar, and S. Choi, "Adaptive Spectral Estimation by the Conjugate Gradient Method (CGM) with Pseudo Frequency Elimination (PFE)," *AEU*, pp. 43–49, Jan. 1990.

[11]   S. Haykin, *Adaptive Filter Theory*, Prentice Hall, Upper Saddle River, NJ, 1996.

[12]   E. M. Hofstetter, "Random Processes," Chap. 3 in H. Margenau and G. M. Murphy (eds.), *The Mathematics of Physics and Chemistry*, Vol. 11, Van Nostrand, Princeton, NJ, 1964.

[13]   W. A. Gardner, *Statistical Spectral Analysis: A Nonprobabilistic Theory*, Prentice Hall, Englewood Cliffs, NJ, 1987.

[14]   W. A. Gardner, *Cyclostationarity in Communications and Signal Processing*, IEEE Press, Piscataway, NJ, 1994.

[15]   J. Capon, "High Resolution Frequency–Wavenumber Spectrum Analysis," *Proceedings of the IEEE*, Vol. 57, pp. 1408–1418, 1969.

[16]   N. L. Owsley, "A Recent Trend in Adaptive Signal Processing for Sensor Arrays: Constrained Adaptation," pp. 551–604 in J. W. R. Griffiths et al. (eds.), *Signal Processing*, Academic Press, New York, 1973.

[17] N. L. Owsley, "Sonar Array Processing," pp. 115–193 in S. Haykin (ed.), *Array Signal Processing*, Prentice Hall, Englewood Cliffs, New Jersey, 1985.

[18] D. H. Johnson and D. E. Dudgeon, *Array Signal Processing*, Prentice Hall, Englewood Cliffs, NJ, 1993.

[19] T. K. Sarkar, S. Nagaraja, and M. C. Wicks, "A Deterministic Direct Data Domain Approach to Signal Estimation Utilizing Uniform 2-D Arrays," *Digital Signal Processing: A Review Journal*, Vol. 2, pp. 114–125, Apr. 1998.

[20] Y. Hua and T. K. Sarkar, "A Note on the Cramer–Rao Bound for 2-D Direction Finding Based on a 2-D Array," *IEEE Transactions on Signal Processing*, Vol. 39, No. 5, pp. 1215–1218, May 1991.

4

DIRECT DATA DOMAIN LEAST SQUARES APPROACHES TO ADAPTIVE PROCESSING BASED ON SINGLE SNAPSHOTS OF DATA

SUMMARY

In this chapter a direct data domain least squares (D^3LS) approach to adaptive processing using a single snapshot of data is presented. In contrast to the conventional adaptive techniques, where at the first step one needs to form the covariance matrix of the data, and then invert it, in the present approaches the signal of interest (SOI) is obtained directly from the solution of a matrix equation. Use of the conjugate gradient and the fast Fourier transform in solving this matrix equation makes this procedure highly suitable to real-time implementation of these algorithms, as the computational time is considerably less than that of the other conventional adaptive techniques. Here a least squares approach is applied directly to the data on a snapshot-by-snapshot basis and hence is computationally quite efficient. A snapshot in this specific case is defined as the phasor voltages measured at the feed points of all the antenna elements in the array at a particular instance of time. Nonstationarity in the data then has little effect for these classes of direct data domain methods, as no assumption is made about the statistics of the environment. Yet the environment is modeled in a realistic fashion. Even though we are processing the data on a snapshot-by-snapshot basis, one can still extract N coherent sources using an antenna array consisting of approximately $1.5N$ elements, which is the minimum possible under any circumstances. However, the direct data domain methods result in a slightly reduced number of degrees of freedom as opposed to a conventional statistical analysis for noncoherent interferers, where one needs to processes a block of data snapshots to form the covariance matrix. In addition, it is shown how to carry out adaptive processing with a prespecifed main lobe width of the adapted beam pattern, thereby preventing signal cancellation in the direct data domain algorithms. Finally, it is shown that the norm of the adapted weights may provide a refined estimate for the actual direction of arrival of the signal of interest when there are uncertainties associated with their initial direction of arrival.

4.1 INTRODUCTION

In this chapter we address the radar problem where the direction of arrival of the signal of interest (SOI) is known *a priori*. For these classes of problems we generally transmit the energy along a certain specific direction and then try to detect targets along that direction from the signal returns. Since we know the direction along which the energy was initially transmitted, one has a reasonably good knowledge of the direction of arrival of the reflected energy from the target, if they exist along that specific direction. Initially, we will develop this methodology based on the assumption that each antenna element is an omnidirectional point radiator and is spaced uniformly along a line. However, in later chapters these assumptions are relaxed to handle realistic antenna elements and we demonstrate how the electromagnetic analysis could be used in the signal-processing framework to achieve an accurate estimation of the SOI, when the array is operating in a near-field environment. In addition, to the SOI contributing to the received voltages at each antenna element, we also have contributions due to jammers, clutter, and thermal noise. The incoming interferers and the clutter may be coherent with the SOI. Consider a uniformly spaced linear array consisting of $N + 1$ isotropic omnidirectional point radiators, as shown in Figure 4.1.

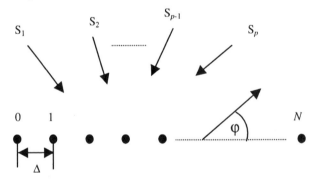

Figure 4.1. Linear uniform array.

The phasor voltage X_n induced at the nth antenna element at a particular instance of time will then be given by

$$X_n = s e^{\frac{j2\pi n \Delta \cos\varphi_s}{\lambda}} + \sum_{p=1}^{P} A_p e^{\frac{j2\pi n \Delta \cos\varphi_p}{\lambda}} + C_n + \xi_n \qquad (4.1)$$

where s =   complex amplitude of the SOI (to be determined)
       φ_s =   direction of arrival of the SOI (assumed to be known)

Δ = spacing between each of the antenna elements

λ = wavelength of transmission (here it is assumed that we are dealing with narrowband signals)

P = total number of interferers

A_p = complex amplitude of pth undesired interferer

φ_p = direction of arrival of the pth interferer

C_n = clutter induced at the nth element (in some cases, this may be diffused electromagnetic radiation from distant mountains, lands, buildings, water, and so on)

ξ_n = thermal noise induced at the nth antenna element

Here we model the clutter as a bunch of reflected/diffracted rays bouncing back from the ground, platforms on which the array is mounted, and from nearby buildings, trees, or uneven terrains. The amplitude and phase of these rays have been determined by two random number generators. Hence, the clutter is modeled by a true physics of an electromagnetic model and not based on some probability distributions which do not satisfy any known electromagnetic phenomenon. As discussed in Appendix A, stochastic models may not even be correct for the specific situation that we are dealing with. For the array shown in Figure 4.1, the measured voltages X_n for $n = 0, 1, \ldots, N$, at the antenna elements are assumed to be known along with θ_s, the direction of arrival of the SOI. The goal is to estimate the complex amplitude s for the SOI. Here we define a single snapshot by the voltages X_n measured at the nth element at a certain instant of time t_m. It is understood that all the SOI, jammers, clutter, and thermal noise vary as a function of time.

In conventional adaptive processing we assume that we have a set of weights W_n, $n = 0, 1, \ldots, N$, connected to each antenna element. Then we generate a block of data corresponding to $M + 1$ snapshots (i.e., X_n^m for $m = 0, 1, \ldots, M$ and $n = 0, 1, \ldots, N$). Here the superscript m on X_n denotes that the voltage X_n^m is induced at element n at a specific time instance m. Then a covariance matrix of this block of data of $(N + 1) \times (M + 1)$ samples is evaluated and the adaptive weights are given by the Wiener solution, which is related to the inverse of the covariance matrix, as explained in Chapter 3. Some of the problems with this conventional covariance matrix-based approach , [1, 2] - are:

1.) The computational load is proportional to $\Theta\{2(N + 1)^3\}$ if we consider $N = M$. Here $\Theta(\cdot)$ represents "on the order of." To form a covariance matrix requires $(N + 1)^3$ operations and a similar number of operations to invert it. Hence, it is difficult to implement it in real time.

2.) In addition, the procedure assumes that the data are stationary over these $(M + 1)$ snapshots (i.e., the environment of the SOI, clutter, and jammer scenarios have not changed over all the $M + 1$ snapshots).

Because of these disadvantages, we propose a direct data domain least squares approach (D^3LS) using a single snapshot of the data. The next section presents this method along with some numerical simulations. The properties of this D^3LS method are also outlined. In this chapter, we develop the methodology using point sources and illustrate in the next chapters how the mutual coupling between the sensors can be incorporated, including the presence of near-field scatterers to carry out reliable adaptive processing.

4.2   DIRECT DATA DOMAIN LEAST SQUARES PROCEDURES

Consider the linear array as shown in Figure 4.1. Here we consider that we have a single snapshot of the voltages measured at the feed point of the antenna elements (i.e., at a time $t = t_m$, we have the voltages X_n, for $n = 0, 1, \ldots, N$ measured at the feed points of all $N + 1$ antenna elements). The superscript m on X has been suppressed for convenience, as we are dealing with a single snapshot of data. Our goal is to estimate s given φ_s. We also know that in order to obtain the SOI, the number of coherent jammers must be $\leq N/2$ in the absence of clutter and noise. It is important to point out that in this procedure we do not make any distinction between coherent or noncoherent interferers. The classical techniques based on the statistical methodology will be able to handle more than $N/2$ noncoherent interferers but no more than $N/2$ coherent interferers. However, the price to be paid for this is that we require at least $N + 1$ snapshots of voltages. Therefore, for these direct data domain methods there may be a significant loss in the number of degrees of freedom (DOFs) for the noncoherent case. But when the interferers are coherent with the signal, the number of degrees of freedom is the same for both the direct and the stochastically based methods even though the latter is using multiple snapshots as opposed to a single snapshot for the former. However, later in this section we show how to go beyond this limitation by using a modification of this technique. But for now, we have this constraint that the DOF $\cong N/2$ in addition to the assumption that all the antenna elements are point radiators.

4.2.1   Eigenvalue Method

In this section we present the eigenvalue method, which is one of the forms of the D^3LS method. Under the assumption that P of (4.1) $\leq N/2$, one can form the matrix pencil $[X] - \alpha[S]$ of dimension $L + 1$ (here, α is the estimate of the complex amplitude for the unknown SOIs to be solved for), where

$$[X] = \begin{bmatrix} X_0 & X_1 & \cdots & X_L \\ X_1 & X_2 & \cdots & X_{L+1} \\ \vdots & \vdots & \cdots & \\ X_L & X_{L+1} & \cdots & X_N \end{bmatrix}_{(L+1)\times(L+1)} \tag{4.2}$$

$$[S] = \begin{bmatrix} S_0 & S_1 & \cdots & S_L \\ S_1 & S_2 & \cdots & S_{L+1} \\ \vdots & \vdots & \cdots & \\ S_L & S_{L+1} & \cdots & S_N \end{bmatrix}_{(L+1)\times(L+1)} \tag{4.3}$$

The variables in (4.2) and (4.3) are defined in such a way that the difference $X_n - \alpha S_n$ at each element represents the contribution due to signal multipaths, jammers, clutter, and thermal noise (i.e., all the undesired components of the signals except the SOI, which is s) since it is assumed that S_n is the voltage induced at the nth element due to a signal arriving from the same direction as the SOI but whose amplitude is unity. Therefore,

$$S_n = e^{\dfrac{j2\pi\, n\Delta\cos\varphi_s}{\lambda}} \tag{4.4}$$

In this procedure we assume that $(N+1) \geq (2L+1)$ and the total number of antenna elements $N+1$ is always odd. For the development of these principles, if we assume that there are P jammers, we have a total of $2P+1$ unknowns to deal with. For each jammer, the direction of arrival and its complex amplitude are unknowns, which accounts for $2P$ terms. For the signal we have the direction of arrival but do not know its strength. Hence the +1 term takes care of the unknown signal strength. Therefore, the total number of unknowns is always $2P+1$ in this procedure, so $N+1$ is an odd number.

Next, in an adaptive processing methodology, the column vector of weights [W] is chosen in such a way that the contribution from the jammers, clutter, and thermal noise are minimized to enhance the output signal to interference plus noise ratio. Hence, if we define the matrix $[U] = \{ [X] - \alpha[S] \}$, we obtain the following generalized eigenvalue problem:

$$[U]_{(L+1,L+1)}\,[W]_{(L+1)} = \{ [X] - \alpha[S] \}_{(L+1)\times(L+1)}\,[W]_{(L+1)\times 1} = 0 \tag{4.5}$$

where α, the estimate of the complex amplitude for SOI in (4.1), is given by its solution (i.e., the generalized eigenvalue, and the weights [W] are given by the generalized eigenvector). Since we have assumed that there is only one signal arriving from θ_s, the matrix [S] is of rank unity, and hence the generalized eigenvalue equation given by (4.5) has only one eigenvalue, and that eigenvalue α provides an estimate for the complex amplitude for the SOI.

Alternatively, one can view the left-hand side of (4.5) as the total noise signal at the output of the adaptive processor due to jammer, clutter, and thermal noise. Hence, the weighted sum of the total interference plus noise voltage is given by

$$\Xi_{\text{out}} = [U][W] = \{[X] - \alpha[S]\}[W] \tag{4.6}$$

Therefore, the total noise power would be given by

$$\Xi_{\text{power}} = [W]^H \{[X] - \alpha[S]\}^H \{[X] - \alpha[S]\}[W] \qquad (4.7)$$

Our objective is to minimize this noise power by selecting [W] for a fixed signal strength α. This is achieved by differentiating Ξ_{power} with respect to each of the individual weights W_i and setting each one of the partial derivatives equal to zero. When each of the individual equations is assembled together, this results in (4.5).

From a computational point of view, an alternative way to solve for α is to make the determinant of the following matrix equal to zero:

$$\det \{[X] - \alpha [S]\} = 0 \qquad (4.8)$$

for a suitable value of α.

Equation (4.5) can be rewritten in terms of the following two generalized eigenvalue equations:

$$[X] [W] = \alpha [S] [W] \qquad (4.9)$$

or alternatively,

$$[S] [W] = 1 / \alpha [X] [W] \qquad (4.10)$$

Next we introduce two different methods for solving the generalized eigenvalue problem.

4.2.1.1 Two Methods for Solving the Generalized Eigenvalue Problem. The QZ algorithm is a well-known method for solving the generalized eigenvalue problem [3]. Consider the generalized eigenvalue problem

$$[X] [W] = \alpha [S] [W] \qquad (4.11)$$

In the QZ algorithm, first the matrices [X] and [S] are reduced to upper Hessenberg triangular forms by means of simultaneous orthogonal equivalence of the orthogonal matrices [Q] and [Z]. Then the Hessenberg triangular pair [X], [S] is further reduced to the generalized Schur form through implicit QR iterations on $[X] [S]^{-1}$. The eigenvalues are easily calculated from the diagonal elements of the generalized Schur form.

A second approach in our case consists of exploiting the Hankel structure of the matrices [X] and [S] to use the singular value decomposition (SVD) to compute the eigenvalues of (4.11). One starts by calculating the SVD of [X] as

$$[X] = [U][\Sigma][V]^H \qquad (4.12)$$

where $[\Sigma]$ is a rectangular matrix whose diagonal elements contain the singular values of $[X]$. The matrices $[U]$ and $[V]$ are two orthogonal matrices consisting of the left and right singular vectors corresponding to all the singular values of $[X]$. The superscript H denotes the conjugate transpose of a matrix. The pseudo-inverse of $[X]$ can then be written as

$$[X]^{-1} = [X]^{\dagger} = [V]\,[\Sigma]^{-1}[U]^{H} \qquad (4.13)$$

where $[\Sigma]^{-1}$ is a diagonal matrix with only diagonal entries, which are the reciprocals of the nonzero singular values of $[X]$ and $[U]^{H} = [U]^{-1}$; $[V]^{H} = [V]^{-1}$. This tool can be very useful using a signal/noise subspace decomposition analysis.

Two useful properties of the pseudo-inverse are

(i) $[X][X]^{\dagger}[X] = [X]$ $\qquad\qquad$ (4.14)

(ii) $[X]^{\dagger}[X][X]^{\dagger} = [X]^{\dagger}$ $\qquad\qquad$ (4.15)

By exploiting the Hankel structure of $[S]$, we can rewrite it as

$$[S] = [u][v] \qquad (4.16)$$

where

$$[u] = \begin{bmatrix} 1 \\ s_1 \\ s_2 \\ \vdots \\ s_L \end{bmatrix} \qquad (4.17)$$

and

$$[v] = [\,1 \quad s_1 \quad s_2 \quad \cdots \quad s_L\,] \qquad (4.18)$$

Note that

$$[v] = [u]^{T} \qquad (4.19)$$

So for a Hankel matrix $[X]$,

(i) $[X]^{T} = [X]$ $\qquad\qquad$ (4.20)

(ii) $\{[X]^{*}\}^{T} = [X]^{H}$ $\qquad\qquad$ (4.21)

where the superscript $*$ denotes a conjugate of the complex quantity. Now multiplying (4.11) by $[X][X]^{\dagger}$ yields

$$[X][X]^{\dagger}[X][W] = [X][X]^{\dagger}\alpha[S][W] \tag{4.22}$$

Using property (i) of the pseudo-inverse, we obtain

$$[X][W] = [X][X]^{\dagger}\alpha[S][W] \tag{4.23}$$

or equivalently,

$$[X]^{\dagger}[X][W] = [X]^{\dagger}\alpha[S][W] \tag{4.24}$$

Now substitute (4.11) into (4.24) to obtain

$$\alpha[S][W] = [X][X]^{\dagger}\alpha[S][W] \tag{4.25}$$

Assuming that $\alpha \neq 0$, we can cancel the scalar from both sides of (4.25). We then use (4.16) to rewrite (4.25) as

$$[u][v][W] = [X][X]^{\dagger}[u][v][W] \tag{4.26}$$

Now note that $[v][W]$ is a scalar. We can then rewrite (4.26) as

$$[u] = [X][X]^{\dagger}[u] \tag{4.27}$$

Using the fact that $[u]$ and $[v]$ are transposes of each other and through properties (i) and (ii) for Hankel matrices, we obtain

$$[v] = [v][X]^{\dagger}[X] \tag{4.28}$$

By premultiplying (4.28) by $[u]$ and using (4.16), we finally get

$$[S] = [S][X]^{\dagger}[X] \tag{4.29}$$

Substitution of (4.29) into (4.11) results in

$$[X][W] = \alpha[S][X]^{\dagger}[X][W] \tag{4.30}$$

Using (4.11) to substitute for $[X][W]$ on both sides of (4.30), we get

$$\alpha[S][W] = \alpha[S][X]^{\dagger}\alpha[S][W] \tag{4.31}$$

Again, canceling the scalar α from both sides of (4.31) and using (4.16) to rewrite $[S]$, we get

$$[u][v][W] = \alpha[u][v][X]^{\dagger}[u][v][W] \tag{4.32}$$

Since $[v][W]$ is a scalar and $[u]$ is a vector of rank one, therefore,

$$\alpha = \frac{1}{[v][X]^{\dagger}[u]} \tag{4.33}$$

In summary, the two procedures for the solution of (4.11) are:

1.) Application of the QZ algorithm, which solves (4.11) through similarity transforms.
2.) The SVD-based algorithm, which computes the pseudo-inverse, $[X]^{\dagger}$, and forms an estimate of the eigenvalue using (4.33).

The QZ algorithm rotates spaces to zero-out subdiagonal elements. When the spaces are close together, a fine enough rotation can still give the desired results. The SVD uses the fixed orthogonal projections of the null space and the range space of one matrix applied to the basis of the other space. When the two spaces are close together, the projection has a harder time to carry out the discrimination than the QZ rotation projection. Adding noise to the data whitens the spaces and moves them closer together. So when using single precision the QZ algorithm provides a better result; however, if we use double precision, both techniques produce similar results, as evidenced in [3]. This is illustrated next through a perturbation analysis for both algorithms.

The QZ algorithm uses unitary similarity transforms to get [S] and [X] into more tractable forms. It is known that if [P] and [Q] are similar matrices, a perturbation in [P] will result in a perturbation in [Q] of the same magnitude [3]:

$$[Q] = [U]^{H}[P][U] \tag{4.34}$$

$$[U]^{H}\{[P] + [\Delta P]\}[U] = [Q] + [\Delta Q] \tag{4.35}$$

$$\|\Delta Q\| = \|\Delta P\| \tag{4.36}$$

where $[\Delta P]$ and $[\Delta Q]$ are the perturbations of the matrices [P] and [Q], respectively. Here, $\|\bullet\|$ denotes the norm of the matrix, which is a multiplication of the matrix by its complex conjugate transpose. It is also known from the Bauer–Filke theorem that an eigenvalue of a perturbed matrix will be magnified by the order of the condition number of the matrix [4]. When the matrix of interest consists of two spaces which are very close together, it becomes nearly singular. So when the angle of the arrival of the jammer is close to the DOA of the SOI or when noise whitens both spaces, the problem becomes ill conditioned.

For the SVD-based algorithm, we have

$$\alpha = \frac{1}{[v][X]^{\dagger}[u]} \tag{4.37}$$

and ignoring second-order effects, we can write

$$[[X] + [\Delta X]]^{-1} \approx [X]^{-1} - [X]^{-1}[\Delta X] [X]^{-1} \tag{4.38}$$

Therefore, for small perturbations in the received matrix, the corresponding perturbation in the eigenvalue is

$$\alpha + \Delta \alpha \approx \frac{1}{[v][X]^{\dagger}[u] - [v][X]^{\dagger}[\Delta X][X]^{\dagger}[u]} \tag{4.39}$$

Subtracting (4.37) from (4.39), we get

$$\Delta \alpha \approx \frac{[u][X]^{\dagger}[\Delta X][X]^{\dagger}[u]}{\{[v][X]^{\dagger}[u]\}^2} \tag{4.40}$$

In summary, when using single precision, the performance of the QZ algorithm is much better than that of the SVD algorithm. However, when double precision is used in the computation, the performance of the two methods is similar. However, the solution of a generalized eigenvalue problem results in a computational complexity $\Theta[(L + 1)^3]$, and in real-time applications it may be difficult to solve this generalized eigenvalue problem in an efficient way, particularly if the dimension L—the number of weights—is large. For that reason we convert the solution of a nonlinear eigenvalue problem defined by (4.5) to the solution of a linear matrix equation.

4.2.2   Forward Method

Note that the (1,1) and (1,2) elements of the interference plus noise matrix, [U], as defined in (4.5), are given by

$$U(1,1) = X_0 - \alpha S_0 \tag{4.41}$$

$$U(1,2) = X_1 - \alpha S_1 \tag{4.42}$$

where X_0 and X_1 are the voltages received at antenna elements 0 and 1 due to the signal, jammer, clutter, and noise whereas S_0 and S_1 are the values of the SOI only at those elements due to a signal of unit strength. Define

$$Z = \exp\left[j2\pi\frac{d}{\lambda}\cos\varphi_s \right] \tag{4.43}$$

Then, $U(1,1) - Z^{-1} U(1,2)$ contains no components of the SOI, as

$$S_0 = \exp\left[j2\pi\frac{(n=0)d}{\lambda}\cos\varphi_s \right] \quad \text{with } n = 0 \tag{4.44}$$

and

$$S_1 = \exp\left[\cdot j 2\pi \frac{(n=1)d}{\lambda} \cos\varphi_s\right] \quad \text{with } n = 1 \tag{4.45}$$

The same is true for $U(1, 2) - Z^{-1}U(1, 3)$ and, in general, for $U(i, j) - Z^{-1}U(i, j+1)$, for $i = 1, ..., L + 1$, $j = 1, ..., L$. Here we have $L = N/2$. Therefore, one can form a reduced rank matrix $[T]_{L \times (L+1)}$, generated from $[U]$ such that

$$[T] = \begin{bmatrix} X_0 - Z^{-1}X_1 & X_1 - Z^{-1}X_2 & \cdots & X_L - Z^{-1}X_{L+1} \\ \vdots & \vdots & \vdots & \vdots \\ X_{N-L-1} - Z^{-1}X_{N-L} & X_{N-L} - Z^{-1}X_{N-L+1} & \cdots & X_{N-1} - Z^{-1}X_N \end{bmatrix}_{L \times (L+1)}$$

$$\times \begin{bmatrix} W_0 \\ \vdots \\ W_L \end{bmatrix}_{(L+1) \times 1} = 0 \tag{4.46}$$

In order to restore the signal component in the adaptive processing, we fix the gain of the subarray formed by the $L + 1$ elements along the direction θ_s and then evaluate a weighted sum of the voltages $\sum_{i=0}^{L} W_i X_i$. Let us say that the gain of the subarray is C along the direction of φ_s. This provides an additional equation, resulting in a square matrix:

$$\begin{bmatrix} 1 & Z & \cdots & Z^L \\ X_0 - Z^{-1}X_1 & & & X_L - Z^{-1}X_{L+1} \\ \vdots & \vdots & & \vdots \\ X_{N-L-1} - Z^{-1}X_{N-L} & X_{N-L} - Z^{-1}X_{N-L+1} & \cdots & X_{N-1} - Z^{-1}X_N \end{bmatrix}_{(L+1) \times (L+1)}$$

$$\times \begin{bmatrix} W_0 \\ W_1 \\ \vdots \\ W_L \end{bmatrix}_{(L+1) \times 1} = \begin{bmatrix} C \\ 0 \\ \vdots \\ 0 \end{bmatrix}_{(L+1) \times 1} \tag{4.47}$$

or equivalently,

$$[F][W] = [Y] \tag{4.48}$$

Once the weights are solved for by using (4.48), the signal component α [estimate for s for SOI in (4.1)], may be estimated by using

$$\alpha = \frac{1}{C} \sum_{i=0}^{L} W_i X_i \qquad (4.49)$$

The proof of (4.49) is available in [4–6].

It is also possible to get an estimate of α from any one of the following $L + 1$ equations:

$$\alpha_v = \frac{1}{CZ^v} \sum_{i=0}^{L} W_i X_{i+v} \qquad \text{for } v = 0, ..., L \qquad (4.50)$$

or by averaging any number of the equations given by the set of $L + 1$ equations in (4.50). However, it is interesting to note that because of (4.46), averaging $L + 1$ estimates of α obtained from (4.50) is no better than using (4.49)!

As noted in [6–9], equation (4.47) can be solved very efficiently by applying the conjugate gradient method, which may be implemented to operate in real time utilizing a digital signal processing chip (the algorithm was actually implemented on a DSP32C chip produced by AT&T) [7, 8].

For the solution of $[F][W] = [Y]$ in (4.48), the conjugate gradient method starts with an initial guess $[W]_0$ for the solution and continues with the calculation of [4–9]

$$[P]_0 = -b_{-1}[F]^H [R]_0 = -b_{-1}[F]^H \{[F][W]_0 - [Y]\} \qquad (4.51)$$

where H denotes the conjugate transpose of a matrix. At the kth iteration the conjugate gradient method develops the following:

$$c_k = \frac{1}{\left\| [F][P]_k \right\|^2} \qquad (4.52)$$

$$[W]_{k+1} = [W]_k + c_k[P]_k \qquad (4.53)$$

$$[R]_{k+1} = [R]_k + c_k[F][P]_k \qquad (4.54)$$

$$b_k = \frac{1}{\left\| [F]^H [R]_{k+1} \right\|^2} \qquad (4.55)$$

$$[P]_{k+1} = [P]_k - b_k[F]^H [R]_{k+1} \qquad (4.56)$$

The norm is defined by

$$\left\| [F][P]_k \right\|^2 = [P]_k^H [F]^H [F][P]_k \qquad (4.57)$$

The equations above are applied in an iterative fashion until the desired error criterion for the residuals $\|[R]_k\|$ is satisfied, where $[R]_k = [F][W]_k - [Y]$. In our case, the error criterion is defined by

$$\frac{\|[R]_k\|}{\|[Y]\|} = \frac{\|[F][W]_k - [Y]\|}{\|[Y]\|} \le 10^{-6} \tag{4.58}$$

The iterative procedure is stopped when the criterion defined above is satisfied. A detailed description of this method along with a few sample computer programs is presented in Appendix B. The strength of the conjugate gradient method is that the final solution is still going to converge to an acceptable one even if the matrix [F] is exactly singular.

The computational bottleneck in the conjugate gradient method is in the evaluation of the matrix-vector products $[F][P]_k$ and $[F]^H[R]_{k+1}$. Typically, matrix vector products in real-time computations can slow down the computational process when they are transported to a digital signal-processing chip. However, in our examples, these computational bottlenecks can be streamlined through exploitation of the block Hankel structure in the matrix [F] as seen from (4.47). A block Hankel structure implies that the elements along any diagonal are equal. Under this special circumstance, that the matrix [F] has a block Hankel structure, the matrix-vector products defined by $[F][P]_k$ or $[F]^H[R]_{k+1}$ can be carried out efficiently through use of the fast Fourier transform (FFT) [7, 8]. This is accomplished as shown in the next paragraph.

Consider the following matrix-vector product, when the matrix has a block Hankel structure so that we have the expression

$$\begin{bmatrix} f_1 & f_2 & f_3 \\ f_2 & f_3 & f_4 \\ f_3 & f_4 & f_5 \end{bmatrix}_{(r \times r)} \times \begin{bmatrix} w_1 \\ w_2 \\ w_3 \end{bmatrix}_{(r \times 1)} \tag{4.59}$$

In (4.59), the value of $r = 3$. A matrix-vector product is usually accomplished in r^2 operations, where r is the dimension of the matrix. However, since the matrix has a Hankel structure, we can rewrite the matrix-vector product as a result of the convolution of the two sequences $\{f\} \; \odot \; \{w\} = \{f_1\, f_2\, f_3\, f_4\, f_5\} \; \odot \; \{w_3\, w_2\, w_1\, 0\, 0\}$, where $\odot$ denotes a convolution operation. We observe that the fourth, fifth, and sixth elements of this convolution provide the correct expression for the matrix-vector product. The convolution actually results in more terms than we require for the matrix-vector product. However, that is not relevant. In fact, convolutions can be carried out very efficiently using the FFT. Here, since we have finite sequences, the FFT will provide the correct solution even though it is periodizing both sequences in carrying out the convolution. We take the FFT of the two sequences $\{f\}$ and $\{w\}$. Next, we multiply the two transformed sequences term by term. Then we take an inverse FFT to obtain the results for the matrix-vector product. In this procedure, the total operation count for the operations $\text{FFT}^{-1}[\text{FFT}\{f\} \times \text{FFT}\{w\}]$

will be $3[2r - 1] \log[2r - 1]$. For a value of r greater than 30, this procedure becomes quite advantageous, as the operation count is on the order of $(r \log r)$ as opposed to r^2 for a conventional matrix-vector product. Also, in this new procedure, there is no need to store an array. Thus, the time spent in accessing the elements of the array in the disk is virtually nonexistent, as everything is now one-dimensional and can be stored in the main memory. This procedure is quite rapid and easy to implement in hardware [7]. Hence this direct data domain least squares method is not only efficient [as the computational count for K iterations is $K \times (L + 1) \log(L + 1)$ as opposed to $2(L + 1)^3$] but can also be implemented efficiently on a DSP chip for accurate solution of the adaptive problem.

Next we see how to increase the degrees of freedom L for free.

4.2.3   Backward Method

Next we reformulate the problem using the same data to obtain a second independent estimate for the solution. This is achieved by reversing the data sequence and then complex conjugating each term of that sequence.

It is well known in the parametric spectral estimation literature that a sampled sequence, which can be represented by a sum of exponentials with purely imaginary argument, can be used in either the forward or reverse direction, resulting in the same value for the exponent. From physical considerations we know that if we solve a polynomial equation with the weights W_i as the coefficients, its roots provide the direction of arrival for all the unwanted signals, including the interferers. Therefore, whether we look at the snapshot as a forward sequence as presented in Section 4.2.2 or by a reverse conjugate of the same sequence, the final results for W_i must be the same. Hence for these classes of problems, we can observe the data in either the forward or reverse direction. This is equivalent to creating a virtual array of the same size but located along a mirror symmetry line. Therefore, if we now conjugate the data and form the reverse sequence, we get an independent set of equations similar to (4.47) for the solution of the weights [W]. This is represented by

$$
\begin{bmatrix}
1 & Z & \cdots & Z^L \\
X_N^* - Z^{-1} X_{N-1}^* & X_{N-1}^* - Z^{-1} X_{N-2}^* & \cdots & X_L^* - Z^{-1} X_{L-1}^* \\
\vdots & \vdots & \cdots & \vdots \\
X_{L+1}^* - Z^{-1} X_L^* & X_L^* - Z^{-1} X_{L-1}^* & \cdots & X_1^* - Z^{-1} X_0^*
\end{bmatrix}_{(L+1)\times(L+1)}
$$

$$
\times
\begin{bmatrix}
W_0 \\
W_1 \\
\vdots \\
W_L
\end{bmatrix}_{(L+1)\times 1}
=
\begin{bmatrix}
C' \\
0 \\
\vdots \\
0
\end{bmatrix}_{(L+1)\times 1}
\tag{4.60}
$$

or equivalently, in matrix form as

$$[B]\,[W] = [Y] \tag{4.61}$$

The signal strength α can now be determined from

$$\alpha_v = \left[\frac{Z^{L+v}}{C'} \sum_{i=0}^{L} W_i X^*_{L-i+v} \right]^* \quad \text{for } v = 0, ..., L$$

Note that for both the forward and backward methods described in Sections 4.2.2 and 4.2.3, we have $L = N/2$. Hence the degrees of freedom are the same for both the forward and backward methods. However, we have two independent solutions for the same adaptive problem. In a real situation when the solution is unknown, two different estimates for the same solution may provide a level of confidence on the quality of the solution.

4.2.4   Forward-Backward Method

Finally, in this section we combine the forward and backward methods to double the given data and thereby increase the number of weights or the degrees of freedom significantly over that of either the forward or backward method alone. In the forward-backward model we double the amount of data not only by considering the data in the forward direction but also conjugating it and reversing the direction of increment of the independent variable. This type of processing can be done as long as the series to be approximated can be fit by exponential functions of purely imaginary argument. This is always true for the adaptive array case. So by considering the data set simultaneously in both the forward and backward directions, X_n and X^*_{-n}, we have essentially doubled the amount of data without any penalty, as these two data sets for our problem are linearly independent.

An additional benefit accrues in this case. For both the forward and backward methods, the maximum number of weights we can consider is given by $N/2$, where $N + 1$ is the number of antenna elements. Hence, even though all the antenna elements are being utilized in the processing, the number of degrees of freedom available for this approach is essentially $N/2$. For the forward-backward method, the number of degrees of freedom can be increased significantly without increasing the number of antenna elements. This is accomplished by considering the forward and backward versions of the array data. For this case, the number of degrees of freedom can reach $N/1.5 + 1$. This is approximately equal to 50% more weights or number of degrees of freedom than the two previous cases. The equation that needs to be solved for the weights is given by combining (4.47) and (4.60), with $C' = C$, into

$$
\begin{bmatrix}
1 & Z & \cdots & Z^Q \\
X_0 - Z^{-1}X_1 & X_1 - Z^{-1}X_2 & \cdots & X_V - Z^{-1}X_{V+1} \\
\vdots & \vdots & \vdots & \vdots \\
X_{N-Q-1} - Z^{-1}X_{N-Q} & X_{N-Q} - Z^{-1}X_{N-Q+1} & \cdots & X_{N-1} - Z^{-1}X_N \\
X_N^* - Z^{-1}X_{N-1}^* & X_{N-1}^* - Z^{-1}X_{N-2}^* & \cdots & X_{N-Q}^* - Z^{-1}X_{N-Q-1}^* \\
\vdots & \vdots & \vdots & \vdots \\
X_{Q+1}^* - Z^{-1}X_Q^* & X_Q^* - Z^{-1}X_{Q-1}^* & \cdots & X_1^* - Z^{-1}X_0^*
\end{bmatrix}_{(Q+1)\times(Q+1)}
$$

$$(4.62)$$

$$
\times \begin{bmatrix} W_0 \\ W_1 \\ \vdots \\ W_V \end{bmatrix}_{(Q+1)\times 1} = \begin{bmatrix} C \\ 0 \\ \vdots \\ 0 \end{bmatrix}_{(Q+1)\times 1}
$$

or equivalently,

$$[FB]\,[W] = [Y] \tag{4.63}$$

The value of Q in equation (4.62) is now much greater than the value of L in equations (4.47) and (4.60). Since in (4.62), the total amount of data is now doubled, the number of degrees of freedom Q in this case will be much greater than L. This increase in the degrees of freedom has been achieved by considering both the forward and reverse forms of the data sequence. In summary, in a conventional adaptive technique where there is a weight attached to each element and the processing is done in time, the number of degrees of freedom is $N + 1$, provided that the environment is stationary in time and the interferers are noncoherent. For coherent interferers, whether one is using the conventional methods or this technique, the maximum number of interferers Q that can be handled is much greater than L or $N/2$. So for the forward-backward method this proposed spatial processing based on a snapshot-by-snapshot analysis will provide the number of degrees of freedom $Q = N/1.5 + 1$. It is important to note that this is the maximum number of degrees of freedom for handling coherent interferers by any method! In addition, this is a least squares-based approach. The advantage of doing snapshot-by-snapshot processing is that the stationarity assumption about the data can be relaxed.

4.2.5   Numerical Simulations

A set of examples has been chosen to illustrate the direct data domain method where use of a conventional stochastic methodology may not yield satisfactory results.

For the first example, we consider the performance of these different methods in estimating the SOI in the presence of jammer, clutter, and thermal noise. We assume a signal of amplitude $1 + j1$ arriving from $\varphi = 95°$ impinging on a 13-element array. The array consists of elements which are omnidirectional point radiators and are considered to be spaced a half wavelength apart. So the half-power beamwidth of the array for broadside processing in this case is approximately 9°. We consider two jammers 47 dB stronger than the signal. One is arriving from 93° and hence is located in the main beam, whereas the other is coming from 47°. In addition, we have thermal noise at each antenna element, which is assumed to be uniformly distributed in amplitude, and the phase is chosen between 0 and 2π so that the signal-to-thermal noise ratio is +26 dB. We also consider clutter arriving at the array from $\varphi = 125°$ to 130° and from $\varphi = 10°$ to 12°. Here clutter is modeled by plane waves with complex amplitudes that are random. So the clutter patches contain many specular electromagnetic reflections, which are arriving in azimuth 0.1° apart, with a complex amplitude determined by two random number generators. The magnitude of the signals is determined by a uniformly distributed random number generator, with values distributed between 0 and 1. The phase is also determined by another uniformly distributed random number generator, with values between 0 and 2π. The signal-to-clutter ratio is −10.5 dB. We consider a single snapshot of the voltages across the array at a particular instance of time. So the data that is available for analysis are:

1.) The complex voltages that consist of the signal, two jammers, clutter, and thermal noise at each of the 13 elements
2.) The direction of arrival of the SOI, which is 95°.

The goal is to estimate the strength of the SOI from the data above.

When one applies the forward method to this data based on a single snapshot, the signal-to-interference ratio jumped from −47 dB at the input to +11 dB at the output. In this case the number of weights is seven. The conjugate gradient method took 14 iterations to arrive at the solution. Similar results are obtained when the backward method is applied to the same data. It is important to note that this is a second independent estimate of the signal. A third independent estimate can be obtained by applying the forward-backward method to the same data sets. In this case, we are doubling the data and so the number of weights is nine. Indeed, in this case the output signal-to-noise ratio is now +14 dB and is better than either the forward or backward method. It is important to note that we have obtained three independent solutions for the same data set. In a real problem, when the solution is unknown, application of three different methods to obtain three independent realizations of the same solution may enhance the level of confidence on the processed results if the variations among the three methods are small. This is an added advantage over the classical stochastic methodology in addition to the computational speed in obtaining the results. The advantage of performing digital beam forming is that an interference in the main lobe can be suppressed. The CPU time taken by the forward-backward method on an Inspiron 3000 laptop computer with a clock speed of 233 MHz was less than 0.3 s. The antenna beam patterns for the forward, backward and the forward-backward methods are given in Figure 4.2. The clutter is better mitigated by the forward-backward method. The interesting part of a digital beam former is that the SOI is always at 0dB of the pattern and that the maximum does not always correspond to the location of the SOI.

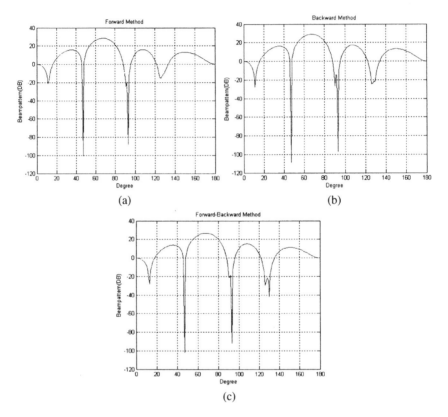

Figure 4.2. Antenna beam pattern for the three methods.

For the next example we consider the performance of the direct data domain least squares method due to clutter and thermal noise. For this example we assume a signal of unity amplitude arriving from $\varphi_s = 90°$ impinging on a 19-element array, where the omnidirectional isotropic antenna elements are assumed to be located a half wavelength apart in free space. So the antenna beamwidth in this case is approximately 5.5°. We consider clutter arriving at the array from $\varphi = 0.1°$ to 85° and from $\varphi = 95°$ to 179°. Here the clutter is modeled by plane waves whose complex amplitudes are random. Thus, the clutter patches contain many specular electromagnetic reflections which are arriving in azimuth 0.1° apart with a complex amplitude determined by two random number generators. The amplitude is determined by a uniformly distributed random number generator distributed between 0 and 1. The phase is also determined by another uniformly distributed random number generator uniformly distributed between 0 and 2π. In addition, we introduce thermal noise to each of the antenna elements. Its magnitude is assumed to be uniformly distributed between 0 and 1, and its phase is chosen between 0 to 2π. The signal-to-total thermal noise power is +23 dB at the array. Figure 4.3a provides the output signal-to-noise ratio yielded by the various methods as a function of the signal-to-clutter ratio in dB. Figure 4.3a illustrates that if the input signal-to-clutter ratio in the array is −10 dB and if we use the forward (FRW) or the backward (BAC) method, namely use either (4.47) or (4.60), to do the processing the processed output signal-to-noise ratio is about +5 dB. The eigenvalue method (EIG) described by equation (4.5) yields a similar value. However, if we utilize the forward-backward (FB) method to do the processing, namely by (4.62), the processed output signal-to-noise ratio is +8.2 dB. The difference between the FRW in the processed output signal-to-noise ratio or the EIG method and the FB method becomes much larger as the signal-to-clutter ratio increases. For the FRW, BAC and EIG methods, the number of degrees of freedom is 9, whereas for the forward-backward method the number of weights is 13. Figure 4.3b provides the antenna beam pattern for the forward-backward method.

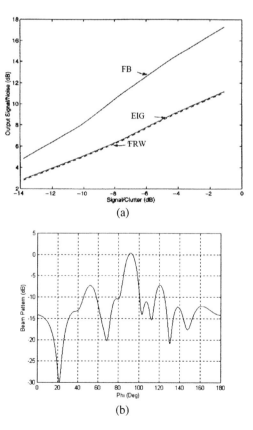

(a)

(b)

Figure 4.3. Plot of (a) Output signal-to-noise ratio as a function of input signal in clutter and thermal noise for the direct domain least squares approach and (b) The radiation Pattern.

As a third example, consider the same 19-element array being arranged so as to receive a signal of 0 dB with respect to $1+j1$ volts from $\varphi = 90°$. In addition, we have a 69-dB jammer coming from $\varphi = 140°$, a 50-dB jammer arriving from $\varphi = 95°$, a 60-dB jammer arriving from $\varphi = 85°$, and a 56.5-dB jammer arriving from $\varphi = 20°$. In addition, we have two clutter patches. The first clutter patch is located from 0.1° to 30° and is modeled by discrete scatterers located every 0.1° apart. The second clutter patch extends from 35° to 59°. The complex amplitudes for the point source clutter returns are generated by two uniformly distributed random number generators as outlined before. The total signal-to-clutter ratio is −13.2 dB. In addition, we have thermal noise at each antenna element. The total signal-to-thermal noise at the array is 23 dB. The beamwidth of the antenna is approximately 5.5°. If we utilize the forward-backward (FB) method to do the processing, with the only a priori information that signal is arriving from $\varphi = 90°$, the processed output signal-to-interference plus noise ratio is 26.6 dB. If we use either the forward (FRW) or backward (BAC) method, the processed output signal-to-noise ratio is 13.4 dB, whereas for the eigenvalue (EIG) method it is 13.41 dB. The number of weights is the same as before: 9 for the forward, backward, and the eigenvalue methods and 13 for the forward-backward method.

For the fourth example consider the same 19-element array receiving a signal of strength 0 dB with respect to $1 + j1$ volts, from $\varphi = 95°$. In addition, we have a 50.5-dB jammer coming from $\varphi = 50°$, a 60-dB jammer arriving from $\varphi = 80°$, a 56.5-dB jammer arriving from $\varphi = 70°$, and a 69-dB jammer arriving from $\varphi = 20°$. In addition, we have two clutter patches. The first clutter patch is located from $\varphi = 15°$ to 50° and is modeled by discrete scatterers separated in azimuth by 0.1° and whose complex amplitudes are considered random and generated by two uniformly distributed random number generators. In addition, we have a clutter patch from $\varphi = 100°$ to 130° modeled by discrete scatterers every 0.1° apart. The total signal-to-clutter ratio of the signals arriving at the array is −13.2 dB. In addition, we have thermal noise at each antenna element, and the total signal-to-thermal noise at the antenna array is 23 dB. If we utilize the FB method to do the processing, the output signal-to-noise ratio at the output is given by 7.4 dB. If the processing is done by the EIG method, the processed output is 1.01 dB, whereas for the FRW or BAC method it is 1.01 dB. Again, the number of weights in this case is the same as in the previous example.

For all the examples it is seen that the FB method given by (4.62) yields a much higher output signal-to-noise ratio than that given by any of the other methods. This is to be expected, as we have more degrees of freedom to null out the undesired signals.

So far we have assumed that the direction of arrival of the signal is known exactly. However, due to atmospheric diffraction or to some errors committed in aligning the antenna elements in the array, it is quite possible that the a priori information about the direction of arrival may be only an approximation. So the direction of arrival of the signal is not exactly φ_s, but it comes from $\varphi_s \pm \Delta\varphi$, where $\Delta\varphi$ is not known a priori. The slight deviation in the direction of arrival can also be due to atmospheric refraction. The processed result may not be correct, as all the methods will not find any signal exactly at φ_s. In fact, there will be signal cancellation. To alleviate such problems of signal cancellation when there is uncertainty in knowing *a priori* the direction of arrival φ_s of the signal, we utilize the main beam constraints as described in the next section.

4.3   MAIN BEAM CONSTRAINTS FOR PREVENTION OF SIGNAL CANCELLATION

So far, we have addressed the problem of eliminating unwanted jammers to extract the signal from an arbitrary look direction. However, in practice the expected signal (target returns) can occur over a finite angle extent. For example, in the radar case the angle extent is established by the main beam of the transmitted wave (usually, between the 3-dB points of the transmit field pattern). Target returns within the angle extent must be processed coherently for detection, and estimates made of target Doppler and angle. Adaptive processing, which impacts these processes, will lead to unacceptable performance. Correction for this effect is accomplished in the least squares procedures by establishing look-direction constraints at multiple angles within the transmitter main beam extent. The multiple constraints are established by using a uniformly weighted array pattern for the same size array as that of the adaptive array under consideration. Multiple points are chosen on the nonadapted array pattern and a row is implemented in the matrix equations (4.47), (4.60), and (4.62) for each of the desired angles, and the corresponding uniform complex antenna gains are placed in the [Y] vector of (4.48), (4.61), and (4.63). Hence, for this problem the size of the matrix [U], for example, is established by the following parameters:

$$Q = \text{number of look-direction constraints}$$
$$L + 1 = \text{number of weights to be calculated}$$
$$L - Q + 1 = \text{number of jammers that can be nulled}$$

The first canceling equation uses data from the $L + 1$ elements, and each successive canceling equation is shifted by one element, and therefore $N - L$ equations are required to use the data from $N + 1$ elements effectively. Thus there are Q-constraint equations and $N - L$ canceling equations for the case of the forward method described by (4.47). The number of equations must equal the number of weights; therefore,

$$L = Q + N - L \tag{4.64}$$

This leads to the relationship between the number of weights, number of constraints, and number of elements:

$$N = 2L - Q \tag{4.65}$$

Similar constraints can be applied to the backward method or the forward-backward method.

4.3.1 Examples

To illustrate the effectiveness of the least squares approach to the adaptive array problem, we consider an array of $N + 1 = 21$ antenna elements and we employ the forward method. For all the examples, the value of N will be fixed. The performance across the main beam will be compared for the cases of one, three, and five look-direction constraints. This leads to the following relationships:

- $N + 1 = 21$, $Q = 1$, and so $L + 1 = 11$ and ten jammers can be canceled.
- $N + 1 = 21$, $Q = 3$, and so $L + 1 = 12$ and nine jammers can be canceled.
- $N + 1 = 21$, $Q = 5$, and so $L + 1 = 13$ and eight jammers can be nulled.

As an example, consider a target at 94° with the main beam look-direction constraint placed at 90°. It is seen from the main beam array pattern depicted in Figure 4.4 that the target at 94° has been nulled out. In Figures 4.5, 4.6, and 4.7 the complex array gain is shown for one, three, and five main beam constraints using the same sets of random noise generated at the 21 elements. For the three cases, the array gain along the target direction (denoted by × in the figures) is reduced more in the one-constraint case (Figure 4.5) than for the three-constraint case (Figure 4.6) or the five-constraint case (Figure 4.7). Also, the 10 vectors for the different simulations of noise are less randomly distributed for the five-constraint case, and hence some coherent integration gain is possible. For the three-constraint case, the constraints are placed at 85°, 90°, and 95°. For the five-constraint case, the main beam constraints are placed at 85°, 87.5°, 90°, 92.5°, and 95°.

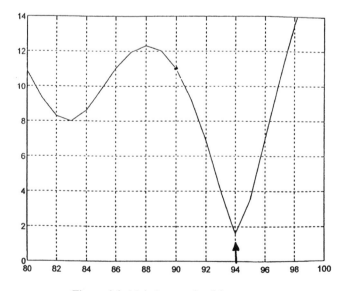

Figure 4.4. Main beam gain of the array.

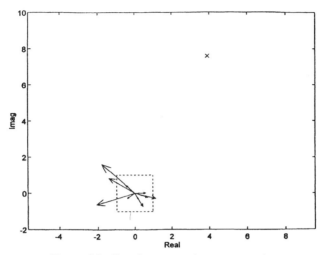

Figure 4.5. Complex array gain: one-constraint case.

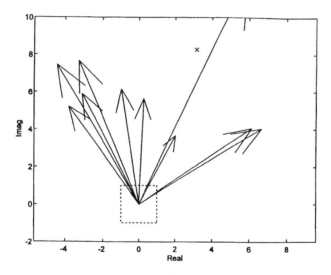

Figure 4.6. Complex array gain: three-constraint case.

Thus either the first three or first five rows of the matrix [F] of equation (4.47) or matrix [B] of equation (4.60) are of the same form as the first row of the matrices defined above but with the appropriate steering vector. The excitation function [Y] would have one, three, or five nonzero elements, respectively, depending on the number of constraints used for the main beam. For the five-constraint case, [Y] would be of the form

$$[Y]^T = [\, 13, 7.72 + j\,8.32, 7.72 - j8.32, -8.16 + j7.149, -8.16 - j7.149, 0, 0, 0, 0, 0, 0, 0, 0]$$

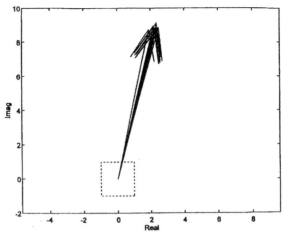

Figure 4.7.   Complex array gain: five-constraint case.

So for the five-constraint case there is practically no loss in the array gain, and the estimated vectors representing the SOI from the 10 different runs are very nearly aligned. The five-constraint approach would permit effective radar processing across the main beam extent with little loss in performance. Figure 4.7 shows the main beam gain in the presence of three jammers, with five constraints in the main beam.

As the signal strength is increased, the distortion of the main beam increases. The results above have been generated utilizing a 20-dB signal-to-noise ratio per channel per pulse. This would be a strong radar return under most circumstances. Simulation results indicate that the five-constraint approach is still effective at a 40-dB signal-to-noise-ratio (SNR), but breaks down at an SNR of 60 dB.

In summary, the main beam constraint allows look-direction constraints to be established over a finite beamwidth while maintaining the ability to null jammers adaptively in the sidelobe region. Although the main beam gain can become degraded if the signal becomes very strong, this does not appear to be a serious limitation for practical radar processing cases.

4.4   MINIMUM NORM PROPERTY OF THE OPTIMUM WEIGHTS

The optimum weights are obtained as a minimum norm solution of (4.47) when the assumed direction of arrival coincides exactly with the actual direction of arrival. One of the open problems is exactly how to estimate the direction of arrival of the signal when there is uncertainty associated with the assumed direction of arrival. It has been our experience that the norm of the weights can be used to estimate the direction of arrival of the signal accurately if that is necessary. Hence this method can be used as a multiple-step approach that can arrive at a good approximation to the optimum weights (i.e., the weights that would be obtained if the arrival angle of the target return were known exactly). This approach will evaluate the weights at multiple values of the angle assumed to have the correct value in the canceling equations but will make decisions only on the values of the weights and will accomplish the detection process only once.

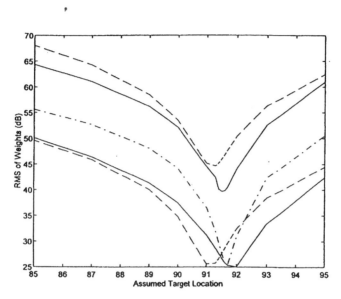

Figure 4.8. RMS of weights as a function of assumed target location: strong target at 91.5° in the presence of jammers and noise.

As a step to further developing this approach, consider the data in Figures 4.8 through 4.10. Again the simulation is implemented for a 21-element array and the number of weights are 13 and we are using the five-constraint algorithm to shape the adapted receive beam pattern. In each case, the simulation is repeated for five samples of noise at each element. With the five constraints in place (at 85°, 87.5°, 90°, 92.5°, and 95°), the assumed location of the target return is varied across the main beam of the array, the adaptive process is implemented and the sum of the absolute values of the weights is calculated. The actual location of the target is 91.5° in all cases. For all these cases (Figures 4.8 through 4.10), in addition to target return and noise, three jammers are present. Figure 4.8 shows the results for a very strong target return (20-dB SNR at each element). The minimum of the sum of the weights is obtained very close to the true target direction in all five samples of the receiver noise. The five different curves in the figures represent five different simulations of the problem. The jammers are effectively nulled in all cases, and the only effect of receiver noise is that the target return angle is estimated to be between 91.4° and 91.6° instead of the true 91.5°. The location of the minimum of the weights is easy to identify with the ratio of the largest sum at any location across the main beam to the minimum of about 50. The multiple-constraint approach is thus successful in identifying the true target direction, and the detection process can be implemented optimally at this angle. Figure 4.9 repeats the process for a moderate target return (10-dB *S/N* at each element). The effects of receiver noise are more

significant in this case. The estimates of the target location have greater spread (the
estimate varies between 91° and 92°). The location of the minimum is still easy to
identify, but the ratio between the maximum and minimum is now only about 20.
Figure 4.10 shows the data for a weak target return 0-dB *S/N*). The estimated target
locations have even greater spread (between 90° and 93°). The plots of the sum of
weights now have a relatively broad null, and the ratio between the maximum and
minimum is now about 5.

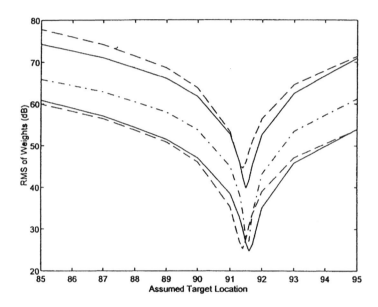

Figure 4.9. RMS of weights as a function of assumed target location: moderate target at
91.5° in the presence of jammers and noise.

The existence of a minimum in the sum of the weights can be used to estimate
the target return angle. It could also be used to perform the detection process as
well. If there is a large ratio between the minimum and maximum values of the
sum of the weights across the main beam, that is an indication that a target is
present. The strongest linear progression of the random noise samples sets a lower
limit on that detection process. That component of the random noise samples that
has a linear progression across the array appropriate to a far-field source in a main
beam direction will detect a weak target.
Therefore, the sum of the adaptive weights varies as the value of the assumed target
direction is varied across the main beam. When a strong target is present, the ratio
between the largest sum of the weights and that at the target direction is large.
When there is no target present, the ratio between the largest and smallest sums is

small. This could lead to a more accurate estimation of the direction of arrival of the signal or on a detection process, when the direction of arrival information is not available *a priori*.

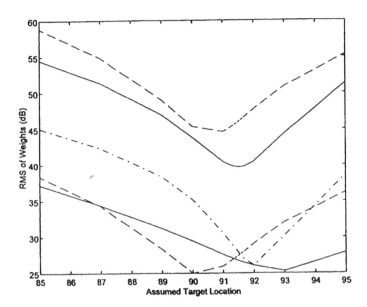

Figure 4.10. RMS of weights as a function of assumed target location: weak target at 91.5° in the presence of jammers and noise.

4.5   CONCLUSION

A direct data domain approach based on the spatial samples of a single snapshot of data is presented. In this approach the adaptive analysis is done on a snapshot-by-snapshot basis, and therefore nonstationary environments can be handled quite easily, including coherent multipaths. Associated with adaptive processing is the same *a priori* knowledge about the nature of the signal, which in this case is the direction of arrival. The assumption that the target signal is coming from an exactly known direction will probably never be met in any real array. In communication systems the location of the transmitter may be known only approximately, or the propagation of the signal through the atmosphere may distort the wavefront such that it appears to be coming from a slightly different direction. For example, diffraction could cause enough error for the determination of an elevation angle to be important for some systems. Or the adaptive receive array may be surveyed into a location with small errors, and thus the angle to the transmitter from the broadside of the array will be in error. Other applications of adaptive arrays will also have at least small errors in the direction of arrival of the desired signal. In this chapter two methodologies are presented to treat this signal cancellation problem. One is through the main beam constraint and the other is through the norm of the weight vectors. It has been illustrated that the norm of the adapted weights is a monotonically increasing function of the separation between the assumed direction of the target and the actual direction of the target. It is thus possible to find the true angle of arrival of the desired signal by minimizing the sum of the absolute values of the weights. Once this true angle is known, an optimum set of weight vectors could be formed using the existing algorithm.

The advantage of this direct data domain least squares approach based on spatial processing of the array data may prove beneficial over conventional adaptive techniques utilizing time averaging of the data. This will be quite relevant in a nonstationary environment.

REFERENCES

[1]   S. Haykin, *Adaptive Filter Theory*, Prentice Hall, Upper Saddle River, NJ, 1996.

[2]   J. Capon, "High Resolution Frequency–Wavenumber Spectrum Analysis," *Proceedings of the IEEE*, Vol. 57, pp. 1408–1418, 1969.

[3]   S. Park, T. K. Sarkar, and Y. Hua, "A Singular Value Decomposition Based Method for Solving a Deterministic Adaptive Problem," *Digital Signal Processing: A Review Journal*, Vol. 9, pp. 57–63, 1999.

[4]   T. K. Sarkar and N. Sangruji, "An Adaptive Nulling System for a Narrowband Signal with a Look Direction Constraint Utilizing the Conjugate Gradient Method," *IEEE Transactions on Antennas and Propagation*, Vol. 37, No. 7, pp. 940–944, 1989.

[5]   T. K. Sarkar, S. Park, J. Koh, and R. A. Schneible, "A Deterministic Least Squares Approach to Adaptive Antennas," *Digital Signal Processing: A Review Journal*, Vol. 6, No. 3, pp. 185–194, 1996.

[6]   S. Park and T. K. Sarkar, "Prevention of Signal Cancellation in Adaptive Nulling Problem," *Digital Signal Processing: A Review Journal*, Vol. 8, No. 2, pp. 95–102, Apr. 1995.

[7]   R. Brown and T. K. Sarkar, "Real Time Deconvolution Utilizing the Fast Fourier Transform and the Conjugate Gradient Method," in *5th Acoustic Speech and Signal Processing Workshop on Spectral Estimation and Modeling*, Rochester, NY, 1990.

[8]   T. K. Sarkar, E. Arvas, and S. M. Rao, "Application of FFT and the Conjugate Gradient Method for the Solution of Electromagnetic Radiation from Electrically Large and Small Conduction Bodies," *IEEE Transactions on Antennas and Propagation*, Vol. 34, No. 5, pp. 635–640, 1986.

[9]   M. G. Bellanger, *Adaptive Digital Filters and Signal Analysis*, Marcel Dekker, New York, 1987.

5

ELIMINATION OF THE EFFECTS OF MUTUAL COUPLING ON ADAPTIVE ANTENNAS

SUMMARY

In this chapter we consider the antenna elements in the phased array to have finite dimensions (i.e., they are not omnidirectional radiators). Hence, the elements sample and reradiate the incident fields, resulting in mutual coupling between the antenna elements. Objects in the near field (near-field scatterers) have the same effect. Mutual coupling not only destroys the linear wavefront assumption for the signal of interest but also for all the interferers and clutter impinging on the array. Thus, the voltages induced at the antenna elements need to be corrected to compensate for the mutual coupling. In this chapter we describe a relatively straightforward method that can be applied to an array of arbitrary elements. The methodology is illustrated for an array of linear dipoles. Numerical results are presented to illustrate the proposed methods, including the direction of arrival estimation in a CDMA environment. However, we still process the signal on a snapshot-by-snapshot basis.

5.1 ACCOUNTING FOR MUTUAL COUPLING AMONG AN ARRAY OF DIPOLES

The assumption of a plane wavefront impinging on an array implies that the fundamental starting equation in an adaptive algorithm relating the received voltages due to the signals incident on the individual antenna elements can be written as a sum of complex exponentials. The presence of mutual coupling among the elements of an antenna array destroys this plane wave assumption. Therefore, unless the effects of mutual coupling are accounted for, adaptive algorithms will be incapable of extracting the solution for the problem. The electromagnetic analysis of the antennas should be coupled to the adaptive algorithm to account for the mutual coupling and/or near-field effects.

To illustrate the elimination of mutual coupling among the antenna elements
and to illustrate how these two diverse methodologies can be coupled, we
consider an array of parallel thin-wire dipole antennas. The dipoles are assumed
to be z-directed of length L and radius a and are placed along the x-axis,
separated by a distance Δx. The port of each antenna element is located at the
center and is loaded with an impedance of Z_L ohms. The array lies in the x–z
plane, as shown in Figure 5.1. The voltages for a single snapshot are measured
across these loads. Each antenna element is now no longer an omnidirectional
radiator but has an electrical length of L/λ and an electrical radius of a/λ. Here λ
is the wavelength corresponding to the operating angular frequency ω.

Consider an incident electric field E^{inc} impinging on an array with N
elements. This results in an induced current flowing along the axis of the wires.
Since the array is composed of thin wires, the following simplifying assumptions
are valid [1, 2]:

1.) The current flows only along the direction of the wire axes (here the z-
    direction) and there is no circumferential variation of the current. Let this
    current be $I(z)$.
2.) The current and charge densities on the wire are approximated by
    filaments of current and the charge distribution on the wire axes (that lie in
    the $y = 0$ plane).
3.) Surface boundary conditions can be applied to the relevant axial
    component on the wire axes.

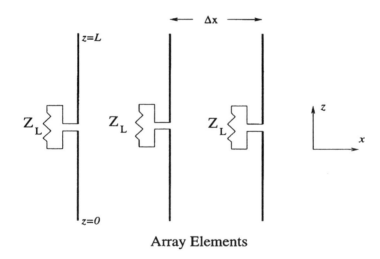

Figure 5.1. Model of the receiving antenna as a linear array.

Based on these assumptions, the integral equation that relates the incident field to the current on the wires and describes the behavior of the array is

$$
E^{\text{inc}}(z) = \frac{1}{j\omega 4\pi\varepsilon_0} \left[(k^2 + \nabla \cdot \nabla) \int_\ell I(z')G(z,z')\,dz' \right]
$$

$$
= -\mu_0 \int_{\text{axes}} I(z') \frac{G(z,z')}{4\pi}\,dz' + \frac{1}{\varepsilon_0} \frac{\partial}{\partial z} \int_{\text{axes}} \frac{\partial I(z')}{\partial z'} \frac{G(z,z')}{4\pi}\,dz'
$$

(5.1)

where axes implies that integration is to be carried out along the axis of the thin wires, and

$$
G(z,z') = \frac{e^{-jkR}}{R}
$$

(5.2)

$$
R = \sqrt{(z-z')^2 + a^2}
$$

(5.3)

with k, the wavenumber in free space, is given by $2\pi/\lambda$. ∇ is the divergence/gradient operator. In addition, μ_0 and ε_0 are the permeability and permittivity of free space, respectively. If we expand the unknown currents in terms of known basis functions with unknown constants we know from [1, 2] that the integrodifferential equation above will be transformed to a matrix equation:

$$
[V] = [Z][I]
$$

(5.4)

$$
I(z) = \sum_{n=1}^{N} \sum_{p=1}^{P} I_{p,n}\, f_{p,n}(z)
$$

(5.5)

where $f_{p,n}$ represents the pth basis function on the nth antenna element. The matrix $[V]$ is related only to the incident field, E^{inc}. The matrix $[Z]$ is related to the array manifold only and $[I]$ corresponds to the unknown current amplitudes on the structure. Specifically, if one assumes a piecewise sinusoidal current distribution on the structure, we have

$$
f_{p,n}(z) = \begin{cases} \dfrac{\sin[k(z - z_{p-1,n})]}{\sin(k\,\Delta z)} & \text{for} \quad z_{p-1,n} < z < z_{p,n} \\[2em] \dfrac{\sin[k(z_{p+1,n} - z)]}{\sin(k\,\Delta z)} & \text{for} \quad z_{p,n} < z < z_{p+1,n} \\[2em] 0 & \text{elsewhere} \end{cases} \tag{5.6}
$$

$$
\Delta z = L/(P + 1) \tag{5.7}
$$

$$
z_{p,n} = z_{0,n} + p\,\Delta z \tag{5.8}
$$

$$
V_i = \int_{z_{q-1,m}}^{z_{q+1,m}} f_{q,m}(z)\, E_z^{\mathrm{inc}}(z)\,dz \tag{5.9}
$$

$$
Z_{i,\ell} = \int_{z_{q-1,m}}^{z_{q+1,m}} f_{q,m}(z) \left\{ \begin{aligned} & j\omega\mu_0 \int_{z_{p-1,n}}^{z_{p+1,n}} f_{p,n}(z') \frac{e^{-jkR}}{4\pi R}\, dz' \\[1em] & - \frac{1}{j\omega\varepsilon_0} \frac{\partial}{\partial z} \int_{z_{p-1,n}}^{z_{p+1,n}} \frac{df_{p,n}(z')}{dz'} \frac{e^{-jkR}}{4\pi R}\, dz' \end{aligned} \right\} dz' \tag{5.10}
$$

where

$$
i = [(m-1)p + q] \tag{5.11}
$$

$$
\ell = [(n-1)p + q] \tag{5.12}
$$

and $z_{0,n}$ is the z-coordinate of the bottom of the nth antenna. Note that several types of basis functions are possible. However, we choose the piecewise sinusoidal functions because it results in closed-form expressions to equation (5.10).

The left-hand side of equation (5.4) is not influenced by the array manifold and is solely dependent on the incident fields. This is what signal-processing algorithms, described in Chapters 3 and 4, require. However, in actuality, what we have are the measured voltages V_L across the loads Z_L of the wire antennas, as shown in Figure 5.2.

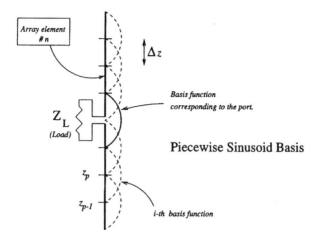

Figure 5.2. Basis functions used in the method of moments.

In (5.4) the matrix [Z] is $NP \times NP$ and [V] is $NP \times 1$. Here the dimension of the matrix equals $NP = P \times N$. This procedure of converting an integrodifferential equation to a matrix equation is popularly known as the *method of moments* (MM). This numerical method is well understood and has been used in numerical electromagnetics for over 30 years. If an incoming electric field is linearly polarized, the z-component of the field is of the functional form [2]

$$E_z = E_0 e^{-j k \cdot r} \tag{5.13}$$

where $k = -k\,[\hat{x} \cos \varphi \sin \theta + \hat{y} \sin \varphi \sin \theta + \hat{z} \cos \theta]$ is the wave vector associated with the direction of arrival of the incident field (θ, φ). Here θ is the elevation angle defined from the z-axis and φ is the azimuth angle defined in the x–y plane with the starting angle from the x-axis. Using (5.13) and (5.10) in (5.5) and (5.6) results in

$$V_i = \frac{E_0\, e^{j k\, x_m\, \cos \varphi \sin \theta}}{k \sin(k\,\Delta z)\sin^2 \theta}\; 2 e^{j k\, z_{q.m} \cos \theta}\; [\cos(k\,\Delta z \cos \theta) - \cos(k\,\Delta z)] \tag{5.14}$$

where x_m is the x-coordinate of the axis of the mth antenna. The term V_i tends to zero as θ tends to the end-fire case, which corresponds to $\theta = 0$ and $\theta = \pi$. The entries of the matrix [Z] are given by

$$Z_{i,\ell} = \frac{j\,30}{\sin^2(k\Delta z)} \int_{z_{q-1,m}}^{z_{q,m}} \sin[k(z - z_{q-1,m})]$$

$$\times \left\{ \frac{e^{-jkR_1}}{R_1} - 2\cos(k\,\Delta z)\frac{e^{-jkR_2}}{R_2} + \frac{e^{-jkR_3}}{R_3} \right\} dz$$

$$+ \frac{j\,30}{\sin^2(k\Delta z)} \int_{z_{q,m}}^{z_{q+1,m}} \sin[k(z_{q+1,m} - z)]$$

$$\times \left\{ \frac{e^{-jkR_1}}{R_1} - 2\cos(k\,\Delta z)\frac{e^{-jkR_2}}{R_2} + \frac{e^{-jkR_3}}{R_3} \right\} dz$$

(5.15)

with

$$R_1 = \sqrt{(x_m - x_n)^2 + (z - z_{p-1,n})^2}$$ (5.16)

$$R_2 = \sqrt{(x_m - x_n)^2 + (z - z_{p,n})^2}$$ (5.17)

$$R_3 = \sqrt{(x_m - x_n)^2 + (z - z_{p+1,n})^2}$$ (5.18)

If $m = n$, the term $(x_m - x_n)$ is set equal to the radius a. A closed-form expression for $Z_{i,\ell}$ is available in [2]. The MM admittance matrix is the inverse of the impedance matrix (i.e., $[Y] = [Z]^{-1}$). To take the load impedance into account, we simply add the load impedance, Z_L, to the entries of the diagonal elements of $[Z]$ that correspond to the port of excitation. Define a new impedance matrix $[T]$ such that

$$T(i, i) = Z(i, i) + Z_L \qquad \text{if } i \text{ corresponds to a port} \qquad (5.19)$$

$$T(i, \ell) = Z(i, \ell) \qquad \text{otherwise} \qquad (5.20)$$

The coefficients of the MM current expansion for the loaded antennas are given by $[I] = [S][V]$, where $[I]$ and $[V]$ are the MM current and voltage vectors of dimension $NP \times 1$, respectively. The new admittance matrix $[S] = [T]^{-1}$ is of size $NP \times NP$. Due the choice of an odd number of basis functions, only a single basis function is nonzero at the port. The measured voltages at the ports are therefore given by

$$[V_{\text{meas}}] = [Z_L][I_{\text{port}}] = [Z_L][S_{\text{port}}][V] = [C][V] \qquad (5.21)$$

with

$$[Z_L] \quad = \quad \text{diag} \, [Z_L \, , \, Z_L \, , \, ..., \, Z_L] \, _{N \times N} \qquad (5.22)$$

where $[V_{meas}]$ is a vector of length N, $[I_{port}]$ is the vector of currents at the N ports, $[Z_L]$ is the diagonal load matrix, and $[S_{port}]$ is the $N \times NP$ rectangular matrix corresponding to the N_w row of $[S]$ that corresponds to the N_w port. The $N \times NP$ matrix $[C] = [Z_L] \, [S_{port}]$ has dimensionless entries.

5.1.1 Compensation Using Open-Circuit Voltages

Research in the past into compensating for the mutual coupling has been based mainly on the idea of using open-circuit voltages, first proposed by Gupta and Ksienski [3]. They argue that due to the lack of a terminal current, the re-radiated fields are reduced and the open-circuit voltages are free of mutual coupling. The principal idea of [3] is to derive the open-circuit voltages from the measured voltages for further signal processing. However, the theory is valid only for half-wavelength dipoles with half-wavelength spacing. For the more general case, one can use the MM analysis in conjunction with the Thévenin and Norton equivalent circuits to obtain the open-circuit voltages. Using the notation developed above, we define a new $N \times N$ matrix $[Y']$ whose entries are those rows and columns of the MM admittance matrix $[Y]$ that correspond to the feed ports, that is,

$$
\begin{aligned}
Y'_{pq} &= Y_{i,\ell} \quad 1 \le p,q \le N \\
i &= (p-1)P + (P+1)/2 \\
\ell &= (q-1)P + (P+1)/2
\end{aligned}
\qquad (5.23)
$$

Note that this matrix $[Y']$ is defined from the admittance matrix that does not include the load impedances. The open-circuit voltages are then related to the short-circuit currents as

$$[V_{oc}] \quad = \quad [Y']^{-1} [I_{sc}] \qquad (5.24)$$

The measured voltages are also related to the short-circuit currents through

$$[V_{meas}] \quad = \quad \left[Y' + [Z_L]^{-1} \right]^{-1} [I_{sc}] \qquad (5.25)$$

Eliminating the short-circuit currents from (5.24) and (5.25) yields the open-circuit voltages as

$$[V_{oc}] \quad = [Y']^{-1} \left[[Y'] + [Z_L]^{-1} \right] [V_{meas}] \qquad (5.26)$$

In the following sections the open-circuit voltages we refer to are obtained from the measured voltages using (5.26).

We have defined four admittance matrices, the MM admittance matrix $[Y] = [Z]^{-1}$, the admittance matrix including the load impedances, $[S] = [T]^{-1}$, $[Y_{port}]$, the $N \times NP$ matrix corresponding to the ports of $[S]$, and finally, $[Y']$, the $N \times N$ matrix whose rows and columns correspond to the N rows and columns of $[Y]$ that correspond to the ports.

5.1.2   Compensation Using the Minimum Norm Formulation

As shown by Adve and Sarkar [4], use of the open-circuit voltages only reduces the effects of mutual coupling. The technique presented in [4] is more effective in suppressing mutual coupling effects. In [4], a part of the MM voltage vector is reconstructed under the assumption of a linear dipole array. The motivation comes from the fact that from (5.14), the MM voltages are directly related to the incident fields and so are free of mutual coupling. A big drawback with the approach of [4] is that it is only applicable for linear arrays of linearly polarized dipole elements. In this section we introduce the use of a *minimum norm technique* for general antenna arrays with arbitrary-shaped elements. The MM technique is used to accurately model the interactions between the antenna elements in an array. In this minimum norm approach, the MM admittance matrix is used to estimate the incident fields, with minimum energy, that would generate the received voltages. Unlike in [5], this technique does not require the solution to the entire MM problem. This new approach is theoretically valid for any kind of array, in any configuration.

For the general case, from (5.21), the equation relating the measured and MM voltages form an underdetermined system of equations and the matrix [V] cannot be reconstructed exactly. However, one can find the *minimum norm solution* to this equation. This solution provides the vector with the minimum norm (minimum energy) that would result in the received voltages. The resulting vector is an estimate of the MM voltage vector. Using (5.21), the minimum norm solution to the MM voltage vector is given by

$$[\tilde{V}] = [C]^H \left\{ [C][C]^H \right\}^{-1} [V]_{meas} \tag{5.27}$$

where the superscript H denotes the conjugate transpose of a matrix. Entries in [V] corresponding to the ports may be used for further signal processing.

Physically, the compensation procedure may be interpreted as finding the signal with minimum energy that results in the measured voltages. Since using the MM analysis, the matrix [C] may be obtained *a priori*, the computational time required to solve (5.27) is no greater than finding the open-circuit voltages or the technique of [4]. Therefore, through (5.27) we have transformed the measured voltages that have been affected by the mutual coupling and/or near-field scatterers to a set of voltages where these effects have been reduced through the use of Maxwell's equations.

5.2   EFFECT OF MUTUAL COUPLING

Two examples demonstrate the effect of mutual coupling between the elements of the array and illustrate the effectiveness of the compensation technique described in Section 5.1. In each example, an array receives a signal corrupted by three jammers. To focus only on the effects of mutual coupling, these examples neglect thermal noise.

For each example two scenarios are compared. In the first scenario, we deal with the ideal case, where we assume that no mutual coupling exists between the elements. Hence, the antenna elements are considered to be point sources. The voltages at the ports of the antenna elements constituting the array are given by a sum of complex exponentials, as defined in (5.14). The antenna array is shown in Figure 5.3. These voltages measured at the respective ports are then passed to the signal recovery subroutine to find the weights using the direct data domain least squares method (D^3LS) presented in Chapter 4.

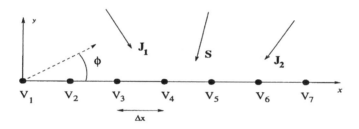

Figure 5.3. Model of an ideal array consisting of omnidirectional isotropic point radiators.

In the second scenario the mutual coupling between the antennas is taken into account as they are not isotropic elements. The antenna array is analyzed using the method of moments (MM). The intensities of the signal and the interferences along with their directions of arrival are used in conjunction with (5.14) to calculate the MM voltage vector. Equation (5.21) is used to find the voltages that are measured across the load at the individual feed ports. These measured voltages then serve as input to the signal recovery subroutine. The signal intensity is then recovered using the D^3LS method. No attempt is made to compensate for mutual coupling.

The details of the array chosen for the second scenario are presented in Table 5.1. The receiving algorithm tries to maintain the gain of the array along the azimuth direction of $\varphi_s = 45°$ while automatically placing nulls along the directions of the interferences. All signals and jammers arrive from the elevation $\theta = 90°$. The signal of interest and the intensities of the jammer along with their directions of arrival are given in Table 5.2.

TABLE 5.1

Parameters defining the elements of the array

Number of elements in array	7
Length of z-directed wires	$\lambda/2$
Radius of wires	$\lambda/200$
Spacing between wires	$\lambda/2$
Loading at the center	$50\,\Omega$

TABLE 5.2

Complex amplitudes of the signals and jammers and their directions of arrival

	Magnitude	Phase	DOA
Signal	1.0 V/m	0.0	45°
Jammer 1	1.0 V/m	0.0	75°
Jammer 2	1.5 V/m	0.0	60°
Jammer 3	2.0 V/m	0.0	30°

In all the simulations the intensities of the jammer along with their directions of arrival and the signal intensity are used only to find the voltages that will be the input to the signal processing algorithm. The adaptive algorithm itself uses information about only the direction of arrival of the signal (i.e., the look direction, φ_s, is considered to be known).

5.2.1   Constant Jammers

For constant jammers, the magnitude of the incident signal is varied from 0 V/m to 10.0 V/m in steps of 0.05 V/m while maintaining the jammer intensities constant, as given in Table 5.2. If the jammers have been nulled correctly and the signal recovered properly, it is expected that the recovered signal will have a linear relationship with respect to the intensity of the incident signal.

Figure 5.4 plots the results of using the D^3LS algorithm presented in Chapter 4 to recover the magnitude and phase of the signal in the presence of jammers and in the absence of mutual coupling. So in this situation, all the antenna elements are considered to be isotropic point radiators, as shown in Figure 5.3. Now if we apply an adaptive algorithm to the voltages measured at these ideal elements, we get a perfect reconstruction of the original signal in both magnitude and phase, as shown in Figure 5.4.

Now we replace the idealized array by the realistic array of Figure 5.1. In this case, for each value of the incident signal intensity, mutual coupling is taken into account and the measured voltages are obtained using the MM. So the effects of mutual coupling have affected the measured voltages, and these values are now used in the adaptive algorithms.

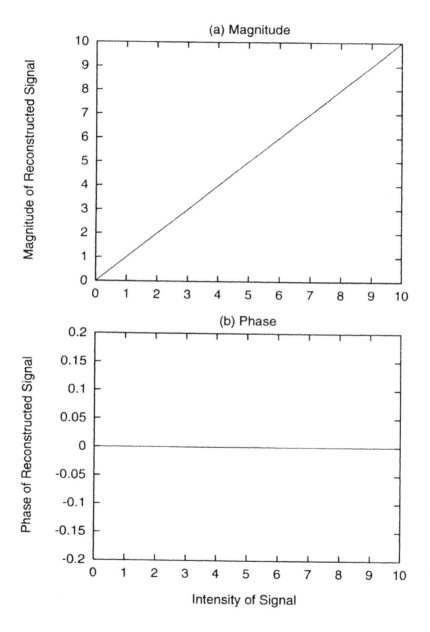

Figure 5.4. Signal recovery using idealized omnidirectional point radiators in the absence of mutual coupling: (a) magnitude; (b) phase.

The results of the reconstruction using the voltages affected by mutual coupling are presented in Figure 5.5a for the magnitude and Figure 5.5b for the phase, using the D³LS procedure of Chapter 4. As can be seen from the figure, in the presence of mutual coupling the reconstruction is completely inaccurate. As the incident signal increases in intensity, the reconstructed signal displays a nonlinear behavior.

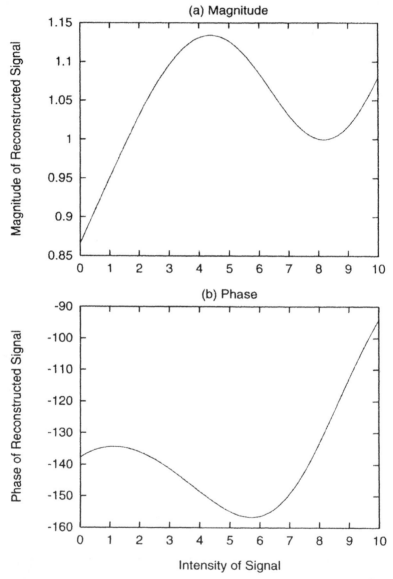

Figure 5.5. Signal recovery in a dipole array in the presence of mutual coupling: (a) magnitude; (b) phase.

In this example, jammers are included in the simulation to illustrate the effects of mutual coupling on interference suppression. Without any interference, the plot of the reconstructed signal versus intensity of incident signal again would have a linear relationship as in Figure 5.4. This is because in the absence of mutual coupling the beam pattern would change little with changing signal strength. Most of the signal is eliminated by the subtraction operation as used in the direct data domain method.

5.2.2   Constant Signal

In the second example, the signal is kept constant at 1.0 V/m, as given in Table 5.2. The intensity of the first jammer arriving from $\varphi = 75°$ is varied from 1.0 V/m (0 dB with respect to the signal) to 1000.0 V/m (60 dB) in steps of 5 V/m. The same antenna array described in Section 5.2.1 is considered. If the jammers are properly nulled, we expect the reconstructed signal to have no residual jammer component. Therefore, as the jammer strength is increased, we expect the reconstructed signal to remain constant.

Figure 5.6 presents the results of using the adaptive algorithm when mutual coupling is absent. The magnitude and phase of the reconstructed signal are indistinguishable from the expected value of 1.0 V/m and 0°. The figure demonstrates that in the absence of mutual coupling, the receiving algorithm is highly accurate and can null a strong jammer.

Figure 5.7 shows the results of using in the adaptive algorithm the measured voltages, which have been affected by the mutual coupling. The magnitude of the reconstructed signal varies approximately linearly with respect to the intensity of the jammer. This is because the strong jamming has not been fully eliminated and the residual jammer component completely overwhelms the signal.

The reason the signal cannot be recovered when mutual coupling is taken into account can be visually understood by comparing the adapted beam patterns in the ideal case of mutual coupling with the case where mutual coupling is present. In Figure 5.8a we see the beam pattern for the ideal case. The antenna pattern clearly displays the three deep nulls along the directions of the interferers. The high sidelobes are in the region where there is no interference. Because of the deep nulls, the strong interference can be completely nulled and the signal recovered correctly. This is an important feature of digital beam forming. The antenna pattern has no clear physical significance as the level of the sidelobes can even greater than mail lobe. However, 0 dB is attained exactly along the look direction and the pattern nulls are placed along the desired directions.

Figure 5.8b shows the beam pattern when the mutual coupling between the antenna elements is taken into account. As is clear, the gain of the antenna along the direction of arrival of the signal is considerably reduced. In addition, the nulls in the antenna array pattern are shallow and are displaced from the desired locations. The shallow nulls result in an inadequate nulling of the interference environment. Hence, the signal cannot be recovered.

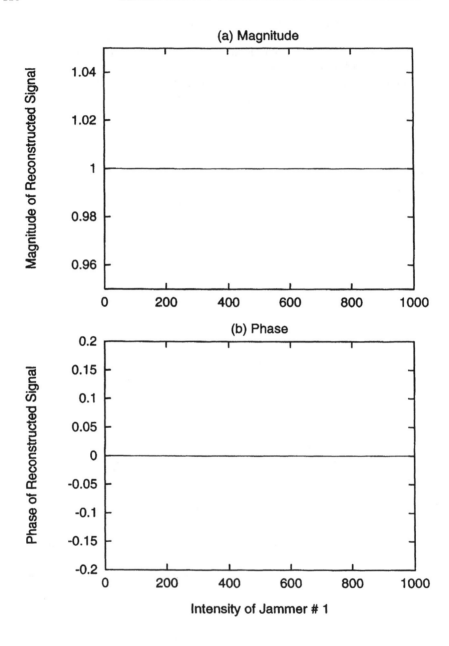

Figure 5.6. Signal recovery using the idealized omnidirectional point radiators in the absence of mutual coupling: (a) magnitude; (b) phase.

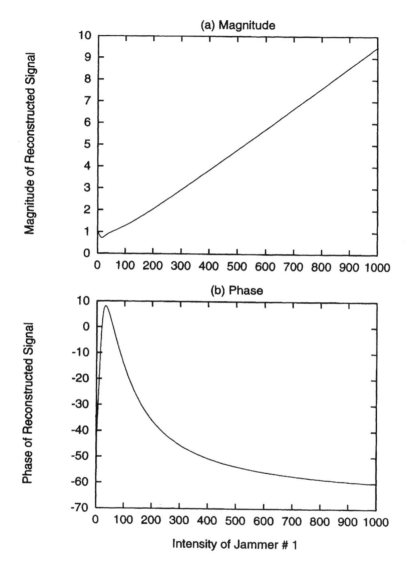

Figure 5.7. Signal recovery in a dipole array in the presence of mutual coupling: (a) magnitude; (b) phase.

The two examples presented here illustrate the importance of the problem at hand. When mutual coupling is taken into account, not only is the main beam of the adaptive array pointed in the wrong direction, but the ability to form deep nulls along the directions of the interference is considerably reduced.

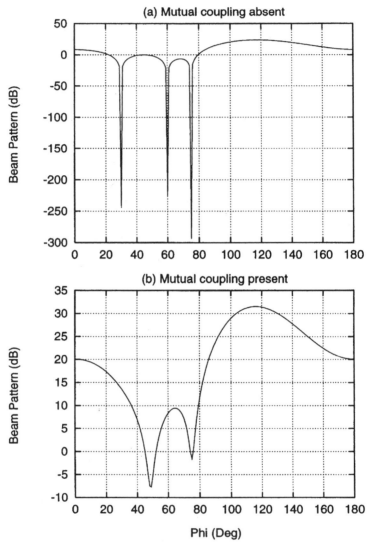

Figure 5.8. Adapted beam patterns in the (a) absence and (b) presence of mutual coupling.

In summary, the direct data domain algorithm [6–8] of Chapter 4 is a promising alternative to the traditional classical statistical adaptive algorithms. However, using the voltages measured at the ports of the array yields incorrect results and the mutual coupling between the elements undermine the ability of the algorithm to suppress interference.

The next section presents examples using (5.27) [4, 9–11] to compensate for the effects of mutual coupling for any type of antenna arrays. This technique is demonstrated to be more effective than the compensation technique of [3, 5, 12].

5.3   COMPENSATION FOR MUTUAL COUPLING

Most adaptive algorithms assume that each element in the array is independent of the other elements in the array. The mutual coupling arises due to the reradiation of the incident fields from the elements themselves. To eliminate the effects of mutual coupling, we begin by realizing that the MM voltages of (5.4) are related directly to the incident fields and are not related to the array manifold, so they are not affected by mutual coupling. In contrast, the measured voltages are related to the currents flowing through the antenna and are affected by the array manifold described by the impedance matrix in (5.4). The approach here will be to recreate some part of the MM voltage vector from the given measured voltages, which are related to the currents on the structure [6–9] through the use of (5.27).

5.3.1   Constant Jammers

The seven-element array defined in Table 5.1 is designed to receive a signal which is corrupted by three jammers. The phase and directions of arrival of the signal and the electrical characteristics of the interferers are given in Table 5.2. The magnitude of the incident signal is varied from 0 V/m to 10.0 V/m in steps of 0.05 V/m while maintaining the jammer intensities constant as given in Table 5.2. For each value of the signal intensity, the MM voltage vector is evaluated to yield the measured voltages. The measured voltages and the signal DOA are treated as the known quantities.

Using the measured voltages and MM admittance matrix, the open-circuit voltages are obtained [3]. These open-circuit voltages are then passed to the D^3LS algorithm described in Chapter 4, and an attempt is made to recover the signal. It is expected that the recovered signal would vary linearly with the intensity of the incident signal. Figure 5.9a presents the results of using the open-circuit voltages. The expected linear relationship is clearly seen, giving the user the false impression that the jammers have been nulled and the signal recovered correctly. However, the actual values of the recovered signals are not correct.

In the second scenario, the measured voltages are used to estimate the vector $[\tilde{V}]$ using (5.27). These voltages are used to recover the signal. Figure 5.9b shows the results of using the voltages $[\tilde{V}]$. Again, the expected linear relationship is clearly visible with the correct scale. Furthermore, Figure 5.10

clearly indicates that both the magnitude (Figure 5.10a) and the phase (Figure 5.10b) of the signal of interest have been recovered correctly using the formulation presented in this section.

This example has shown that open-circuit voltages do provide some compensation for mutual coupling. The use of open-circuit voltages provides significantly better signal recovery than using the measured voltages directly, as shown in Figure 5.5. The technique to eliminate the effects of mutual coupling introduced in Section 5.1.2 also proves that it actually compensates for the mutual coupling between the antenna elements.

In this example, however, the interference has been relatively weak. A more stringent test for both compensation techniques is to check their ability to suppress strong interference.

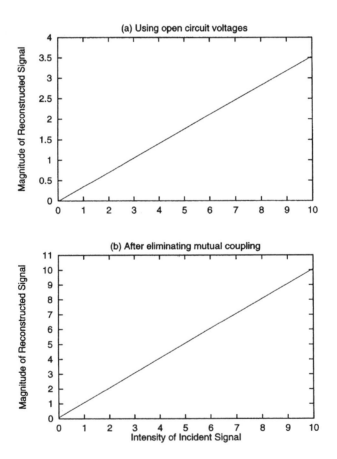

Figure 5.9. Signal recovery: (a) using open-circuit voltages; (b) after eliminating mutual coupling.

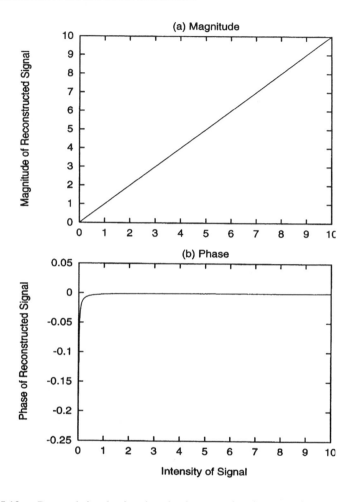

Figure 5.10. Restored signal using the adaptive procedure incorporating proper mutual coupling analysis: (a) magnitude; (b) phase.

5.3.2 Constant Signal

In this example, the intensity of the incident signal is held constant at 1.0 V/m. The intensity of the first jammer is varied from 1.0 V/m to 1000 V/m (60 dB above the signal) in steps of 5 V/m. The rest of the parameters of signal and interferers are given by Table 5.2. For each value of the jammer intensity, the MM voltage vector is calculated and the measured voltages are computed. In the first scenario, the measured voltages are used to find the open-circuit voltages. The open-circuit voltages are then passed on to the direct data domain algorithm of [6–8]. In the second scenario (5.27) is used to find the voltage vector [$\tilde{V}$].

These voltages are used to recover the signal and null the jammers using the same algorithm. If the jammers are properly nulled, the reconstructed signal magnitude should remain constant as a function of jammer strength.

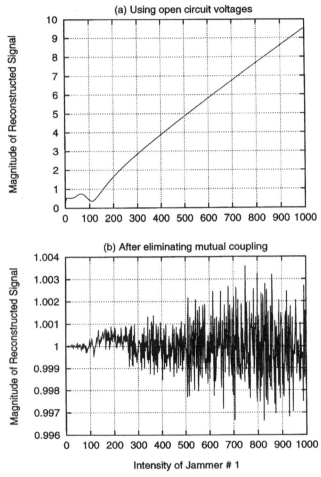

Figure 5.11. Signal recovery: (a) using open-circuit voltages; (b) after eliminating mutual coupling.

Figure 5.11a presents the results when the open-circuit voltages are used to recover the signal. As can be seen, the recovered signal shows a near-linear relationship as a function of jammer strength. This indicates that the jammer has not been adequately nulled and the residual jammer strength has overwhelmed the signal.

The results of compensating for the mutual coupling using the technique presented in Section 5.1.2 are shown in Figure 5.11b. The magnitude of the reconstructed signal varies between 0.996 and 1.004 V/m (i.e., the error in the signal recovery is very small). This figure shows that the strong jammer has been effectively nulled and the signal can be reconstructed. As shown in Figure 5.12, when the mutual couplings between the antenna elements are taken care of properly, the magnitude (Figure 5.12a) and phase (Figure 5.12b) of the actual signal can be recovered with great accuracy.

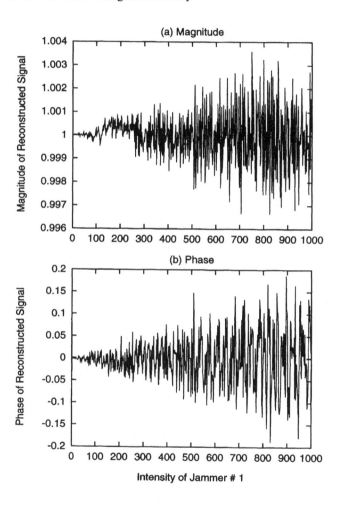

Figure 5.12. Restored signal using the adaptive procedure incorporating proper mutual coupling analysis: (a) magnitude; (b) phase.

The reasons that using the open-circuit voltages are inadequate to compensate for the mutual coupling while the technique presented in Section 5.1.2 is adequate are illustrated through the adapted beam patterns in the two cases. The adapted beam pattern associated with using the open-circuit voltages is shown in Figure 5.13a. The nulls are placed in the correct locations. However, they are shallow, resulting in the inadequate nulling of the interference.

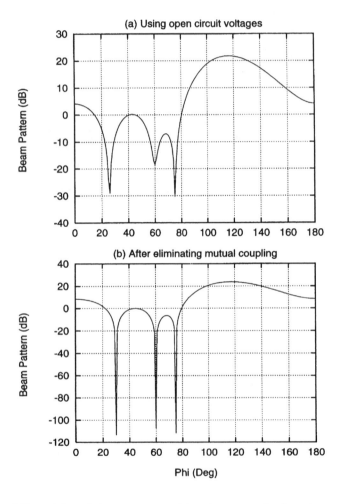

Figure 5.13. Adapted beam pattern: (a) using open-circuit voltages; (b) after compensation for mutual coupling.

The beam pattern associated with proper compensating for mutual coupling using the technique presented in this chapter is shown in Figure 5.13b. The nulls are deep and placed in the correct directions. This demonstrates that the mutual coupling has been suppressed enough so as to null even a strong jammer. It is important to observe in Figure 5.13 that the maximum of the beam pattern does not always occur at the direction of arrival of the signal. This is an important distinction between digital and analog beam forming. In an analog beamformer the magnitude of the weights can never be greater than unity, as that will imply the existence of an amplifier, whereas in digital beam forming the weights are mere numbers and they can be greater than 1 and the beam pattern has no physical meaning, unlike its analog counter part. The only constraint that the digital beamformer guarantees is that 0 dB of the beam pattern scale will align with the direction of arrival of the signal of interest. The beam pattern may exceed 0 dB along directions where there are no incident signals and in this way preserve the integrity of the adaptive process. It is important to note that the sidelobes can exceed the mainlobe without interfering with the desired constraints.

In [3] the objective has been to compensate for the effects of mutual coupling by relating the open-circuit voltages (voltages at the ports of the array if all were open circuited) with the voltages measured at the ports. The stated assumption is that the open-circuit voltages are free of mutual coupling. This assumption is valid only in a limited sense. The open-circuit voltages are the voltages in the presence of the other open-circuited elements. This implies that the effects of mutual coupling have been reduced but not eliminated. The work of [3] remains the only published effort analyzing the effects of and compensating for mutual coupling in adaptive antenna arrays used for signal recovery. Many authors have used this formulation to analyze and eliminate the effects for mutual coupling on DOA estimation algorithms [10, 13, 14]. Pasala and Friel [5, 12] use the MM to quantize and eliminate the effects of mutual coupling on DOA estimation algorithms. However, the authors solve the entire MM problem, requiring knowledge of the incident fields. In practice, the information about the interferers is not available. Therefore, the complete problem cannot be solved.

In the methodology presented in Section 5.1.2, the problem of signal recovery by a linear array of equispaced thin half-wavelength dipoles is analyzed accurately by the MM. In addition, the compensation of mutual coupling is done not only for the signal but also for the interferers. This short analysis allows us to conclude that using the open-circuit voltages does reduce the effect of mutual coupling somewhat. However, the reduction is inadequate to suppress strong interference. This is because the open-circuit voltage at an array element is the voltage in the presence of the other open-circuited elements. The technique presented in this chapter proves to be far superior in compensating for mutual coupling. This is because by using multiple basis functions per antenna element, the mutual coupling information has been represented accurately.

5.3.3   Results for Different Elevation Angles

In this example, we deal with interfering signals arriving from an elevation other than that of the signal of interest. The same seven-element array of the earlier examples has been used. The description of the array is given in Table 5.1 and the angle of arrival of the signal and the jammers is given in Table 5.3.

TABLE 5.3

Complex amplitudes of the signals and jammers and their azimuth and elevation angles of arrival

	Magnitude (V/m)	Phase	DOA (φ)	DOA (θ)
Signal	1.0	0.0	80°	90°
Jammer 1	1.0 – 1000.0	0.0	130°	80°
Jammer 2	1.5	0.0	60°	95°
Jammer 3	2.0	0.0	100°	90°

Here, the intensity of the signal is kept constant while the intensity of jammer 1 is varied. The jammer strength is increased from 1 V/m to 1000 V/m in steps of 5 V/m. For each value of the jammer intensity, the measured voltages are obtained. Because the interference does not arrive from the same elevation as the signal, a voltage relationship similar to (5.27) is approximately satisfied. Therefore, the jammers are not completely nulled. The results of reconstructing the signal under this situation are shown in Figure 5.14a for the magnitude and the phase in degrees in Figure 5.14b. The reconstructed magnitude is approximately linear with respect to the jammer intensity. This is because the jammer has not been completely nulled and the residual jammer strength interferes with the signal. However, the jammer leakage has been significantly reduced. The effect of the 60-dB jammer has been reduced to less than 10% of the signal value (i.e., about 20 dB below the signal). This amounts to an 80-dB effective null at the jammer. In certain applications, this may be an effective nulling capability.

The reason why the strong jammer has been effectively nulled in the example of Section 5.3.2 but not in this case can be understood by looking at the beam pattern of the weighted array. Figure 5.15 shows the beam patterns as a function of the azimuth angles, with the jammer intensity at 995 V/m. Figure 5.15a shows the beam pattern for the broadside case (i.e., the elevation angle corresponds to $\theta = 90°$). Figure 5.15b shows the beam pattern for the elevation corresponding to $\theta = 80°$. Jammer 1 arrives from this elevation. The 80-dB null along the direction of the first jammer ($\varphi = 130°$ and $\theta = 80°$) can be seen clearly. The problem here is that we assume that the signal is arriving from a particular elevation angle, but in reality it is coming from a different angle. Hence, these uncertainties in the *a priori* assumption of the direction of arrival of the signal may introduce errors in the final solution.

5.3.4   Effect of Noise

The examples presented above illustrate the effects of mutual coupling and ignored the inclusion of additive noise at each antenna element. This example presents the effect of thermal noise on the adaptive algorithm. The thermal noise is additive and is modeled as a Gaussian random variable. The noise at any antenna element is assumed to be independent of the noise at the other elements. Since the noise introduces a random component to the data, comparisons will be made in terms of averages over many random samples using a Monte Carlo simulation.

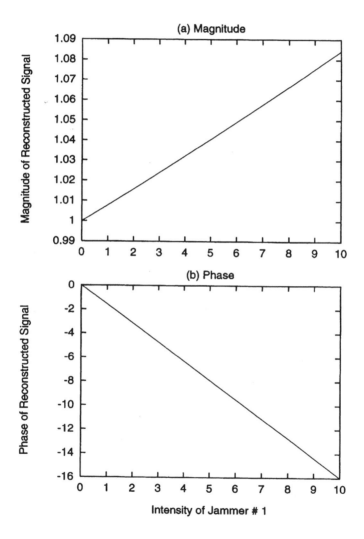

Figure 5.14. Restored signal using the adaptive procedure incorporating proper mutual coupling analysis: (a) magnitude; (b) phase.

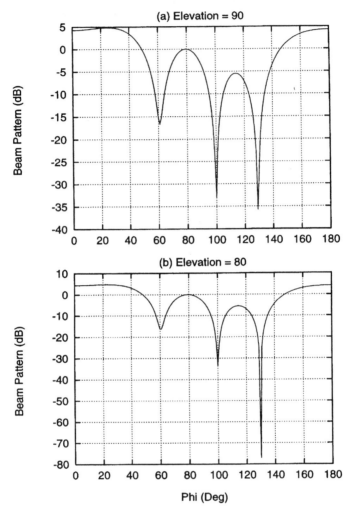

Figure 5.15. Adapted beam patterns for different elevation angles: (a) $\theta = 90°$; (b) $\theta = 80°$.

   In this example, a 13-element array of thin half-wavelength-long wire dipoles receives a signal corrupted by three jammers as given in Table 5.4. The z-directed dipoles each have radius $\lambda/200$ and are spaced a half wavelength apart. Each wire is centrally loaded with a 50 Ω resistance. Seven unknowns per wire are used in the MM analysis, leading to a total of 91 unknowns. The signal-to-noise ratio was set at 13 dB. Note that jammer 1 is a strong jammer (66 dB with respect to the signal).

TABLE 5.4
Complex amplitudes for the signals and jammers and their respective DOAs

	Magnitude (V/m)	Phase	DOA
Signal	1.0	0.0	85°
Jammer 1	2000.0	0.0	135°
Jammer 2	1.5	0.0	60°
Jammer 3	2.0	0.0	100°

For each of the 13 channels, a complex Gaussian random variable is added to the measured voltages due to the signal and jammers. This set of voltages, affected by noise, is passed to the signal recovery routine described in Chapter 4. This procedure is repeated 500 times with different noise samples. These 500 samples are used to compute the average and the variance for the estimated signal components. The results for this simulation after correction for mutual coupling can be seen in Figure 5.16. The output signal-to-interference plus noise ratio (SINR) in decibels is defined as

$$SINR_{out} = 10.0 \; \log \left[\frac{|S|^2}{|\text{bias}|^2 + \text{var}} \right] \qquad (5.28)$$

Here S is the value of the signal, which is 1V, the bias is the deviation of the estimated average value of the signal from S, and the variance represents a spread of the different estimated values. The results of the simulation above are presented in Table 5.5.

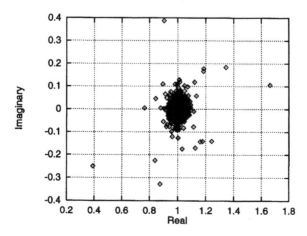

Figure 5.16. Estimate for the adapted signal for individual noise simulations.

TABLE 5.5
Summary of the results of 500 simulations with different noise samples

	Without Compensating for Mutual Coupling	After Compensating for Mutual Coupling
Input signal-to-noise ratio	13 dB	13 dB
Number of samples	500	500
True value	(1.0, 0.0) V/m	(1.0, 0.0) V/m
Mean of 500 estimates	(0.033, 0.493) V/m	(1.004, −0.003) V/m
Bias of estimate	(−0.067, 0.493) V/m	(0.004, −0.003) V/m
Variance of estimates	0.010	0.010
Output SINR	6.355 dB	19.866 dB

When the measured voltages are used directly to recover the signal mainly due to the high bias in the estimate of the signal, the output SINR is only 6.355 dB. The high bias can be directly attributed to the inadequate nulling of the strong jammer. However, when the mutual coupling is eliminated using the technique presented in this chapter, the jammers are completely nulled, yielding accurate estimates of the signal. The total interference power is suppressed to nearly 20 dB below the signal. An adapted sample beam pattern is shown in Figure 5.17.

Next, we look at a different application of the direction of arrival (DOA) estimation with particular emphasis on a code division multiple access (CDMA) environment.

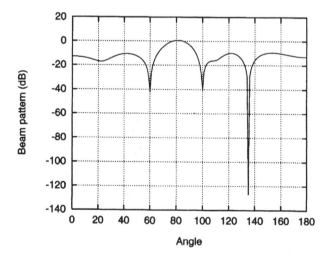

Figure 5.17. Sample adapted beam pattern in the presence of Gaussian noise.

5.4   BEARING ESTIMATION BY COMBINED CDMA AND MATRIX PENCIL WITH MUTUAL COUPLING COMPENSATION

In this section we describe a practical algorithm for direction of arrival (DOA) estimation in code division multiple access (CDMA) multiuser settings in the presence of mutual coupling in an antenna array. The class of noise subspace algorithms for DOA estimation, such as MUSIC, contains some of the most popular algorithms used in the past [15]. These algorithms separate the noise and signal subspaces based on an eigenvalue decomposition of a covariance matrix estimated by averaging over several samples in time (snapshots). In the case of Root-MUSIC, the eigenvectors of the noise subspace are then used to form a complex polynomial whose roots correspond to signal directions. Here we address both the signal processing and electromagnetics aspects of a practical implementation. We use the Matrix Pencil [16–18] method, which does not require formation of a covariance matrix nor the estimation of the roots of a polynomial. Its only computationally intensive step is an eigen decomposition. The Matrix Pencil algorithm is described in Appendix C. This algorithm yields accurate results for the DOA using only a *single* time snapshot. This reduces the computation time drastically, making it very attractive for real-time applications. The only penalty paid is that given an antenna array of N elements, Matrix Pencil can only estimate the directions of $(N + 1)/2$ sources, in contrast to the $(N - 1)$ sources possible using MUSIC. However, the algorithm can distinguish between coherent sources, that is, in-phase or out-of-phase multipath components, while MUSIC requires all the signals to be uncorrelated or must use spatial smoothing to decorrelate the incoming signals. Both Matrix Pencil and MUSIC estimate the complex exponentials. The DOA of the signal can be obtained using the relation

$$\phi_p = \cos^{-1}\left[\frac{\text{Im}(\ln(\gamma_p))}{k\,\Delta_x}\right] \tag{5.29}$$

where γ_p is the estimate of the pth complex exponential.

To overcome the issue of number of resolvable sources, in a CDMA setting we can use the CDMA processing gain to enhance the SNR of the signal of interest (SOI), reducing the amplitudes of all the interfering signals. The problem reduces to effectively estimating the DOA of a single SOI with interference. Matrix Pencil, which has lower computational complexity than Root-MUSIC, serves as the basis of the direction of arrival (DOA) estimation. In addition, CDMA processing gain enhances the SNR of the signal of interest (SOI) and reduces the effects of mutual coupling: mutual coupling compensation, based on the minimum norm formulation of (5.27). This new CDMA/MP combined technique is able to estimate angles of arrival with a low root mean-squared error (RMSE) using only a single time snapshot of the voltages at the elements of the array in a multiuser setting. The RMSE of CDMA/MP is comparable or better than the popular Root-MUSIC algorithm, which takes approximately twice the CPU time as that of the Matrix Pencil method. When combined with the minimum norm compensation of (5.27), the algorithm is able to perform well even with small element spacing when using realistic arrays. This compensation procedure is valid for all types of arrays and elements.

First we present the CDMA/MP method of estimating the DOA and then we illustrate the enhancement in the results when accounting for mutual coupling.

5.4.1   CDMA/MP DOA Estimation

In a realistic communication setting, the number of users will surely exceed the number of antenna elements in the array. In addition, each user's signal may arrive over multiple paths. In such a case, estimating the directions of all users appears to be difficult. However, in the case of a CDMA signal, each user's data is spread by a pseudo-noise (PN) code. The receiver-matched filter enhances the signal-to-interference ratio (SIR) by a factor equal to the spreading gain.

In a CDMA setting, we can write this situation mathematically at the nth antenna element as

$$y_n(t) = \sum_{q=1}^{Q} K_q \, b_q(t) c_q(t) \, e^{j\vartheta_q} \, e^{jnk\Delta_x \cos\varphi_q} + \xi_n(t) \qquad (5.30)$$

where K_q is the signal amplitude incorporating the data and the spreading, $b_q(t)$ is the symbol stream (the data), $c_q(t)$ is the chip stream, and $e^{j\vartheta_q}$ is the random phase of the qth signal. Here $k = 2\pi/\lambda$ is the wavenumber, $\xi_n(t)$ represents additive noise corrupting the signal, Δ_x is the inter element spacing, and φ_q is the angle of arrival of the qth signal. Note that by including the relative delay between multipaths into the time variable, each multipath signal is treated as a separate signal to be included in Q. The individual users' codes may be long or short.

Without loss of generality, we can designate one path of one user as the SOI and the others as interference. Here we choose the first path of the first user. The matched filter in the receiver is tuned to the delay and code of this signal. The output of the matched filter at the nth element for a particular symbol is given by

$$y_n = G K_1 b_1 e^{j\vartheta_1} z_1^n$$
$$+ \sum_{q=2}^{Q} K_q e^{j\vartheta_q} \gamma_q^n \int_{t=0}^{T_s} b_q(t) \, c_q(t) \, c_1(t) \, dt + \int_{t=0}^{T_s} \xi_n(t) c_1(t) \, dt \qquad (5.31)$$

where T_s is the symbol duration and G is the CDMA processing gain. Here $\gamma_q = \exp(jk\Delta_x \cos\varphi_q)$. As one can see from (5.30), due to the use of pseudo-noise (PN) sequences, the SINR is effectively increased by a factor of G. We apply

DOA estimation to the output of the matched filter looking to estimate the direction of a single signal ($Q = 1$).

As mentioned earlier, Matrix Pencil requires only a single snapshot of data, that is, the matched filter outputs from a single symbol suffice. The actual symbol is not required. In the case that multiple symbols are available (a training sequence), the output signal-to-noise ratio (SNR) can be further enhanced by averaging over the training symbols. In this case Matrix Pencil is applied to the averaged output, given by

$$\hat{y}_n = \sum_{k=1}^{K} b^*[k] \, y_n[k] \tag{5.32}$$

where K is the number of training symbols available, $b^*[k]$ is the conjugate of the kth training symbol, and $y_n[k]$ is the output of the matched filter at the nth element for the kth symbol.

Note that we do not assume any data processing beyond what is already available in commercial CDMA systems. As we treat the residual interference as additional noise, any use of multiuser detection to further suppress interference will only improve the performance. The need for a training sequence to further enhance the SNR is one drawback with the Matrix Pencil method proposed here. Correlation-based techniques do not require a training sequence. However, as we shall show in the examples below, this does not present a serious difficulty, as Matrix Pencil provides excellent accuracy with a few training symbols.

5.4.1.1 Numeral Examples.

We now present some examples illustrating the performance of the Matrix Pencil and combined CDMA/MP algorithm in terms of accuracy, computation speed, and need for input data. These examples deal with an asynchronous CDMA environment with multipath. Each source uses BPSK modulation, and each multipath component has a uniformly distributed random phase and an amplitude that defines the SNR.

The performance of the CDMA/Matrix Pencil is compared to the Root-MUSIC algorithm, a popular approach within the class of subspace techniques. Like Matrix Pencil, Root-MUSIC also estimates the parameters γ_q in (5.29) and serves as a fair comparison [15].

Tables 5.6 and 5.7 summarize the parameters in these examples. The delay is in terms of number of chips, with all values as stated unless otherwise specified. Path 1 of user 1 acts as the SOI.

Due to the four samples per chip and the spreading gain, the effective SNR, after the matched filter, of the SOI is $-20 + 10 \log_{10}[4 \times 128] = 7.1$ dB. Note that the seven-element array receives six signals and Matrix Pencil should not be able to isolate any signals.

5.4.1.2 Accuracy and Computational Efficiency of the CDMA/MP Algorithm.

Table 5.8 summarizes the results, including the cases of using a training sequence to improve accuracy. The table presents the mean, standard deviation,

TABLE 5.6
Summary of the parameters

Parameters	Symbols	Values
No. of elements	N	7
Spreading gain	G	128
PN sequence period	PN	$2^{15} - 1$
Element spacing	Δ_x/λ	0.5
Samples per chip	s/c	4
Pencil parameter	L	4

TABLE 5.7
Mobile user characteristics

User	Path	SNR (dB)	DOA	Delay (no. chips)
1	1	−20	85°	0
	2	−27	80°	8
	3	−23	90°	20
2	1	−20	50°	24
	2	−20	100°	4
	3	−20	120°	28

TABLE 5.8
Performance of Matrix Pencil and Root-MUSIC algorithms (signal is arriving from 85°)

	One Snapshot			10 Snapshots		
	Mean	Std. Dev.	RMSE	Mean	Std. Dev.	RMSE
CDMA/Matrix Pencil	85.07°	1.35°	1.36°	85.00°	0.39°	0.39°
CDMA/Root-MUSIC	85.40°	8.07°	8.08°	85.01°	0.37°	0.37°

and RMSE of 1000 independent trials when using just 1 bit (a single snapshot) and 10 bits (averaging over 10 snapshots).

As can be seen from the results, with a RMSE of 1.36°, the Matrix Pencil approach is extremely accurate, when using only a single snapshot. For a single snapshot, Root-MUSIC fails, with an extremely large RMSE of approximately 8°. This is despite the fact that Root-MUSIC is estimating the direction of only a single SOI. Figures 5.18 and 5.19 are the histograms of 1000 estimates generated by the Matrix Pencil and Root-MUSIC algorithms, respectively, using a single snapshot. The accuracy of the Matrix Pencil is clearly visible, as are several erroneous estimates by Root-MUSIC. Using multiple snapshots improves the estimate of the noise subspace and so the performance of Root-MUSIC. But Matrix Pencil and Root-MUSIC provide similar accuracy.

In a more realistic wireless communication setting, each multipath component would undergo fading. Table 5.9 and Figures 5.20 and 5.21 provide a performance comparison between Matrix Pencil and Root-MUSIC when each multipath undergoes slow Rayleigh fading. When using a single snapshot of the data, the Matrix Pencil method, with an RMSE of 6.67°, is approximately twice as accurate as Root-MUSIC, which has an RMSE of 12.86°.

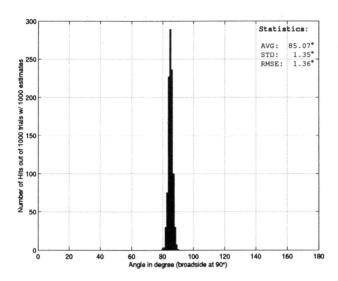

Figure 5.18. Results for Matrix Pencil with the pencil parameter $L = 4$ and one time snapshot.

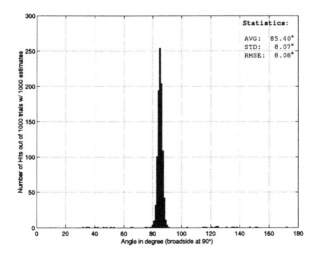

Figure 5.19. Results for Root-MUSIC using one time snapshot.

TABLE 5.9

Performance of Matrix Pencil and Root-MUSIC algorithms during fading.

	One Snapshot			10 Snapshots		
	Mean	Std. Dev.	RMSE	Mean	Std. Dev.	RMSE
CDMA/Matrix Pencil	84.82°	6.67°	6.67°	85.03°	3.53°	3.53°
CDMA/Root-MUSIC	86.49°	12.86°	12.95°	85.22°	5.96°	5.96°

Using 10 time samples or snapshots, Matrix Pencil achieves a mean and RMSE of 85.03° and 3.53°, respectively, while Root-MUSIC has a mean and RMSE of 85.22° and 5.96°, respectively. Note that the accuracy of the two techniques does not improve as quickly with the number of bits when there is fading. This is to be expected, as the signal levels fluctuate. This is consistent with results reported in [19]. Again the performance of the Matrix Pencil algorithm is significantly more accurate than Root-MUSIC.

A major advantage of the Matrix Pencil algorithm over Root-MUSIC is its computation efficiency. Figure 5.22 plots the average time to finish computing the 1000 estimates as a function of the number of snapshots. It illustrates a crucial advantage of using the CDMA/MP for direction finding. The execution time of Matrix Pencil stays about the same as the number of snapshots increases. It is because, as explained in Appendix C, the computation load depends mostly on the size of the antenna array and the pencil parameter, L, rather than the number of snapshots. Due to the computation load involved in averaging over multiple snapshots, the execution time of Root-MUSIC increases with number of snapshots. The computation load of Matrix Pencil is significantly lower than Root-MUSIC. Matrix Pencil is about 1.7 times faster than the Root-MUSIC algorithm initially (based on the MATLAB Profile function). The advantage in execution time increases with the number of snapshots. This savings in execution time does not come at the expense of accuracy.

In summary, the Matrix Pencil algorithm shows some clear advantages over traditional approaches such as Root-MUSIC. Matrix Pencil does not require the estimation of a covariance matrix and can be applied with remarkable accuracy with only a single time sample. Because it does not need to estimate a covariance matrix, unlike MUSIC-based techniques, it can distinguish between coherent sources. During Rayleigh fading, Matrix Pencil yields greater accuracy than Root-MUSIC. In a CDMA application, the limitation on the number of sources is overcome by using the CDMA spreading gain. The only drawback is that to further increase the SNR, a training sequence is required.

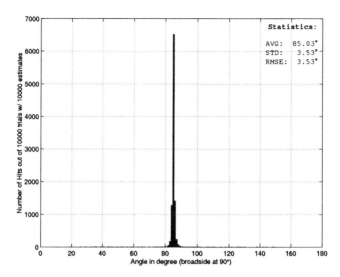

Figure 5.20. Results for Matrix Pencil with the pencil parameter $L = 4$, 10 snapshots, slow Rayleigh fading.

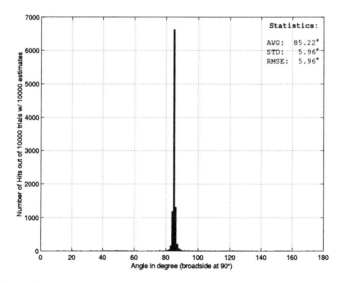

Figure 5.21. Results for Root-MUSIC, 10 snapshots, slow Rayleigh fading.

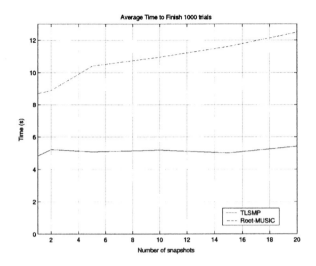

Figure 5.22. Computation time required by Matrix Pencil (TLSMP) and Root-MUSIC.

5.4.2   Compensation of Mutual Coupling in CDMA/MP

As with other signal-processing techniques, Matrix Pencil assumes an ideal linear array. Here we investigate the impact of mutual coupling and the performance of the compensation techniques described in Section 5.1.

5.4.2.1 Numerical Examples to Illustrate Compensation for Mutual Coupling.
This example uses a seven-element array with interelement spacing of 0.3λ. The MM analysis uses seven unknowns per element (i.e., a total of 49 unknowns are used). The array receives three signals, from 40°, 90°, and 140°. Each signal has a nominal SNR of 1 dB. The Matrix Pencil algorithm uses only a single snapshot. The plots shown here use the results of 1000 independent trials. Figure 5.23 shows a histogram of the results of using Matrix Pencil without any compensation for mutual coupling. Thirty-eight times the estimation procedure fails completely by resulting in imaginary angles. This happens when the argument to the $\cos^{-1}$ function in (5.24) becomes greater than 1. As is clearly seen, the DOA estimation is very poor with very large errors.

Figures 5.24 and 5.25 plot the performance after compensation for mutual coupling. Figure 5.24 plots the use of open-circuit voltages, while Figure 5.25 plots the results of using the minimum norm technique. In both figures, the hugely improved performance over the uncompensated case is very clear. Neither technique results in any imaginary angles. Note that because of the accurate performance, we can estimate a standard deviation, which for all cases is approximately 3.5°.

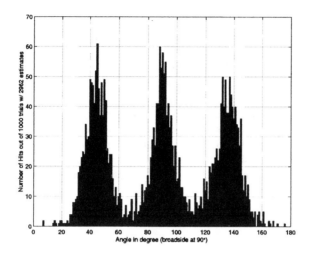

Figure 5.23. Results for Matrix Pencil for an element spacing of $d/\lambda = 0.3$ and no compensation for mutual coupling between the elements.

TABLE 5.10
Comparing open-circuit and minimum norm compensation techniques

	Open Circuit				Minimum Norm			
	Mean	Bias	Std. Dev.	RMSE	Mean	Bias	Std. Dev.	RMSE
Signal 1	37.55°	2.45°	3.88°	4.59°	40.84°	0.84°	3.57°	3.57°
Signal 2	90.05°	0.05°	3.29°	3.29°	90.05°	0.05°	3.13°	3.13°
Signal 3	142.61°	2.61°	3.67°	4.51°	139.28°	0.72°	3.43°	3.50°

As Table 5.10 shows, the crucial difference between the two compensation techniques is in the bias. The bias resulting from using the minimum norm compensation approach is significantly smaller than when using the open-circuit voltages. This is because using the open-circuit voltages only reduces mutual coupling. The currents induced on the arms of the dipole, though zero at the port, still reradiate, resulting in a residual effect of mutual coupling.

In applying the Matrix Pencil technique to a practical array in a CDMA-based communication setting, the CDMA processing gain provides some resistance to mutual coupling. This is so because, after all, the matched filter is effectively only one signal with a relatively weak residual interference. With only one signal impinging on the array, the linear phase front is not fatally corrupted and it is possible to estimate the DOA. This is true particularly of arrays with moderate mutual coupling.

To illustrate this effect, we use the same example as just outlined. For a fair comparison, the power of each signal is reduced by the processing gain of 128×4 (the 4 corresponds to the number of samples per chip). Using the filter matched to the first signal, two of three signals are suppressed. Matrix Pencil is applied *without compensation for mutual coupling*. Figure 5.26 plots the histogram of the resulting estimates. In comparison with Figure 5.23, the accuracy of the final result is greatly improved. No estimates resulted in imaginary angles. In fact, the accuracy is comparable with using the open-circuit voltages as in Figure 5.24. However, as shown in the next example, with stronger mutual coupling, not compensating for mutual coupling can still lead to significantly degraded performance.

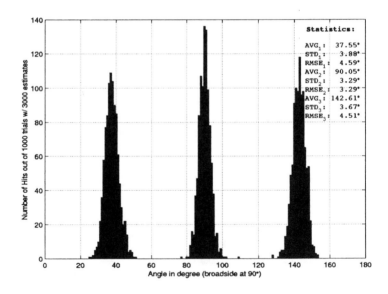

Figure 5.24. Results for Matrix Pencil for an element spacing of $d/\lambda = 0.3$ and using the open-circuit voltages.

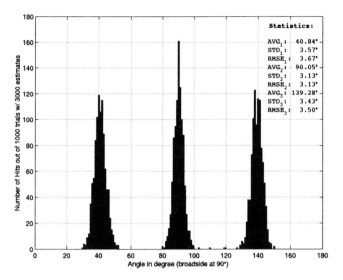

Figure 5.25. Results for Matrix Pencil for an element spacing of $d/\lambda = 0.3$ and using the minimum norm compensation matrix.

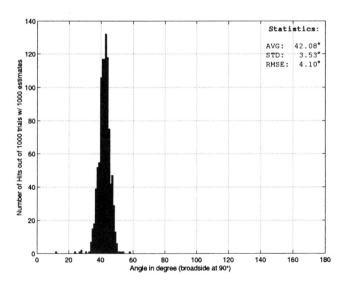

Figure 5.26. Results for the CDMA/MP for an element spacing of $d/\lambda = 0.3$ and no compensation for mutual coupling between the elements.

5.4.3   TLSMP/CDMA with Smaller Element Spacing

This example uses the data of Table 5.7. To increase the mutual coupling between the elements, the interelement spacing is reduced to 0.2λ. Matrix Pencil is applied after the matched filter, matched to the third multipath of user 2. The true angle of arrival is 120°. Figure 5.27 plots the histogram of 1000 independent trials without mutual coupling compensation, while Figures 5.28 and 5.29 plot the histograms after using open-circuit voltages and minimum norm compensation, respectively. As can be seen, the compensation leads to a significantly more accurate estimate. This is because, in comparison to Figure 5.26, with an interelement spacing of 0.2λ the mutual coupling has significantly increased. Consequently, the linear phase front of even a single signal is corrupted.

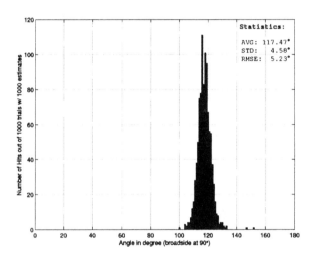

Figure 5.27. Results for the CDMA/MP for an element spacing of $d/\lambda = 0.2$ and no compensation for mutual coupling between the elements.

5.5   CONCLUSION

This chapter demonstrates that for the development of practical direct data domain algorithms, the electromagnetic nature of the array must be taken into account. We have shown that the mutual coupling between the elements of the array causes adaptive algorithms to fail. This problem is associated with both covariance matrix approaches (stated earlier in [3]) and direct data domain approaches (investigated here).

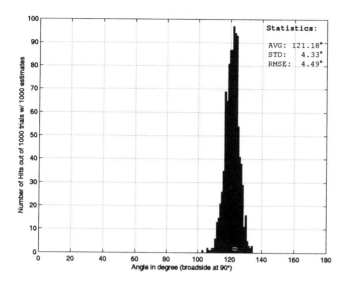

Figure 5.28. Results for CDMA/MP for an element spacing of $d/\lambda = 0.2$ and using the open-circuit voltages.

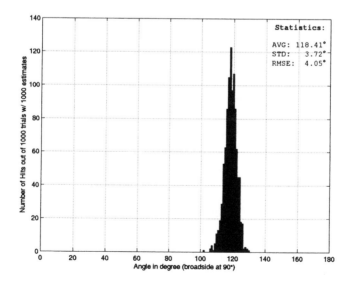

Figure 5.29. Results for CDMA/MP for an element spacing of $d/\lambda = 0.2$ and using the minimum norm compensation matrix.

To properly characterize the antenna, the MM is used. Previously published work [3, 5, 12] in this area has used only one basis function per element. However, this is usually inadequate for an accurate antenna analysis. The use of multiple basis functions per element in a practical manner is a major advance over previously published methods.

Recognizing that the MM voltage vector is free from mutual coupling eliminates the mutual coupling. We estimate this MM voltage vector by using the minimum norm solution to an underdetermined system of equations. It is shown that this method works very well in the presence of strong interfering sources. Furthermore, it is shown that the proposed technique is superior to the earlier suggested method of using the open-circuit voltages.

Next, this chapter presents a comprehensive approach to DOA estimation using a combination of signal processing and electromagnetic issues. The application chosen here is DOA estimation in CDMA wireless communications. On the signal processing side, a combined Matrix Pencil/CDMA algorithm has several attractive features. The algorithm is accurate, does not require multiple data snapshots, can handle a coherent multipath, and requires about half the execution time of other popular techniques, such as Root-MUSIC. This gain in speed and reduced data requirements is due to Matrix Pencil not requiring an estimate of a covariance matrix. The only drawback is that an N-element array can estimate the DOA of $(N + 1)/2$ signals.

Combining Matrix Pencil with CDMA eliminates this drawback. The CDMA processing gain enhances the SNR of the SOI while suppressing the other signals. Matrix Pencil can then be applied as if only one signal were impinging on the array. The examples show this combined CDMA/MP algorithm to be fast, accurate, and effective in multipath fading situations. The accuracy can be further improved by using a training sequence to increase the SINR. We also introduce a very effective technique to compensate for the mutual coupling based on the minimum norm solution to an underdetermined system of equations. The approach is to find the uncoupled signals, with minimum energy, that would produce the mutually coupled measured signals. This technique proves to be significantly more accurate than the classical open-circuit approach.

In applying CDMA/MP to a practical array, a curious result emerges. Because the CDMA processing gain suppresses all signals other than the SOI, leaving only one effective signal, the effect of mutual coupling is also significantly reduced. In a practical implementation in CDMA-based communications, therefore, it must be determined what impact mutual coupling has when combined with CDMA. In the case of moderate mutual coupling, mutual coupling compensation may not be required. In more tightly packed arrays, mutual coupling compensation plays an important role, leading to significantly more accurate estimates. The impact of mutual coupling is a function of the type of antenna elements, interelement spacing, and CDMA processing gain.

In summary, this chapter describes a topic that is very important to the development of practical implementation of adaptive algorithms. The proposed method is easy to implement and does not add an inordinate computational burden on the adaptive process. In the next chapter we describe a more flexible methodology that can not only handle nonuniformity in the array but can also deal with the presence of near-field scatterers in the performance of an adaptive array.

REFERENCES

[1]   A. R. Djordjevic, M. B. Bazdar, T. K. Sarkar, and R. F. Harrington, *Analysis of Wire Antennas and Scatterers: Software and User's Manual,* Artech House, Norwood, MA, 1995.

[2]   B. J. Strait, T. K. Sarkar, and D. C. Kuo, *Special Programs for Analysis of Radiation by Wire Antennas,* Syracuse University, Technical Report AFCRL-TR-73-0399, June 1973.

[3]   I. J. Gupta and A. A. Ksienski, "Effect of Mutual Coupling on the Performance of Adaptive Arrays," *IEEE Transactions on Antennas and Propagation,* Vol. 31, pp. 785–791, Sept. 1983.

[4]   R. S. Adve and T. K. Sarkar, "Compensation for the Effects of Mutual Coupling in Adaptive Algorithms," *IEEE Transactions on Antennas and Propagation,* Vol. 48, No. 1, pp. 86–94, Jan. 2000.

[5]   K. M. Pasala and E. M. Friel, "Mutual Coupling Effects and Their Reduction in Wideband Direction of Arrival Estimation," *IEEE Transactioins on Aerospace and Electronic Systems,* Vol. 30, pp. 1116–1122, Apr. 1994.

[6]   S. Schneible, "A Least Squares Approach to Radar Array Adaptive Nulling," Ph.D. dissertation, Syracuse University, Syracuse, NY, 1996.

[7]   T. K. Sarkar, S. Park, J. Koh, and R. A. Schneible, "A Deterministic Least Squares Approach to Adaptive Antennas," *Digital Signal Processing: A Review Journal,* Vol. 6, No. 3, pp. 185–194, July 1996.

[8]   S. Park, T. K. Sarkar, and Y. Hua, "A Singular Value Decomposition-Based Method for Solving a Deterministic Adaptive Problem," *Digital Signal Processing: A Review Journal,* Vol. 9, No. 1, pp. 57–63, 1999.

[9]   R. S. Adve, "Elimination of Effects of Mutual Coupling in Adaptive Thin Wire Antennas," Ph.D. dissertation, Syracuse University, Syracuse, NY, Dec. 1996.

[10]   B. Himed, "Application of the Matrix Pencil Approach to Direction Finding," Ph.D. dissertation, Syracuse University, Syracuse, NY, May 1990.

[11]   D. H. Shau, "A Study of the Effects of Mutual Coupling on the Direction-Finding Performance of a Linear Array Using the Method of Moments," Ph.D. dissertation, Syracuse University, Syracuse, NY, 1990.

[12]   M. Friel, "Direction Finding with Compensation for Electromagnetic Effects," Ph.D. dissertation, University of Dayton, Dayton, OH, Dec. 1995.

[13]   M. L. Leou, C. C. Yeh, and D. R. Ucci, "Bearing Estimations with Mutual Coupling Present," *IEEE Transactions on Antennas and Propagation*, Vol. 37, pp. 1332–1335, Oct. 1989.

[14]   B. Friedlander and A. J. Weiss, "Direction Finding in the Presence of Mutual Coupling," *IEEE Transactions on Antennas and Propagation*, Vol. 39, pp. 273–284, Mar. 1991.

[15]   J. C. Liberti, Jr. and T. S. Rappaport, *Smart Antennas for Wireless Communications: IS-95 and Third-Generation CDMA Applications*, Prentice Hall, Upper Saddle River, NJ, 1999.

[16]   Y. Hua and T. K. Sarkar, "Matrix Pencil Method for Estimation Parameters of Exponentially Damped/Undamped Sinusoids in Noise," *IEEE Transactions on Acoustics, Speech, and Signal Processing*, Vol. 38, pp. 814–824, May 1990.

[17]   J. E. Fernandez del Rio and T. K. Sarkar, "Comparison between the Matrix Pencil Method and the Fourier Transform for High-Resolution Spectral Estimation," *Digital Signal Processing: A Review Journal*, Vol. 6, pp. 108–125, 1996.

[18]   R. S. Adve, T. K. Sarkar, O. M. Pereira-Filho, and S. M. Rao, "Extrapolation of Time Domain Responses from Three-Dimensional Objects Utilizing the Matrix Pencil Technique," *IEEE Transactions on Antennas and Propagation*, Vol. 45, pp. 147–156, Jan. 1997.

[19]   J. Joutsensalo and T. Ristaniemi, "Delay Estimation in CDMA System by Differentially Coherent Eigenanalysis," in *Proceedings of the 6th International Conference on Electronics, Circuits and Systems*, Vol. 3, pp. 1279–1282, 1999.

6

DIRECTION OF ARRIVAL ESTIMATION AND ADAPTIVE PROCESSING USING A NONUNIFORMLY SPACED ARRAY FROM A SINGLE SNAPSHOT

SUMMARY

In this chapter a very general methodology is presented for direction of arrival (DOA) estimation and adaptive processing using a single snapshot of the voltages measured at the feed points of a nonuniformly spaced antenna elements in an array. The array may be operating in the presence of mutual coupling between the elements and other near-field scatterers. Here, a snapshot is defined as the phasor voltages measured at the feed point of all the antenna elements in the array at a particular instance of time. First, the voltages induced in the antenna elements of the array due to all the signals, including the signal of interest (SOI), clutter, and interferers, which have been affected by the mutual coupling between the elements and the presence of near-field scatterers, are preprocessed by applying a transformation matrix to the measured snapshot of voltages. The transformation matrix is obtained by using a rigorous electromagnetic analysis tool. This electromagnetic preprocessing technique transforms the voltages that are induced in the nonuniformly spaced array containing real antenna elements due to all the incoming signals to an equivalent set of voltages that will be produced in a uniform linear virtual array (ULVA) containing omnidirectional isotropic point radiators by the same set of incident signals. The preprocessing is carried out using a transformation matrix which includes various electromagnetic effects, such as mutual coupling between the antenna elements, presence of near-field scatterers, and the platform effects on which the antenna array is mounted. This transformation matrix when applied to the actual measured snapshot of voltages yields an equivalent set of voltages that will be induced in the ULVA under the same incoming signal scenario. If the problem is to estimate the DOA, a direct data domain method such as the Matrix Pencil method is then applied to the preprocessed set of voltages to obtain the DOA of the various signals impinging at the array in the presence of noise, as well as their complex amplitudes. If the objective is to extract the SOI from the various electromagnetic signals impinging on the array, including clutter, then again the direct data domain least squares method (D^3LS) described in Chapter 4 can be used to estimate the complex signal amplitude. Limited numerical examples are presented to illustrate the novelty of the proposed method for these two classes of problems.

6.1   PROBLEM FORMULATION

Consider an array composed of $N+1$ nonequally spaced antenna elements located in an array. Assume that $P+1$ narrowband sources impinge at the array from $P+1$ distinct azimuthal directions φ_0, ..., φ_P. So in addition to the SOI there are P undesired signals. Here we assume that the azimuthal angle φ is defined from the end-fire direction of the array as seen in Figure 4.1. For sake of simplicity we assume that the incident fields and the antenna elements are coplanar and that the sources are located in the far field of the array. However, this methodology can easily be extended to the noncoplanar case without any problem, as will be illustrated later.

Using the complex envelope representation, the snapshot represents a $(N+1) \times 1$ vector of phasor voltages $[x(m)]$ received by the elements of the actual array at a particular time instance m and can be expressed by

$$[x(m)] = \begin{bmatrix} x_0(m) \\ x_1(m) \\ \vdots \\ x_N(m) \end{bmatrix} = s_d(m)[a(\varphi_d)] + \sum_{p=1}^{P} s_p(m)[a(\varphi_p)] + [\xi(m)] \qquad (6.1)$$

where $s_d(m)$ is the amplitude of the desired signal and $s_p(m)$ denotes the amplitude of the interference signal at the elements of the array from the pth source, for $p = 1, 2, ..., P$. $[a(\varphi)]$ denotes the field pattern of the array (also called the *steering vector*) toward the azimuth direction φ and $[\xi(m)]$ denotes the *noise vector* associated with each of the voltages measured across the loads of all the antenna elements. By using a matrix representation, (6.1) becomes

$$[x(m)]_{(N+1)\times1} = [A(\varphi)]_{(N+1)\times(P+1)} [s(m)]_{(P+1)\times1} + [\xi(m)]_{(N+1)\times1} \qquad (6.2)$$

where $[A(\varphi)]$ is the $(N+1) \times (P+1)$ matrix of the steering vectors, referred to as the *array manifold* corresponding to each one of the incident signals, and is represented by

$$[A(\varphi)] = [a(\varphi_d), a(\varphi_{1)}, ..., a(\varphi_p)] \qquad (6.3)$$

In a typical array calibration methodology, a far-field source $s_p(m)$ is placed along the angular direction φ_p and then $[x(m)]$ is the voltage measured at the loads of the antenna elements. Here $[s(m)]$ is a $(P + 1) \times 1$ vector representing the complex amplitudes at a particular time instance m that are associated with all the incident waves arriving from each of the directions, say φ_p. Equation (6.2) is applicable to antenna arrays which consist of isotropic omnidirectional point radiators, located in free space. However, this equation will be modified for a real array. There will not be any linear relationship between the incoming signals and the voltages corresponding to the measured snapshot. In practice, the array manifold information in a real array related to these induced voltages is contaminated by the effects of the nonuniformity in the spacing and mutual coupling between the elements of the array, and the presence of near-field scatterers, which undermine the performance of a conventional adaptive signal processing algorithm.

Hence, our problem can be stated as follows: Given the complex array manifold matrix $[A(\varphi)]$ of a nonuniformly spaced array in the presence of mutual coupling between the elements of the array and near-field coupling effects between the platform and other electromagnetic obstacles, our first objective will be to transform the voltages induced in an actual array to a set of voltages that would be induced in ULVA consisting of omni-directional isotropic point radiators radiating in free space. This is numerically carried out by using a transformation matrix which when operating on $[A(\varphi)]$, produces numerically a modified manifold $[A_v(\varphi)]$ which is due to a ULVA manifold matrix. The elements of this virtual array are isotropic omnidirectional point radiators radiating in free space. Thus, we compensate not only for the lack of nonuniformity but also the presence of mutual coupling between the elements in the real array in addition to near-field coupling effects. To estimate the DOA we apply the Matrix Pencil method described in Appendix C to these processed voltages. To carry out adaptive processing we apply the D^3LS to the processed snapshot described in Chapter 4 to estimate the complex amplitude of the SOI in the presence of interferers, clutter, and thermal noise. Moreover, the various signals impinging at the array can be coherent.

6.2   TRANSFORMATION MATRIX TO COMPENSATE FOR UNDESIRED ELECTROMAGNETIC EFFECTS

A preprocessing technique is used to compensate for the lack of nonuniformity in a real array contaminated by the mutual coupling effects. It is based on transforming the nonuniformly spaced array into a ULVA operating in the absence of mutual coupling and other undesired electromagnetic effects. Such a transformation is achieved through the use of a compensation matrix. Hence, our goal is to select the best-fit transformation, $[\mathfrak{I}]$, between the real array manifold, $[A(\varphi)]$, and the array manifold corresponding to a ULVA, $[A_v(\varphi)]$, such that

$$[\Im] [A(\varphi)] = [A_v(\varphi)] \qquad (6.4)$$

for all azimuth angles φ within a predefined sector. Since such a transformation matrix is defined within a predefined sector, the various undesired electromagnetic effects such as nonuniformity in spacing and mutual coupling between the elements and near-field obstacles for an array are made independent of the angular dependence of all the signals, including the SOI and all the interferers. This procedure is similar in concept to the procedure described in [1–4] with the difference that all the electromagnetic effects, including the presence of near-field obstacles, are characterized in an accurate fashion satisfying the real physics through Maxwell's equations. An important point to note is that here we are carrying out the processing using a single snapshot of the data.

The following is a step-by-step description of what needs to be done to obtain the transformation matrix $[\Im]$, which will provide the necessary transformation so as to obtain the array manifold for the ULVA. This transformation will hold even when the antenna elements are nonuniformly spaced and the induced voltages are perturbed by various undesired electromagnetic effects, such as mutual coupling and the presence of various other near field scatterers:

1.) The first step in designing this transformation matrix is to divide the field of view of the array into Q sectors. If the field of view is 180°, it can be divided into six sectors of 30° each, for example. Then, each Q sector is defined by the interval $[\varphi_q, \varphi_{q+1}]$ for $q = 0, 1, 2, ..., Q - 1$. Or one could choose one sector of 180° width. In that case $Q = 1$. So the width of the sector needs to be defined *a priori*.

2.) Next we define a set of uniformly defined angles to cover each sector:

$$\left[\Phi_q\right] = \left[\varphi_q, \ \varphi_q + \phi, \ \varphi_q + 2\phi, \ ..., \ \varphi_{q+1}\right] \qquad (6.5)$$

where the angle ϕ represents the step size.

3.) We measure/compute the steering vectors associated with the set $[\Phi_q]$ of the real array. This is done by placing a signal in the far field corresponding to each of the angles of arrival $\varphi_q, \ \varphi_q + \phi, \ \varphi_q + 2\phi, \ \varphi_q + 1$. The measured/computed vector for each of the signal sources is different from the ideal steering vector, which is devoid of any undesired electromagnetic effects such as the presence of the mutual coupling between the nonuniformly spaced elements and other near-field coupling effects. The measured manifold matrix is defined by

$$\left[A_q(\Phi_q)\right] = \left[a(\varphi_q), a(\varphi_q + \phi), \ . \ . \ ., a(\varphi_{q+1})\right] \qquad (6.6)$$

This can either be actually measured or simulated and includes all the undesired electromagnetic coupling effects. Hence, each row of $[A_q(\Phi_q)]$

represents the relative signal strength received at the antenna elements. They are a function of only the incident angle of an incoming plane wave within the predefined sector.

4.) Next we fix the virtual elements of the interpolated array. In this chapter we always assume that the virtual array is a uniformly spaced linear array consisting of point sources. We denote by $[A_V(\varphi)]$ the array manifold of the virtual array corresponding to the same set of angles $[\Phi_q]$:

$$\left[A_V(\Theta_q)\right] = \left[a_v(\theta_q), a_v(\theta_q + \Delta\theta), \ldots, a_v(\theta_{q+1})\right] \quad (6.7)$$

where $[a_v(\theta)]$ is a set of theoretical steering vectors corresponding to the uniformly spaced virtual linear array.

5.) We now compute the transformation matrix $[\mathfrak{I}_q]$ for each one of the sector q such that $[\mathfrak{I}_q] [A(\Phi_q)] = [A_v(\Phi_q)]$ using the least squares method. This is achieved by minimizing the functional

$$\min_{[\mathfrak{I}_q]} \left\| \left[A_v(\Phi_q)\right] - \left[\mathfrak{I}_q\right]\left[A(\Phi_q)\right] \right\| \quad (6.8)$$

In order to have a unique solution for (6.8), the number of direction vectors in a given sector must be greater than or equal to the number of the antenna elements in the array. The least squares solution to (6.8) is given by

$$\left[\mathfrak{I}_q\right] = \left[A_v(\Phi_q)\right]\left[A(\Phi_q)\right]^H \left\{ \left[A(\Phi_q)\right]\left[A(\Phi_q)\right]^H \right\}^{-1} \quad (6.9)$$

where the superscript H represents the conjugate transpose of a complex matrix.

This transformation matrix needs to be computed only once *a priori* for each sector, and this computation can be done off-line. Hence, once $[\mathfrak{I}_q]$ is known, we can compensate for the various undesired electromagnetic effects such as mutual coupling between the antenna elements, including the effects of near-field scatterers, as well as nonuniformity in the spacing of the elements in the real array simultaneously. Since the transformation matrix $[\mathfrak{I}_q]$ is defined within the predefined angular sector, we can eliminate both the nonplanar effects and the mutual coupling effects independently within each sector.

6.) Finally, using (6.9), we can obtain the processed input voltages in which the effects of non-uniformity in the spacings and the mutual coupling effects, including the presence of near-field scatterers, have been eliminated from the actual voltages. Let us denote the corrected voltages $[x_c(m)]$ by

$$[x_c(m)] = [\mathfrak{I}_q][x(m)] \qquad (6.10)$$

Now, once (6.10) is obtained, we can apply the direct data domain algorithms to these preprocessed set of voltages $[x_c(m)]$ without any significant loss of accuracy.

Next we present two different applications of this electromagnetic compensation technique to eliminate the mutual coupling between antenna elements and presence of near-field scatterers by solving two types of problems. First, we deal with the DOA estimation, and second, we focus on an adaptive problem to extract the SOI in the presence of strong interferers and noise which may be coherent with the signal using a single snapshot of the data.

6.3   DIRECTION OF ARRIVAL ESTIMATION

In this section, first we compensate for the various undesired electromagnetic effects that destroy the fundamental assumptions of the use of omnidirectional isotropic point sources radiating in free space in all signal-processing algorithms. The objective of this preprocessing is to transform the voltages induced in a realistic array to a set of voltages that is induced in an array of omnidirectional isotropic point radiators. Then, we employ the Matrix Pencil technique [5, 6] to fit these transformed voltages obtained for the ULVA by a sum of complex exponentials. The fit is applied directly to the data set. Hence, in this procedure we can deal with coherent signals impinging at the array simultaneously. The imaginary part of the exponents provides the DOA directly and the residues at the poles provide the signal strengths. Hence, it is possible to implement this procedure in real time on a signal-processing chip such as a DSP32C [7, 8]. We observe the performance, accuracy, and feasibility of the above procedure through simulated numerical examples. The important point to make here is that we use a single snapshot of data, unlike the other methods [9–12]. We now illustrate the forgoing principles through examples.

6.3.1   Semicircular Array

For the first example, we consider a semicircular array (SCA) consisting of 24 antenna elements as shown in Figure 6.1. The radius of the semicircular array is 3.82 wavelengths. The elements of the semicircular array are composed of electrically thin dipoles spaced at equal angles as shown in Figure 6.1. Each element of the array is identically point loaded by 50 Ω at the center. The dipoles are z-directed, of length $L = \lambda/2$ and radius $r = \lambda/200$, where λ is the wavelength of operation. The electrical characteristics of the elements in the chosen array are summarized in Table 6.1.

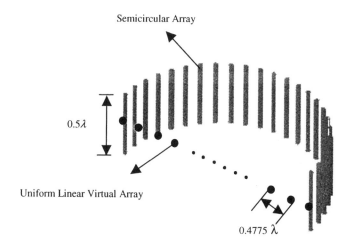

Figure 6.1. Geometry of a semicircular array and a ULVA representing the SCA.

Then the semicircular array is interpolated into a ULVA consisting of 17 isotropic omnidirectional point radiators which are spaced uniformly across the diameter of the SCA and are radiating in free space. The elements of the ULVA are equally spaced and their interelement spacing is given by 0.4775λ. This is also shown in Figure 6.1. By choosing the reference point at the center of the real array, the steering vectors associated with the ULVA are given by

$$[a_v(\theta)] = \left\{ \left[e^{-\frac{j2\pi k\Delta}{\lambda}\cos\varphi}, ..., e^{-\frac{j2\pi 2\Delta}{\lambda}\cos\varphi}, e^{-\frac{j2\pi \Delta}{\lambda}\cos\varphi}, 1, \right. \right.$$
$$\left. \left. e^{-\frac{j2\pi \Delta}{\lambda}\cos\varphi}, e^{-\frac{j2\pi 2\Delta}{\lambda}\cos\varphi}, ..., e^{-\frac{j2\pi k\Delta}{\lambda}\cos\varphi} \right]^T \right\}_{(2k+1)\times 1}$$

$$(6.11)$$

TABLE 6.1
Physical sizes for the antenna elements in the semicircular array

Number of elements in the semicircular array	24
Length of the z-directed wires	$\lambda/2$
Radius of the wires	$\lambda/200$
Loading at the center of the element	$50\,\Omega$

where $(2k + 1)$ is the number of the elements of the ULVA. Δ is the distance between the elements in the ULVA, which is 0.4775λ in this case. The incremental angular size ϕ in the interpolation region, $[\Phi_q] = [\varphi_q, \varphi_{q+1}]$, is chosen to be $1°$. Then, a set of real steering vectors are measured/computed for each of the incident sources located at each of the angles $\varphi_q, \varphi_q +, \varphi_q + 2\varphi, \ldots, \varphi_{q+1}$. The real measured/computed vector $[a(\varphi)]$ will be distorted from the ideal steering vector, due to the presence of mutual coupling between the elements of the real array. These voltages are computed using the electromagnetic analysis code WIPL-D [13]. Then, using (6.9), we obtain the transformation matrix to compensate for the effects of nonuniformity in the spacing and the presence of mutual coupling between the elements of the real array.

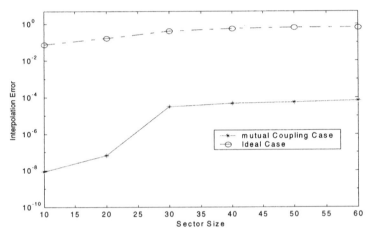

Figure 6.2. Interpolation error of the transformation matrix when using the least squares method as a function of the sector size both without and with mutual coupling effects.

We next define an interpolation error to check the accuracy of this interpolation methodology. The error is defined by

$$\text{interpolation error} = \sqrt{\frac{1}{IJ} \sum_{i=1}^{I} \sum_{i=1}^{J} \left| [A_v(\Phi_q)]_{i,j} - [\Im_q][A(\Phi_q)]_{i,j} \right|^2} \quad (6.12)$$

where I and J represent the number of the columns and rows in matrix $[A_v(\varphi)]$. Figure 6.2 shows the interpolation error as a function of the sector size both with and without mutual coupling effects. The plot labeled "ideal case" deals with the transformation of a set of voltages that are obtained for a SCA consisting of omnidirectional isotropic point radiators to a ULVA consisting of omnidirectional isotropic point radiators spanning the diameter of the SCA. In this case, of course, there is no mutual coupling between the elements of the array. For the plot-marked mutual coupling case, we transform the set of voltages that are induced at the loads of the elements of the SCA, consisting of linear electrically thin dipoles that are centrally point loaded by 50 Ω to that of an ULVA consisting of isotropic omnidirectional point radiators spanning the diameter of the SCA. The interpolation error for both cases is a monotonically increasing function of the sector width. It is interesting to note that the interpolation error is smaller when mutual coupling is present than in the ideal case. Our limited numerical experiments illustrate that the presence of mutual coupling reduces the condition number of the transformation matrix $[\Im_q]$, and therefore it performs better numerically in generating a lower interpolation error. Now, the Matrix Pencil algorithm can be applied directly to the set of complex voltages induced at each of the elements of the uniformly spaced virtual array to obtain the directions of arrival (DOAs) of the various signals and their strengths.

We can use the plots of the interpolation error to select the sector size needed to achieve a desired quality of fit so as to compensate for the lack of non-uniformity and presence of mutual coupling in a real array with prespecified degree of accuracy. For example, for a 80° sector, Figure 6.3 shows the local interpolation error when using the transformation matrix within $[\Phi] = [50°, 130°]$ in the presence of mutual coupling. Also, Figures 6.4 and 6.5 show the local interpolation error when using the transformation matrix within $[\Phi] = [30°, 150°]$ and $[\Phi] = [10°, 170°]$ in the presence of mutual coupling. In both of these cases, the data were interpolated to a 17-element uniformly spaced virtual array of omnidirectional point sources.

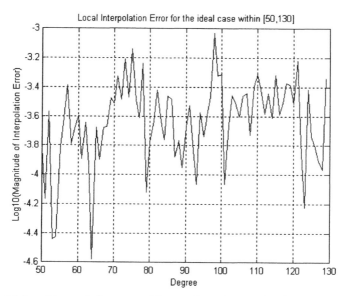

Figure 6.3. Local interpolation error of the transformation matrix within $[\Phi] = [50°, 130°]$ when mutual coupling is present.

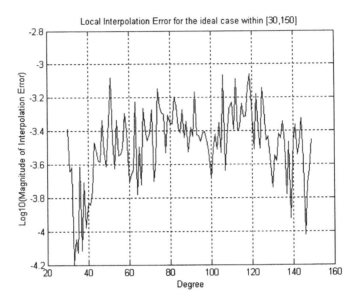

Figure 6.4. Local interpolation error of the transformation matrix within $[\Phi] = [30°, 150°]$ when mutual coupling is present.

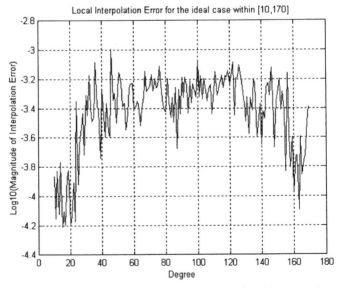

Figure 6.5. Local interpolation error of the transformation matrix within $[\Phi] = [10°, 170°]$ when mutual coupling is present.

An important consideration in the design of the virtual arrays is that the interpolation matrix $[\Im_q]$ must be well conditioned. If $[\Im_q]$ turns out to be ill conditioned, the virtual array must be redesigned. Table 6.2 presents the condition number of the matrix $[\Im_q]$ associated with the number of elements of the virtual array within the predefined angles in the presence and in the absence of mutual coupling. The condition number for the matrix $[\Im_q]$ is defined as the ratio of the largest to the smallest singular values. The improvement in the condition number as the number of elements of the virtual array is reduced is quite clear. If the interpolation region is large, the condition number also improves, as seen from the table. Finally, comparing Table 6.2, it is clear that the presence of mutual coupling makes the condition number of the transformation matrix much less, resulting in a reduced interpolation error. Thus, the presence of mutual coupling actually is beneficial in this case, as it improves the stability of the numerical solution and therefore its performance.

TABLE 6.2

Condition number of $[\Im_q]$ versus the number of elements of the virtual array for the semicircular array within a predefined region:

(a) in the presence of mutual coupling

	$N + 1 = 17$	$N + 1 = 21$	$N + 1 = 25$
$[\Phi] = [70°, 110°]$	6.8850×10^5	1.7074×10^8	4.3518×10^9
$[\Phi] = [50°, 130°]$	1.1714×10^3	6.0434×10^5	1.2701×10^9
$[\Phi] = [30°, 150°]$	344.6679	1.9910×10^5	9.0285×10^7

(b) in the absence of mutual coupling

	$N + 1 = 17$	$N + 1 = 21$	$N + 1 = 25$
$[\Phi] = [70°, 110°]$	1.3391×10^{11}	4.0663×10^{11}	1.0022×10^{12}
$[\Phi] = [50°, 130°]$	4.5788×10^{10}	8.3075×10^{10}	2.7647×10^{11}
$[\Phi] = [30°, 150°]$	3.8131×10^9	1.0881×10^{10}	6.0409×10^{10}

TABLE 6.3.

Complex amplitudes of all the signals along with their DOA

	Magnitude	Phase	DOA
Signal 1	1.0 V/m	0.0	110°
Signal 2	1.0 V/m	0.0	80°
Signal 3	1.0 V/m	0.0	50°
Signal 4	1.0 V/m	0.0	40°

Next, we consider four signals arriving from the azimuth directions 110°, 80°, 50°, and 40° impinging on the SCA, consisting of linear thin-wire dipoles. The intensities of the four signals and their associated DOAs are summarized in Table 6.3.

Noise is introduced to each of the measured antenna voltages so that the received signal-to-noise ratio (SNR) is set to 20 dB at each of the antenna elements. The noise is additive and is modeled as a Gaussian random variable. The interpolation region is defined by $[\Phi] = [30°, 150°]$ and the incremental size ϕ in the interpolation region is chosen to be 1°. After we compensate for the various electromagnetic effects, such as the presence of mutual coupling between the elements of the array using (6.9) and (6.10), we transform the voltages measured at the feed points of the SCA to an equivalent set of voltages that will be induced in a ULVA operating in free space. We then estimate the DOAs along with the complex amplitudes of the various signals by applying the Matrix Pencil method to the voltages induced in the ULVA. The Matrix Pencil method is described in Appendix C. The results for this simulation are presented in Table 6.4.

As can be seen, the technique presented can compute the DOAs of all the signals using a single snapshot of data in the presence of mutual coupling between the elements of the semicircular array. Also, this procedure can resolve signals, which are within the main beam of the array. In this example, the beam width of the array is approximately 35°.

TABLE 6.4
Estimated DOAs of all the signals and their estimated complex amplitudes

	Magnitude	Phase	DOA
Signal 1	1.030 V/m	$-0.12°$	111.15°
Signal 2	1.001 V/m	$-0.4°$	81.06°
Signal 3	0.976 V/m	$0.94°$	51.04°
Signal 4	1.002 V/m	$0.002°$	41.04°

6.3.2   Semicircular Array with a Near-Field Scatterer

In the next example we consider the effects of scatterers located electrically close to the semicircular array. As shown in Figure 6.6, a large near field scatterer is located within a region, which is five times the radius of the semicircular array and is situated along the direction of 110°, which happens to be the DOA of one of the signals. The length and the width of the scatter are both 7.26λ. The height of the scatter is 15.28λ. We again consider the case of four incoming signals arriving from the directions of 110°, 80°, 50°, and 40°. The 24-element semicircular array consists of thin-wire dipole antennas as described in the previous example. All the signal intensities and their directions of arrival are summarized in Table 6.3. The signal-to-noise ratio (SNR) at each of the antenna elements is again set to be 20 dB. The interpolation region is $[\Phi] = [30°, 150°]$ and the incremental size ϕ in the interpolation region is chosen to be 1°.

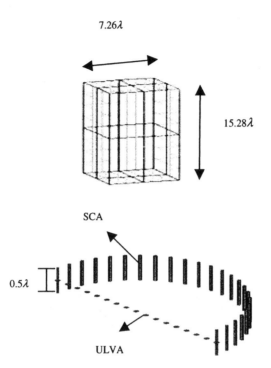

Figure 6.6. Geometry of a SCA and a ULVA representing the SCA along with a near-field scatterer.

After we compensate for the various electromagnetic effects, such as the presence of mutual coupling between the elements of the array including the effects of the near-field scatterer using (6.9), we estimate the DOAs through the Matrix Pencil approach. The estimated results obtained by the Matrix Pencil technique are given in Table 6.5.

TABLE 6.5
Estimated DOAs and the strengths of the four signals in the presence of a near-field scatterer

	Magnitude	Phase	DOA
Signal 1	0.997 V/m	0.028°	111.02°
Signal 2	1.003 V/m	0.0°	81.09°
Signal 3	0.996 V/m	− 0.002°	51.02°
Signal 4	1.003 V/m	0.005°	41.02°

It is interesting to note that even though the near-field scatter is blocking the direct path of the signal arriving from 110°, it can still be picked up by this two-step algorithm. Therefore, the technique presented in this section can be used not only in estimating DOAs using nonuniformly spaced arrays using a single snapshot of data, but the procedure can also simultaneously take into account mutual coupling effects between the antenna elements and the presence of near-field scatterers located close to the array. Algorithms like MUSIC can indeed estimate the DOAs using a nonuniformly spaced array, but have problems when dealing with coherent signals. In addition, MUSIC requires a number of snapshots of the data equal to or greater than the number of antenna elements. Here we not only deal with nonuniformly spaced elements operating in the presence of mutual coupling and near-field scatterers but can also deal with coherent signals. The last point is that all these can be achieved using a single snapshot of data.

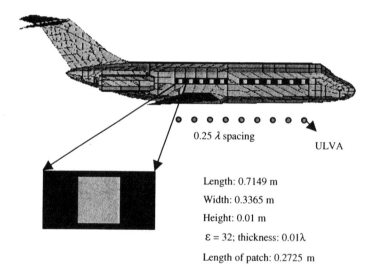

0.25 λ spacing

ULVA

Length: 0.7149 m

Width: 0.3365 m

Height: 0.01 m

$\varepsilon = 32$; thickness: 0.01λ

Length of patch: 0.2725 m

Figure 6.7. Geometry of a microstrip patch array on the side of an aircraft and an equivalent ULVA representing the microstrip patch array mounted on a conformal surface.

6.3.3   DOA Estimation Using a Conformal Microstrip Patch Array on the Side of a Fokker Aircraft

In this example, we estimate the DOA of the various signals impinging on a conformal microstrip patch array on the side of a Fokker aircraft. As seen in Figure 6.7, the elements of the 11-element microstrip patch array are not placed uniformly, nor do they lie on one flat surface. A detailed view of the microstrip

patch antenna element is also shown in Figure 6.7. The rectangular microstrip
patch element is fed by a probe and is situated on a high-dielectric-constant
substrate so that it resonates at an operating frequency of 100 MHz. The
thickness of the dielectric substrate is 0.01λ. There is a strong mutual coupling
between the antenna elements, including the wings and fuselage of the aircraft.
The number of elements in the patch array is 11 and it is interpolated into a
virtual array consisting of 11 uniformly spaced omnidirectional point sources
separated by 0.25λ at 100 MHz. We consider three signals arriving from 139°,
99°, and 60°, impinging at the microstrip patch array on the side of the aircraft.
The intensities and associated DOAs of the various signals are summarized in
Table 6.6 .

TABLE 6.6
Complex amplitudes of all the signals and their DOAs incident on the aircraft

	Magnitude	Phase	DOA
Signal 1	1 V/m	0	139°
Signal 2	1 V/m	0	99°
Signal 3	1 V/m	0	60°

After we compensate for the strong electromagnetic coupling effects
between the elements of the microstrip patch array and the aircraft at 100 MHz,
like the effects of nonuniformly spaced elements on a conformal nonplanar
surface and the effects of mutual coupling between the elements of the patch
array and their interactions with the wings and fuselage by using (6.9), we
estimate the DOAs and their complex amplitudes through use of the Matrix
Pencil approach to the preprocessed voltages induced on the 11-element ULVA.
The simulation results are presented in Table 6.7. From the table it is seen that
not only the DOAs but also the complex amplitudes of the coherent signals have
been recovered with engineering accuracy using a conformal phased array
consisting of nonuniformly spaced microstrip patch elements.

TABLE 6.7
Estimated DOAs and the strengths of the three signals incident on the phased array

	Magnitude	Phase	DOA
Signal 1	1.0444 V/m	−0.0545	139.71°
Signal 2	0.9052 V/m	0.0621	110.86°
Signal 3	1.0838 V/m	−0.0002	61.75°

6.4 ADAPTIVE PROCESSING USING A SINGLE SNAPSHOT FROM A NONUNIFORMLY SPACED ARRAY OPERATING IN THE PRESENCE OF MUTUAL COUPLING AND NEAR-FIELD SCATTERERS

In this section we illustrate how to employ the interpolation technique of the preceding section to carry out adaptive processing using a single snapshot of the data in a nonuniformly spaced array in the presence of mutual coupling and near-field scatterers. Numerical examples are now presented to illustrate the procedure.

In all the examples, we are basically interested in three different kinds of antenna configurations. The three different arrangements we have used are as follows:

1.) A nonuniformly spaced linear array (as shown in Figure 6.8)
2.) A semicircular array (as shown in Figure 6.1)
3.) A spatially sinusoidally modulated array (as shown in Figure 6.9)

In all the antenna configurations, as shown in Figures 6.1, 6.8, and 6.9, the antenna elements of the real array are considered to be half-wavelength-long thin-wire dipoles. Each element of the array is identically point loaded by 50 Ω at the center. The dipoles are z-directed and are of length $L = \lambda/2$ and radius $r = \lambda/200$. The details of the chosen array elements are presented in Table 6.1.

Then, as described in Section 6.3, all the real arrays operating in the three different configurations mentioned above are interpolated into a similar ULVA consisting of 17 uniformly spaced omnidirectional point sources separated by a distance d/λ. Typically d is chosen to be close to $\lambda/2$. By choosing the reference point at the center of the real array in configuration 2 for the SCA and configuration 3 for the sinusoidally modulated array, the steering vectors associated with the virtual array are given by (6.11).

For configuration 1 (a nonuniformly spaced array), we choose the first element in the array as a reference point. Then the steering vectors for the ULVA are given by

$$[a_v(\theta)] = \left\{ \left[1, \, e^{\frac{j2\pi\Delta}{\lambda}\cos\varphi}, \, e^{\frac{j2\pi 2\Delta}{\lambda}\cos\varphi} \, \cdots, e^{\frac{j2\pi(2k)\Delta}{\lambda}\cos\varphi} \right]^T \right\}_{(2k+1)\times 1}$$

(6.13)

Here the number $(2k + 1)$ of elements of the virtual array in both (6.11) and (6.12) is considered to be 17, and λ is the wavelength of the signal located in the far-field region of the array, and the distance, Δ, between the elements in the virtual array is 0.4775λ. The incremental size ϕ in the interpolation region, $[\Phi] = [\varphi_q, \, \varphi_{q+1}] = [30, \, 150°]$, is chosen to be 1°.

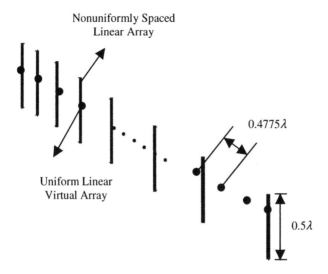

Figure 6.8. Geometry of a nonuniformly spaced array (NLA) and a ULVA representing the NLA.

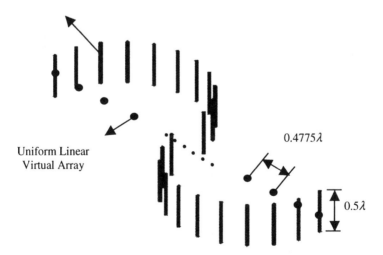

Figure 6.9. Geometry of a sinusoidally spaced array (SSA) and a ULVA representing the SSA.

The sector chosen here, for example, is of width 120° symmetrically located around the broadside. Then, a set of real steering vectors are measured/computed for sources located at each of the angles φ_q, $\varphi_q + \phi$, $\varphi_q + 2\phi$, . . . , ϕ_{q+1}. The measured/computed vector $\bar{a}(\varphi)$ is then distorted from the ideal steering vector, due to the presence of mutual coupling between the elements of the real array. The actual steering vectors having all the undesired electromagnetic effects are computed using the electromagnetic analysis code WIPL-D [13]. Then, using (6.9), we obtain the transformation matrix to compensate for the effects of non-uniformity in spacing and the presence of mutual coupling between the elements of the real array. Finally, using (6.10), we can obtain the corrected input voltage in which the nonuniformity in spacing and mutual coupling effects are eliminated from the actual voltage. We then apply the direct data domain least squares approach described in Chapter 4 to estimate the complex amplitude of the SOI given its DOA.

In the three examples to be discussed, this compensated input voltages are passed through the direct data domain algorithms described in Chapter 4 to recover the magnitude of the desired signal while simultaneously rejecting all other interferences.

6.4.1   Constant Signal

First, we consider a case where we have two jammers arriving at the array along with the SOI. The interference-to-signal ratio for the jammers impinging on the array from an angle of 60° is varied from 0 to 54 dB. The other jammer arrives at the array at an angle of 70° with respect to the x-axis. The SOI is incident at an angle of 80° at a frequency of 300 MHz. The values of the signals are summarized in Table 6.8. The received signal-to-noise ratio at the antenna elements is set at 20 dB in this example. For this input, the signal strength is estimated while rejecting jammers. This estimation is applied to all three antenna configurations.

TABLE 6.8
Complex amplitudes of the signals and their DOAs

	Magnitude	Phase	DOA
Signal	1.0 V/m	0.0	80°
Jammer 1	1.0 V/m	0.0	70°
Jammer 2	1.0–500.0 V/m	0.0	60°

The output signal-to-interference plus noise ratio is an indicator of the accuracy of our estimate. It is defined as

$$\text{SINR}_{\text{out}} = 20 \log \left| \frac{\alpha}{\alpha - \alpha_{\text{est}}} \right| \tag{6.14}$$

where α is the amplitude of the desired signal and α_{est} is the estimate of the amplitude of the reconstructed signal.

The results are shown in Figures 6.10 and 6.11 for all three configurations. The thermal noise is ignored in Figure 6.10 while the signal-to-noise ratio is set at 20 dB in Figure 6.11. The noise is additive and is modeled as a Gaussian random variable. The x-axis of the graph corresponds to the jammer 2-to-SOI power level, while the y-axis corresponds to the output signal-to-interference plus noise ratio, as defined in (6.14). It can be observed in these figures that the interpolation technique using measured/computed steering vector shows proper compensation for all the electromagnetic effects of the real arrays to estimate the magnitude of the SOI based on direct data domain approach.

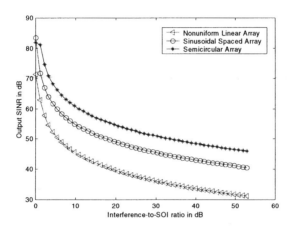

Figure 6.10. Output SINR as a function of interference-to-SOI ratio in dB without noise.

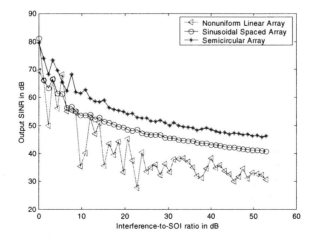

Figure 6.11. Output SINR as a function of interference-to-SOI ratio in dB with noise.

6.4.2 Effect of Variation of the Angular Separation between Signal and Jammer

In the next example we consider the fidelity in the performance for various angular separations between a signal and a jammer. We consider a case where we have one jammer signal arriving at the array along with the SOI. The DOA of the jammer is varied from 101° to 110° while SOI is arriving from 100° with respect to the *x*-axis. The signal-to-interference ratio is fixed at 0 dB, as shown in Table 6.9.

TABLE 6.9
Complex amplitudes of the various signals and their DOAs

	Magnitude	Phase	DOA
Signal	1.0 V/m	0.0	100°
Jammer	1.0 V/m	0.0	101° − 110°

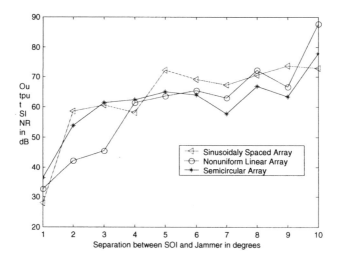

Figure 6.12. Output SINR as a function of separation between SOI and jammer.

The results are shown in Figure 6.12 for the three antenna configurations. The *x*-axis corresponds to the angular separation between the SOI and the jammer, while the *y*-axis corresponds to the output signal-to-interference plus noise ratio. These plots indicate that, after compensation of the various unwanted electromagnetic effects, good estimates for the amplitude of the SOI based on a direct data domain approach have been obtained. It is seen from the figures that the performance of a semicircular array and a spatially sinusoidally modulated array in estimating the DOA is better than that for a nonuniform linearly spaced array.

6.4.3   Effects of Blockage Produced by Near-Field Scatterers

Next, we consider the effects of a near-field scatterer located near each of the array configurations. The near-field scatterer is located within a distance is twice the radius of the array and is located along the direction of 60° in such a way that this near-field scatterer blocks the direct line of sight of the SOI. The length and width of the scatterer are 1.26 wavelengths and the height of the scatter is 3.28 wavelengths. The desired signal and the jammers are as summarized in Table 6.9. The results are shown in Figure 6.13 for the three different array configurations. It can be observed in these figures that the output signal-to-interference plus noise ratio is much lower than where there was no blockage of the SOI, due to a near-field scatterer. However, one can still   obtain a   proper estimate   for the complex amplitude of the SOI after compensating for the various unwanted effects. It is seen that the estimated amplitude of the SOI for a SCA and a sinusoidally modulated spaced array is better than that for a nonuniformly spaced linear array.

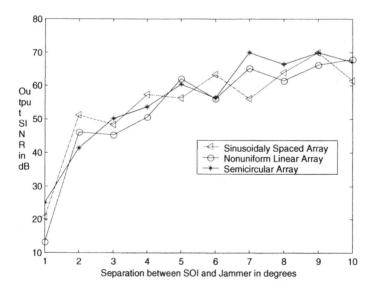

Figure 6.13. Output SINR as a function of separation between SOI and jammer with blockage.

Since a SCA has better performance than the other two configurations, in the next section we do a more detailed simulation using the SCA.

6.4.4   Recovery of a Varying Signal in the Presence of Strong Jammers Using a Semicircular Array

In this example, the intensity of the desired signal is varied from 1 to 10.0 V/m in steps of 0.01 V/m from snapshot to snapshot, while the strong jammer intensities remain constant. All the values of the various signals and their directions of arrival are summarized in Table 6.10.

TABLE 6.10
Complex amplitudes of the various signals and their DOAs

	Magnitude	Phase	DOA
Signal	1.0–10.0 V/m	0.0	100°
Jammer 1	1000.0 V/m	0.0	70°
Jammer 2	500.0 V/m	0.0	130°
Jammer 3	100.0 V/m	0.0	140°

The measured voltages at the loads of the dipoles in the SCA are compensated using (6.10), and then the processed voltages are used to recover the signal and null the jammers using the direct data domain algorithm least squares (D^3LS) method. If the jammers are properly nulled, the reconstructed signal magnitude should remain linear as a function of the incident signal.

Figure 6.14 shows the results of using the D^3LS after compensating for the effects of nonuniformity in spacing and the mutual coupling between the elements of the SCA. The magnitude displays the expected linear relationship and the phase is again close to zero. The beam pattern associated with compensating for the nonuniformity in spacing and the mutual coupling effects using the direct data domain is shown in Figure 6.15. In particular, it can be seen that, the nulls are very deep and placed along the correct directions of the strong jammers. This demonstrates that the nonuniformity of the SCA and the mutual coupling effects has been suppressed.

6.4.5   Effects of Noise in the Received Voltages in a Semicircular Array

This example demonstrates the effects of thermal noise on the performance of the adaptive technique. The noise at any element of the SCA is assumed to be independent of the noise at other elements. The noise in this case is zero mean additive Gaussian noise. In this example, the desired signal is corrupted by three jammers, as given in Table 6.11. The signal-to-noise ratio at each element is set at 20 dB. Note that one of the jammers is strong (100 V/m).

The set of induced voltages, affected by noise, is at first compensated for the various electromagnetic effects using (6.10) and then is passed to the adaptive processor. This procedure is repeated 500 times with different noise samples. The results for the SOI obtained from the 500 samples are then used to find the mean and variance of the estimated signal. The output signal-to-interference plus noise ratio (SINR) in decibels has been defined in (6.14).

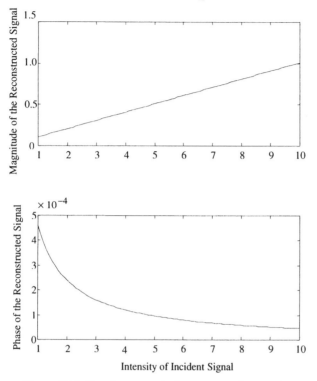

Figure 6.14. Signal recovery in the presence of strong jammers.

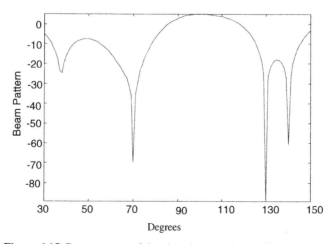

Figure 6.15. Beam pattern of the phased array when nulling strong jammers.

As seen from Table 6.12, after the nonuniformity in spacing and the mutual coupling between the array elements have been compensated properly using the technique presented above, the jammers have been nulled, yielding an accurate estimate for the signal. The total interference power is suppressed to nearly 26 dB below the signal.

TABLE 6.11
Values for the signal and jammer

	Magnitude	Phase	DOA
Signal	1.0 V/m	0.0	100°
Jammer 1	100.0 V/m	0.0	70°
Jammer 2	1.0 V/m	0.0	130°
Jammer 3	1.0 V/m	0.0	140°

TABLE 6.12
Simulation results in the presence of noise

Input signal-to-noise ratio	20 dB
Number of samples	500
True value of the signal	(1.0, 0.0) V/m
Mean of 1000 estimates	(1.029410, − 0.03816)
Variance of the estimate	0.00136
Output signal–to-interference-plus-noise ratio	26.34271 dB

6.4.6   Effect of Large Near-Field Scatterers on the Performance of an SCA

Finally, we consider the effects of a large near-field scatterer located close to the SCA, as shown in Figure 6.6, where it is oriented along the direction of 110°. The length and width of the scatterer are 7.26λ, and the height of the scatterer is 15.28λ. Hence, the semicircular array and the scatterer have strong coupling effects in addition to the presence of mutual coupling between the elements. We again consider the case of four incoming signals from 110°, 80°, 50°, and 40°. The strengths of the desired signal and the jammers along with their DOA have been summarized in Table 6.11.

After we compensate for the nonuniformity in spacing and the presence of mutual coupling between the elements of the array and the scatterer using (6.10), we estimate the amplitude of the desired signal through use of the D^3LS. Figure 6.16 plots the results of using the D^3LS method to recover the signal in the presence of mutual coupling between the elements of the array and a scatterer located close to the array. The magnitude and phase of the signal recovered are shown in Figure 6.16. The expected linear relationship between the intensity of the incident signal and the magnitude of the reconstructed signal is clearly seen, implying that the jammers have been nulled and the signal recovered correctly. The phase of the recovered signal varies within a very small value of zero.

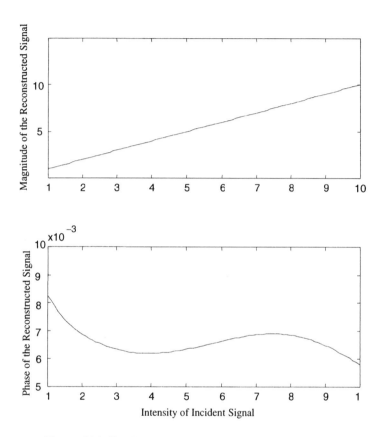

Figure 6.16. Signal recovery in the presence of a near-field scattererer.

The beam pattern associated with this example is shown in Figure 6.17. The nulls are deep and have been placed along the correct directions, implying that the effects of nonuniformity in spacing, mutual coupling between the array elements and near-field scatterers, have been suppressed so as to even null a strong jammer. By comparing the plots of Figures 6.15 and 6.17, one can observe that the presence of a near-field scatterer has influenced the adapted beam pattern.

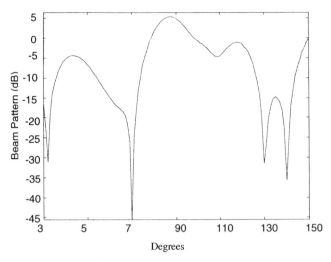

Figure 6.17. Adaptive beam pattern of the SCA in the presence of a near-field scatterer.

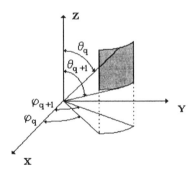

Figure 6.18. Parameters of the sector.

6.5   DOA ESTIMATION USING A PHASED ARRAY LOCATED ON A CONFORMAL HEMISPHERICAL SURFACE

In this section we describe how to carry out direction of arrival estimation using directive antenna elements over a conformal surface like a hemisphere. We consider the use of three directive element, such as shorted dual-patch antennas, dielectric resonator antennas, and horn antennas as elements in a conformal array. The surface selected for the implementation is a hemisphere. The procedure to carry out the analysis to obtain the steering vectors over a conical scan subtended by the azimuth angles $[\varphi_q, \varphi_{q+1}]$ and from elevation angles spanning $[\theta_q, \theta_{q+1}]$. This is shown in Figure 6.18. As outlined earlier in this chapter, we now transform the real array situated over a hemispherical surface into a two-dimensional uniform linear virtual array (2DULVA) lying in the x–y plane containing the projection of the hemisphere. The 2DULVA can either be used in a cross, an L-shaped, or a two-dimensional grid configuration. Here, in this methodology, the one-dimensional transformation matrix, for azimuth angles only given by (6.9) and (6.10) is extended to two dimensions to handle both elevation and azimuth angles. Through this 2D transformation matrix, which is similar to (6.10), we map the voltages that are induced in the feed point of these real antenna elements operating in the presence of mutual coupling and other near-field scatterers to a planar 2DULVA, as shown in Figure 6.19.

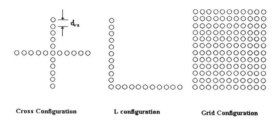

Cross Configuration   L configuration   Grid Configuration

Figure 6.19. Different configurations of two-dimensional uniform virtual arrays (d_{va}: distance between virtual elements).

6.5.1   Shorted Dual-Patch Antenna on a Hemispherical Surface

The first structure to be considered in this technique is the application of a shorted dual-patch antenna (SDPA) to form an array over a hemispherical surface. An antenna element representing a SDPA is shown in Figure 6.20, where the specific dimensions are given in Figure 6.21 [14]. A SDPA consists of two layers, in which the upper patch, which is of trapezoidal shape, is connected along one edge to the ground by a vertical metal wall and the other edge is connected to the lower patch by another vertical wall. These novel features produce a significant reduction in the resonant frequency of the antenna. The numerical results obtained by using a dynamic electromagnetic simulator like WIPL-D [13] are very similar to the experimental results described in [14]. The actual geometry of the SDPA is shown is Figure 6.21, and the input impedance for this geometry as a function of frequency is given in Figure 6.22. For this structure the resonant frequency occurs at about 2.41 GHz, as shown in Figure 6.22. Finally, the radiation pattern of the SDPA along the E-plane is shown in Figure 6.23, illustrating that the element pattern has some directivity and the front lobe is larger than the back lobe. These results are close to the measured ones given in [14]. The SDPA is considered to be situated over a finite square ground plane of size G centimeters on each side.

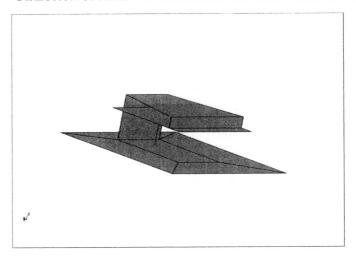

Figure 6.20. Shorted dual-patch antenna.

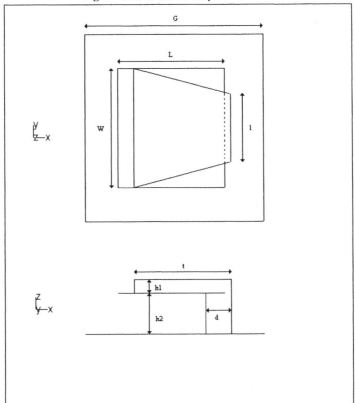

Figure 6.21. Geometry of the shorted dual-patch antenna. $h^1 = 2$mm; $h^2 = 6$ mm; $d = 4$ mm; $t = 15$ mm; $l = 10$ mm; $L = 16.5$ mm; $W = 17.5$ mm; $G = 27.4$ mm.

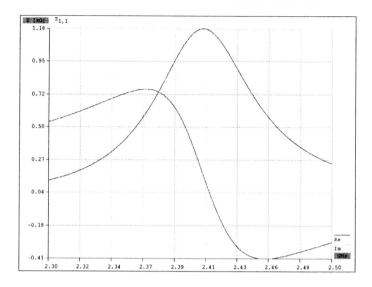

Figure 6.22. Impedance versus frequency of a SDPA. Resonance frequency = 2.41 GHz.

We now consider a 48-element array consisting of the SDPA which is distributed over a hemisphere of radius 1.4λ. The elements are placed in a star configuration as seen in Figure 6.24. We now transform the voltages that are induced at the feed point of these elements to a 16-element omnidirectional point radiator array which is situated on the x–y plane with $z = 0$. The spacing between the virtual elements is 0.5λ at 2.41 GHz, and all the omnidirectional point sources are placed on a 4×4 array located at the center of the circle. For all the examples that we are going to present, the azimuth scan angle, φ, varies from $1°$ to $360°$ with an angular step of $1°$ (i.e., $\Delta\varphi = 1$). The step size for the elevation angle is also $1°$ (i.e., $\Delta\theta = 1$). This information is now used to calculate the transformation matrix [3].

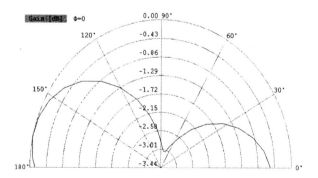

Figure 6.23. Radiation pattern of the shorted dual-patch antenna. Normalized. E-plane.

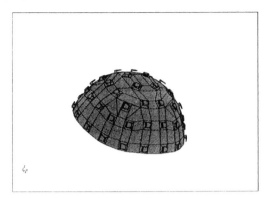

Figure 6.24. Shorted dual-patch antenna array.

The interpolation error and the condition number of the transformation matrix $[\Im_q]$ for different sectors q are given in Table 6.13. The interpolation error has been defined in (6.12). The condition number is defined as the ratio of the largest singular value to the smallest singular value of the matrix. As observed from Table 6.13, the error in the transformation to a ULVA remains more or less constant until $0° \le \theta \le 50°$. The azimuth scan is the entire $360°$. We do not present results beyond $\theta \le 60°$ because all the SDPA elements on the hemisphere do not see the signal and hence the transformation matrix $[\Im_q]$ becomes undefined. In that case we restrict the azimuth angle scan, as will be seen later on in this section.

We now use the Matrix Pencil method in two dimensions to estimate the DOA of three signals impinging on the 48-element array, which has been interpolated to a 16-element 2DULVA as shown in Figure 6.25. The distance between the elements of the virtual array is 0.5λ and is centered at a distance $R/2$, where R is the radius of the hemisphere and is oriented along the direction shown in Figure 6.25. The goal here is to estimate the directions of arrival and the amplitudes of three signals that are incident on the array, from $(\varphi, \theta)_1 = (50°, 5°)$, $(\varphi, \theta)_2 = (80°, 20°)$, and $(\varphi, \theta)_3 = (140°, 30°)$. The amplitudes of the signals are 1 V. The scan for the transformation matrix is carried out spanning over all the azimuth angles $(\varphi_q, \varphi_{qq}) = (30°, 150°)$ and the elevation angles $(\theta_q, \theta_{qq}) = (0°, 40°)$.

First we generate the transformation matrix that electromagnetically maps the voltages induced on a 16-element SDPA as shaded in Figure 6.26 situated over a quadrant of a hemispherical surface to a 16-element 2DULVA as shown by Figure 6.25. Here the real elements are shorted dual-patch antennas placed over the hemispherical array in a star configuration. The estimated DOAs of all three signals given by the Matrix Pencil method using the voltages induced in the 2DULVA are given in Table 6.14. The estimated amplitudes of all three signals are also shown in Table 6.14.

TABLE 6.13

Interpolation error and condition number for the transformation matrix [ℑ] when interpolating to a 16-element ULVA

Elevation Scan Angle (θ)	Interpolation Error (%)	Condition Number
0 → 10	0.2	2.7×10^8
0 → 20	0.36	5×10^5
0 → 30	2.4	8×10^5
0 → 40	5.9	17,191
0 → 50	10.2	15,368
0 → 60	19.7	8,898
0 → 70	35.3	11,820
0 → 80	48.5	4,694
0 → 90	58.9	9,557

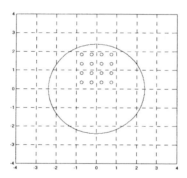

Figure 6.25. Location of the 16-element virtual array when the scanned sector is defined by the azimuth angle $(\varphi_q, \varphi_{qq}) = (30°, 150°)$ and elevation angle $(\theta_q, \theta_{qq}) = (0°, 40°)$.

TABLE 6.14

Estimation of the DOA of three signals: $(\varphi, \theta)_1 = (50°, 5°)$, $(\varphi, \theta)_2 = (80°, 20°)$, $(\varphi, \theta)_3 = (140°, 30°)$

φ	θ	Estimation of φ	Estimation of θ	Estimation of the Amplitude
50	5	51.9152	5.8354	1.0328
80	20	78.8039	18.8433	0.8658
140	30	141.3827	30.5080	0.9275

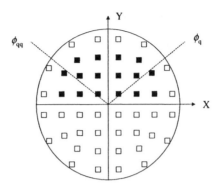

Figure 6.26. Top view of the hemispherical array. The black squares represent the 16 active elements for processing, and the white squares represent the rest of the real elements mounted over the hemispherical surface.

6.5.2   Dielectric Resonator Antennas on a Hemispherical Surface

The antenna array consists of dielectric resonator antennas (RDRs) as elements on a hemispherical surface. The reason that dielectric resonators are chosen is because they have a high radiation efficiency, typically greater than 98%.

An antenna element representing an RDR is shown in Figure 6.27, where the specific dimensions of the RDR are given in Figure 6.28 [15]. The coaxial probe extends into a dielectric of relative permittivity of 8.9. The antenna element is situated over a square finite ground plane of dimension G = 6.0 cm along each side. The input impedance, shown in Figure 6.29, has been obtained using WIPL-D and is similar to the results of [15]. The actual resonant frequency of this RDR occurs at 3.07 GHz, as in Figure 6.29. Finally, the radiation pattern along the *E*-plane is shown in Figure 6.30.

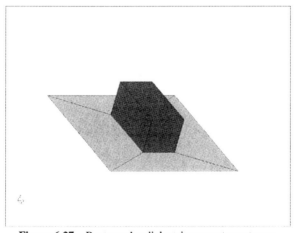

Figure 6.27.   Rectangular dielectric resonator antenna.

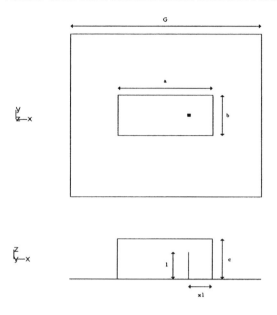

Figure 6.28. Geometry of the rectangular dielectric resonator antenna.
$a = 30$ mm; $b = 15$ mm; $c = 15$ mm; $l = 10.3$ mm; $x_1 = 7.5$ mm; $G = 60$ mm.

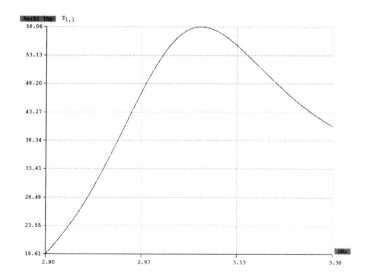

Figure 6.29. Impedance versus frequency resonant frequency = 3.07 GHz.

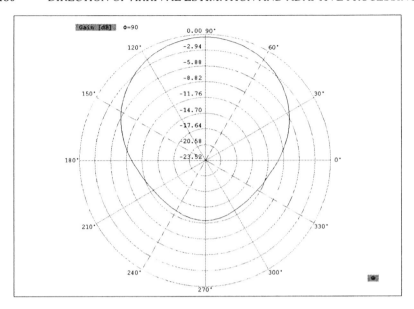

Figure 6.30. Radiation pattern of the rectangular dielectric resonator. Normalized.

A total of 48 RDR elements are placed on a hemisphere of radius 2.1λ, as shown in Figure 6.31, in order to have a RDR-hemispherical array (RDR-HA). We now transform the voltages induced in the real elements to the virtual array. The 18 elements that are a part of this transformation are represented by black squares in Figure 6.32. The sector over which the transformation is carried out is covered by the azimuth angle of (φ_q, φ_{qq}) = (30°, 150°) and the elevation scan angle varies from 10° to 50°, as shown in Table 6.15. The actual voltages induced in the 18 real elements are mapped to the voltages induced that will be induced in the 16-element 2DULVA, as shown in Figure 6.25. For the ULVA the elements are point radiators radiating in free space. These virtual elements are separated by 0.5λ, as shown in Figure 6.25. The results of the transformation are given in Table 6.15.To carry out an estimation of DOA for three signals impinging on this 48-RDR elements array, we transform the voltages induced in the real 18-element array to a 16-element ULVA. The goal here is to estimate the directions of arrival and the amplitudes of three signals that are incident on the array from (φ, θ)$_1$ = (50°, 5°), (φ, θ)$_2$ = (80°, 20°), and (φ, θ)$_3$ = (140°, 30°). The amplitude of all the signals is 1 V. The scan to generate the transformation matrix is carried out over the azimuth angles of (φ_q, φ_{qq}) = (30°, 150°) and the elevation angles of (θ_q, θ_{qq}) = (0°, 40°).

The estimated DOAs of all three signals given by the Matrix Pencil method using the voltages induced in the 2DULVA is given in Table 6.16. The estimated amplitudes of all three signals are also shown in Table 6.16.

TABLE 6.15

Interpolation error and condition number for a RDR-HA using a 16-grid virtual array $(d_{va} = 0.5)$

Sector (θ)	Interpolation Error (%)	Condition Number
$0 \rightarrow 10$	0.2	1.3×10^7
$0 \rightarrow 20$	0.22	5.1×10^5
$0 \rightarrow 30$	0.4	22,763
$0 \rightarrow 40$	1.1	7,029
$0 \rightarrow 50$	4.2	2,970

It is seen that it is possible to carry out DOA estimation by using directive antenna elements located on a hemispherical surface. The 2D Matrix Pencil method is applied to the outputs from the 2DULVA to estimate the DOAs of the various signals. The use of directive elements is made possible through this methodology which increases the efficiency of the system, as we now have a larger signal-to-noise ratio. In addition, it reduces the cost, as fewer of A/Ds are required to carry out digital beam forming.

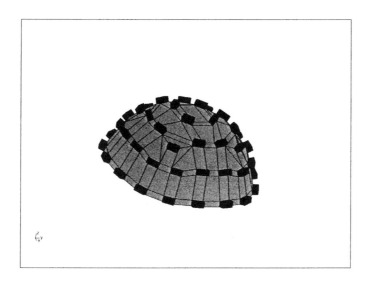

Figure 6.31. Rectangular dielectric resonator–hemispherical array.

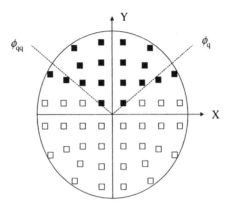

Figure 6.32. Top view of the hemispherical array. The black squares represent the 18 active elements in the processing, and the white squares represent the rest of the real elements mounted over the hemispherical surface.

TABLE 6.16
Estimation of the DOA of three signals: $(\varphi, \theta)_1 = (50°, 5°)$, $(\varphi, \theta)_2 = (80°, 20°)$, $(\varphi, \theta)_3 = (140°, 30°)$

φ	θ	Estimation of φ	Estimation of θ	Estimation of the Amplitude
50	5	51.7	5.9	1.2
80	20	78.3	18.3	1.3
140	30	140.8	29.2	1.0

6.5.3   Horn Antennas on a Hemispherical Surface

Finally, we consider an array of horn antennas located on a hemispherical surface. The probe-fed horn antenna is shown in Figure 6.33, and its dimensions are shown in Figure 6.34. The input impedance at the probe feed is shown in Figure 6.35, demonstrating that the resonant frequency of the horn occurs at 2.41 GHz. At that frequency the element radiation pattern is given by Figure 6.36.

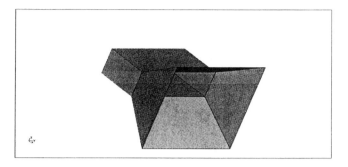

Figure 6.33. Horn antenna

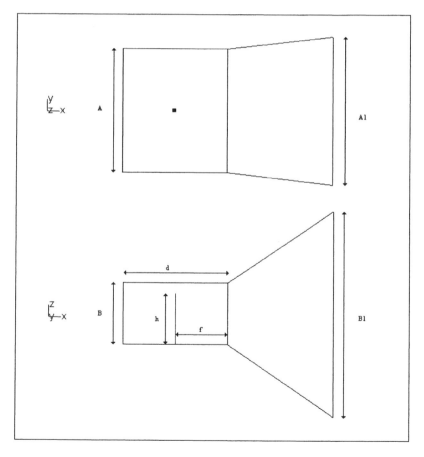

Figure 6.34. Geometry of the horn antenna. $A = 6$ cm; $d = 5$ cm; $f = 2.5$ cm; $h = 2.5$ cm; $B = 3$ cm ; $A_1 = 7.2$ cm; $B_1 = 10$ cm.

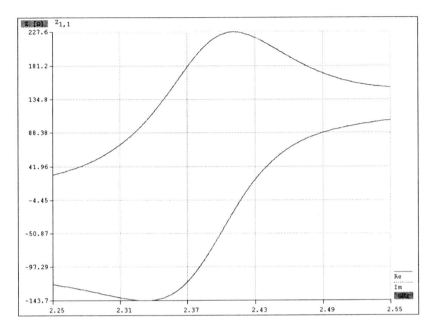

Figure 6.35. Impedance response of the horn antenna.

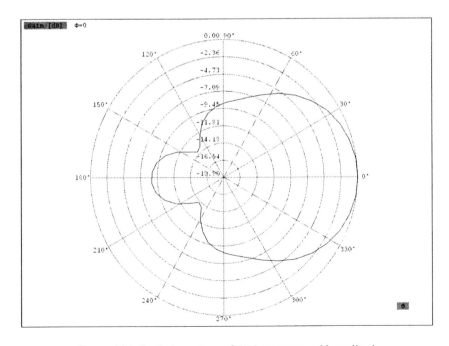

Figure 6.36. Radiation pattern of the horn antenna. Normalized.

A nonplanar conformal array on a hemispherical surface is formed by using 48 horn antennas placed on a hemispherical surface of radius 2.4λ as shown in Figure 6.37. The properties of the transformation matrix when we transform the voltages induced in the actual array consisting of 18 elements (see Figure 6.32) and transform it to a 16-element virtual array (see Figure 6.25) is given in Table 6.17. The sector scan in azimuth and in elevation is given by $(\varphi_q, \varphi_{qq}) = (30°, 150°)$ and $(\theta_q, \theta_{qq}) = (0°, 40°)$, respectively. The interpolation error and the condition number of the transformation matrix $[\Im_q]$ for different sectors q are shown in Table 6.17.

TABLE 6.17
Interpolation error and condition number for the transformation matrix $[\Im]$ when interpolating to a 16-element ULVA (star configuration with a conformal shape)

Elevation Scan Angle (θ)	Error Interpolation (%)	Condition Number
$0 \rightarrow 10$	0.2	5.7×10^7
$0 \rightarrow 20$	0.27	2.7×10^7
$0 \rightarrow 30$	0.73	20,512
$0 \rightarrow 40$	2.4	4,767
$0 \rightarrow 50$	7.1	2,513
$0 \rightarrow 60$	15.2	1,470
$0 \rightarrow 70$	26.4	1,420
$0 \rightarrow 80$	36.2	1,349
$0 \rightarrow 90$	46.3	893

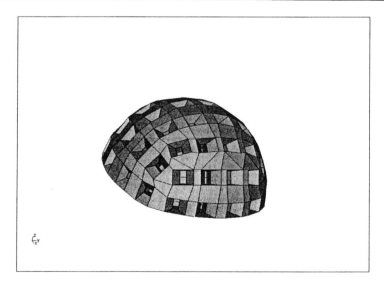

Figure 6.37. Conformal horn antenna aligned on the hemispherical surface.

To carry out DOA estimation of the three signals $(\varphi, \theta)_1 = (50°, 5°)$, $(\varphi, \theta)_2 = (80°, 20°)$, and $(\varphi, \theta)_3 = (140°, 30°)$ impinging on the hemispherical array, we transform the voltages that are induced on the shaded 18-horn antenna elements shown in Figure 6.32 to the 2DULVA of Figure 6.25. The amplitude of all the signals is 1 V. The scan to generate the transformation matrix is carried out over the azimuth angles of $(\varphi_q, \varphi_{qq}) = (30°, 150°)$ and the elevation angles of $(\theta_q, \theta_{qq}) = (0°, 40°)$.

The estimated DOAs of all three signals given by the Matrix Pencil method using the voltages induced in the 2DULVA are given in Table 6.18. The estimated amplitudes of all three signals are also shown in Table 6.18. The estimates using a single snapshot of the data voltages are quite accurate.

It is important to note that the error in the transformation matrix increases as we increase the azimuth angle θ toward 90° because the increase in degrees of freedom is not commensurate with the number of antenna elements. The error can be controlled either by adding additional arrays or by reducing the scan angle for the group of active elements.

TABLE 6.18
Estimation of the DOA of three signals: $(\varphi, \theta)_1 = (50°, 5°)$, $(\varphi, \theta)_2 = (80°, 20°)$, and $(\varphi, \theta)_3 = (140°, 30°)$

φ	θ	Estimation of φ	Estimation of θ	Estimation of the Amplitude
50	5	47.5	4.4	1.1
80	20	80.4	18.1	1.0
140	30	141.6	29.5	1.1

6.6   CONCLUSION

This chapter has presented a preprocessing technique that transforms a nonuniformly spaced array operating in the presence of mutual coupling between the elements of the array and near-field scatterers into a virtual array of omni-directional isotropic point elements that is amenable to the application of a direct data domain algorithm. Through such a transformation obtained using an interpolation technique, we have shown that not only can we compensate for the effects of mutual coupling in a nonuniformly spaced array but the effects of strong near-field scatterers can also be eliminated. Since the transformed output voltages are those of a uniformly spaced linear array consisting of omnidirectional point radiators, a conventional adaptive algorithm can easily be applied to extract the SOI in the presence of jammers. Finally, it is shown how to use directive antenna elements located on a conformal hemispherical surface to perform DOA estimation using a single snapshot of the data.

REFERENCES

[1]   B. Friedlander, "The Root-MUSIC Algorithm for Direction Finding with Interpolated Arrays," *Signal Processing*, Vol. 30, pp. 15–29, 1993.

[2]   T. S. Lee and T. T. Lin, "Adaptive Beamforming with Interpolation Arrays for Multiple Coherent Interferes," *Signal Processing*, Vol. 7, pp. 177–194, 1997.

[3]   B. Friedlander and A. J. Weiss, "Direction Finding Using Spatial Smoothing with Interpolated Arrays," *IEEE Transactions on Aerospace and Electronics Systems*, Vol. 28, No. 2, Apr. 1992.

[4]   B. Friedlander and A. J. Weiss, "Direction Finding in the Presence of Mutual Coupling," *IEEE Transactions on Antennas and Propagation*, Vol. 39, No. 3, Mar. 1991.

[5]   T. K. Sarkar and O. Pereira, "Using Matrix Pencil for Estimating Parameters of Exponentially Damped/Undamped Sinusoids in Noise," *IEEE Antennas and Propagation Magazine*, Vol. 37, No. 1, pp. 48–55, Feb. 1995.

[6]   Y. Hua and T. K. Sarkar, "Generalized Pencil of Functions Method for Extracting Poles of an EM System from Its Transient Response," *IEEE Transaction Antennas and Propagation*, Vol. 37, No. 2, 1989.

[7]   R. Brown and T. K. Sarkar, "Real Time Deconvolution Utilizing the Fast Fourier Transform and the Conjugate Gradient Method," in *5th Acoustic Speech and Signal Processing Workshop on Spectral Estimation and Modeling*, Rochester, NY, 1990.

[8]   S. Choi, H. M. Son, and T. K. Sarkar, "Implementation of a Smart Antenna System on a General Purpose Digital Signal Processor utilizing a Linearized CGM," *Signal Processing*, Vol. 7, No. 2, pp. 105–119, Apr. 1997.

[9]   M. D. Zoltowski and C. P. Mathews, "Direction Finding with Uniform Circular Arrays via Phase Mode Excitation and Beamspace Root-MUSIC," *Proceedings of ICASSP*, 1992.

[10] M. Wax and J. Sheinvald, "Direction Finding of Coherent Signals via Spatial Smoothing for Uniform Circular Arrays," *IEEE Transactions on Antennas and Propagation*, Vol. 42, No. 5, May 1994.

[11] P. N. Fletcher and P. Darwood, "Beamforming for Circular and Semicircular Array Antennas for Low-cost Wireless LAN Data Communication Systems," *IEEE Proceedings on Microwave Antennas and Propagation*, Vol. 145, No. 2, Apr. 1998.

[12] J. Pierre and M. Kaveh, "Experimental Performance of Calibration and Direction-finding Algorithms," *Proceedings of ICASSP*, 1991.

[13] B. M. Kolundzija, J. S. Ognjanovic, and T. K. Sarkar, *WIPL-D: Electromagnetic Modeling of Composite Metallic and Dielectric Structures*, Artech House, Norwood, MA, 2000, (*http://wipl-d.com*).

[14]   R. Chair, K. M. Luk, and K. F. Lee, "Miniature Shorted Dual-Patch Antenna," *IEEE Proceedings on Microwave Antennas and Propagation*, Vol. 147, No. 4, pp. 273–276, Aug. 2000.

[15]   M. W. McAllister, S. A. Long and G. L. Conway, "Rectangular Dielectric Resonator Antenna," *Electronics Letters*, Vol. 19, pp. 218–219, Mar. 1983.

7

ESTIMATING DIRECTION OF ARRIVALS BY EXPLOITING CYCLOSTATIONARITY USING A REAL ANTENNA ARRAY

SUMMARY

In an adaptive process, the goal is to extract the signal of interest (SOI) embedded in other interfering signals and noise. Generally, this adaptive process is carried out by using the available information for the direction of arrival (DOA) of the SOI. This is true for radar applications where that information is available, as we know *a priori* along which direction a beam was transmitted. However, in a mobile communication when the information about the DOA of the SOI is not available, the problem is then how to implement an adaptive process. However, to implement an adaptive process, we first need to estimate the DOA of the SOI. Hence, we propose to use the concept of cyclostationarity to achieve that goal. The term *cyclostationarity* implies that the signal displays characteristic spectral properties, and this property is shared by almost all man-made communication signals. We implement a direct data domain approach to carry out the DOA estimation exploiting cyclostationarity. Thus we avoid the formation of a covariance matrix, which is always problematical when we have short data lengths or the environment is quite dynamic. By exploiting the temporal properties of the signal instead of the spatial characteristics, it is possible to deal with an environment where the number of noncoherent interferers is greater than the number of antenna elements. In this chapter we consider the problem of estimating the DOA for the SOI embedded in noise and other interfering signals whose characteristics are unknown. The DOA of the SOI may coincide with the DOA of one of the interferers. In this approach, the various signals impinging at the array can have different carrier frequencies, which are unknown. In the proposed algorithm, while estimation of the cyclic array covariance matrix is avoided, we develop a new matrix form using extremely short data samples. As a result, the computational load in the proposed approach is relatively reduced and the robustness of the estimation of SOI is significantly improved when the number of available snapshots is extremely limited. We also consider the problem of elimination of multipath using the principle of cyclostationarity. A post-processing technique is developed for identification of multipath for signals having the same cycle frequency of interest. The number of multipaths is less than $N/2$. In this approach, one can eliminate interferers whose number can be greater than that of the number of antenna elements. Finally, we incorporate the concept of electromagnetic analysis to eliminate the problem of mutual coupling between the elements and take into account the effect of near-field scatterers. Numerical results are presented in each section to illustrate the various concepts and how to implement these principles in a simulated system.

7.1   INTRODUCTION

Arrays of sensors are useful in the process of estimating the DOA of propagating signals in areas such as radar, sonar, mobile communications, and others. The information on the DOA is essential to extract the SOI in the presence of interferers and noise. Hence, in adaptive systems where the DOA of the SOI is not known *a priori*, the first step in the computation process is to estimate the DOA. The DOA of the SOI may coincide with the DOA of one of the interferers. The class of problems where one tries to estimate the SOI in the absence of DOA information is called *blind algorithms*. Hence, the method that will be presented in this section is a blind adaptive procedure exploiting cyclostationarity utilizing a direct data domain approach.

For estimating the DOA of multiple narrowband signals impinging at an array, several existing algorithms, such as MUSIC, ESPRIT, and maximum likelihood (ML) methods, have been applied. All of these methods have been shown to be efficient under the following assumptions: (1) the number of snapshots is sufficiently large to obtain an estimate of the covariance matrix of the data, and (2) the overall number of incident signals is smaller than or equal to the number of sensors (a *snapshot* is defined as the voltages measured at the antenna elements at a particular instance of time). However, in some practical applications, there may be situations when the number of available snapshots is limited, and the overall number of incident signals is greater than the number of elements of the array. For these scenarios, the performance of most of the conventional methods will be degraded. On the other hand, even though the Matrix Pencil approach can be applied to a single snapshot of the data, the number of signals it can estimate is always much less than the number of antennas.

To overcome the second limitation of conventional methods in [1–4], authors have investigated other approaches relying on the temporal features of the SOI, especially, cyclostationarity, which is exhibited by almost all man-made communication signals. Cyclostationarity implies that when the temporal signal amplitudes are squared, some spectral peaks are displayed, which can be known *a priori*. An example of this concept can be seen in the demodulation of single- or double-sideband suppressed carrier systems, where the carrier frequency is introduced or generated at the receiver through a similar transformation of the received signals. The signals exhibiting cyclostationary characteristics fluctuate periodically with time. These fluctuating frequencies are usually obtained from knowledge of the transmission baud rate, carrier frequency, or other frequencies, which are related to the periodic phenomena associated with the desired signals. Most known attempts at DOA estimation utilizing the cyclostationarity of SOI are cyclic-MUSIC and cyclic-ESPRIT methods. In these two methods [1, 2] the correlation matrix estimates are replaced by cyclic correlation matrix estimates reflecting the cyclostationarity of SOI, which may be the baud rate or the carrier frequency, as would be the case for radar and in mobile communications. First, we explain the principles of cyclostationarity and then illustrate how it can be applied in practice.

According to Gardner, cyclostationarity is a nonlinear transformation which operating on a signal, will generate finite-strength additive sine-wave components, which result in spectral lines. So, for example, a signal $x(t)$ is assumed to have a cyclostationarity property with cycle frequency η if and only if the product $x(t) x(t - \tau)$ for some delay τ exhibits a spectral line at frequency η. Let $a(t)$ be a real low-pass signal with a power spectral density $S_a(f)$ as shown in Figure 7.1a. It has no identifiable spectral lines. If $a(t)$ is used to modulate a carrier at frequency f_c, we obtain a double-sideband (DSB) amplitude-modulated signal of the form

$$x(t) = a(t)\cos(2\pi f_c t) \qquad (7.1)$$

The power spectral density (PSD) of this amplitude-modulated signal $x(t)$ is $S_x(f)$ and it is plotted in Figure 7.1b. Mathematically, the PSD of the modulated signal can be expressed as

$$S_x(f) = 0.25 \times S_a(f + f_c) + 0.25 \times S_a(f - f_c) \qquad (7.2)$$

As shown in Figure 7.1b, even though the PSD of the DSB signal is centered at $\pm f_c$ there is no spectral line at $\pm f_c$. Now if we use a second-order nonlinear transformation, which is equivalent to multiplying the signal with its shifted version or simply squaring the signal if there is no shift, we get

$$y(t) = x^2(t) = a^2(t)\cos^2(2\pi f_c t) = 0.5 \times [b(t) + b(t)\cos(4\pi f_c t)] \qquad (7.3)$$

where $b(t) = a^2(t)$. The PSD of $y(t)$ is then given by

$$S_y(f) = 0.25 \times [S_b(f) + 0.25 \times S_b(f + 2f_c) + 0.25 \times S_b(f - 2f_c)]$$
$$(7.4)$$

The PSD for $b(t)$, $S_b(f)$, given by Figure 7.1c, contains a line spectrum at the dc value. This can be visualized by making a Fourier series expansion of the function $b(t)$ and then squaring it. This will introduce a spectral line at the dc value. The PSD for $S_y(f)$ is then given by Figure 7.1d. This contains two additional spectral lines at $f = \pm f_c$ as well as a line at $f = 0$. Thus, by employing a second-order nonlinear transformation on the signal, the hidden periodicities have been enhanced. The thinking here is that if we have any DSB, AM, FM, BPSK, or QPSK, or any type of modulated signals, then if we square that signal, a spectral line will be introduced at $\eta = 2f_c$, irrespective of the nature of the modulation.

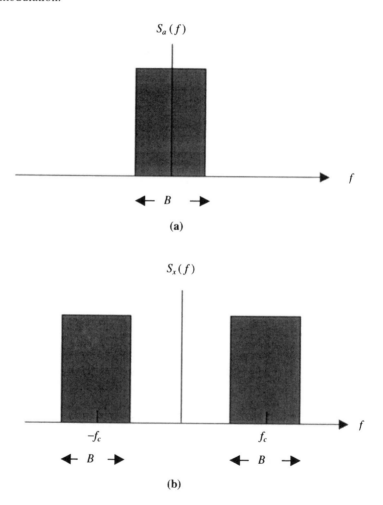

$S_a(f)$

$\leftarrow B \rightarrow$

(a)

$S_x(f)$

$-f_c$ f_c

$\leftarrow B \rightarrow$ $\leftarrow B \rightarrow$

(b)

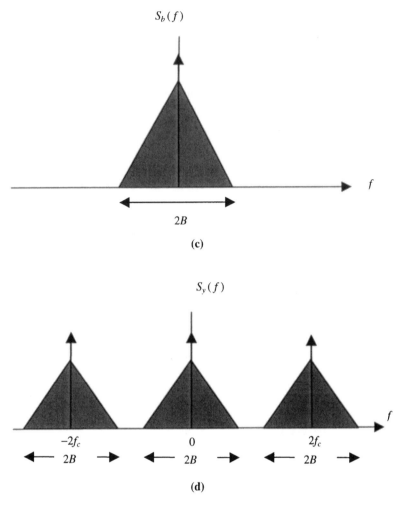

Figure 7.1. PSD of (a) a low-pass signal; (b) an amplitude = modulated (DBS) signal; (c, d) a squared low-pass signal.

Hence cyclostationarity implies that when temporal signal amplitudes, for example, are squared, some spectral peaks are displayed which can be known *a priori*. An example of this concept can be seen in the demodulation of single- or double-sideband suppressed carrier systems, where the carrier frequency is introduced or generated at the receiver through a similar transformation of the received signals. All man-made signals display the characteristics properties of cyclostationarity.

Although the approaches that have been presented in the literature using the concepts of cyclostationarity, so far, have significant advantages over the freedom of choice of the number of elements of an array, they still are a statistics-based approach. Since they require an estimate of the covariance matrix of the signal, they therefore suffer from the effect of finite lengths of data, when the available short data snapshot contains a very limited number of samples. Specifically, statistics-based methods require more snapshots in order to obtain the necessary decorrelation properties between the interference signals and the noise components in the data. This estimation of the covariance matrix in a statistical methodology can be a serious problem if the environment changes from snapshot to snapshot.

Hence, in this chapter, we overcome the limitations of short data records of a statistics-based methodology. In this new method, computation of the covariance matrix is avoided and we propose a direct data domain approach (nonstochastic in nature), which is very effective in dealing with short data sequences. A direct data domain approach is quite useful when the environment changes rapidly, while preserving the advantages associated with the freedom to choose the number of elements in an array. In other words, the number of elements can be less than the number of interfering signals. Here, we extend the direct data domain approach presented in [5, 6] for adaptive processing and use the Matrix Pencil method [7] for DOA estimation in the present applications.

The *a priori* information required to estimate the DOA of the SOI is only the cycle frequency of the desired signal. Hence, its performance is sensitive to the accuracy of the presumed cycle frequency. However, the estimated cycle frequency may not be known very well in a real application. Therefore, it is worthwhile to evaluate the performance of the proposed method in the presence of cyclic frequency error and to develop techniques to compensate for that problem. Here, we propose the multiple cyclic frequency constraint to alleviate the performance degradation due to the error in the *a priori* knowledge of the cyclic frequency. The direct data domain approach is thus different from other [8, 9] stochastic-based approaches.

Finally, a post-processing technique based on the Matrix Pencil method is developed for identification of multipath or signals having the same cycle frequency of interest.

7.2   PROBLEM STATEMENT

For the sake of simplicity we assume that the sources and sensors are coplanar and that the sources are located in the far field of the array. However, this methodology can easily be extended to the noncoplanar case or when there are near-field scatterers.

Signals which display the cyclostationary property are often modulated digital signals where the existence of two periodic phenomena inevitably appears to lead to periodic fluctuations of the SOI. The two periodic phenomena are:

1.) The presence of a sinusoidal carrier
2.) The repetition of a pulse waveform

In this section we consider only periodic fluctuations due to the carrier frequency associated with the SOI, and ignore the second property.

Most digitally modulated signals at a time instance m can be represented analytically by

$$s(m) = \left(\sum_r I_m \, g_b(m - r\zeta) \right) e^{j2\pi f_c m} \tag{7.5}$$

where I_m is the information-bearing sequence, f_c is the carrier frequency associated with the modulated signal, and $g_b(m)$ is a real function representing the pulse waveform. ζ is the duration of the pulse. Here, we assume that the desired source has the following properties:

1.) It exhibits a second-order cyclostationarity with a cyclic frequency $\eta = 2f_c$.
2.) It is cyclically uncorrelated with signals not having this cycle frequency.

The desired signal bearing a cycle frequency η is assumed to be located in the far field of the antenna array and is assumed to impinge at the array from an incident azimuth angle of φ_d.

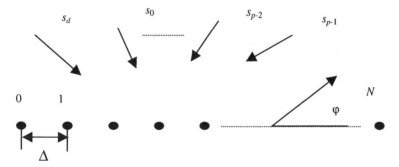

Figure 7.2. Array configuration.

We consider a uniform linear virtual array (ULVA) manifold consisting of $N + 1$ identical and omnidirectional elements with a spacing Δ, as shown in Figure 7.2. Later we consider realistic antenna elements. Furthermore, we assume that additional P narrowband signals (where the carrier frequencies of the p-signals are quite different) are impinging on the array from P distinct directions φ_p for $p = 0, 1, \ldots, P - 1$. It is important to note that the DOA of one of the interferers may coincide with the DOA of the SOI. The received signal $[x(m)]$ at the elements of the array can then be represented by

$$[x(m)] = \begin{bmatrix} x_0(m) \\ x_1(m) \\ \vdots \\ x_N(m) \end{bmatrix} = s_d(m) \left[a(\varphi_d) \right] + \sum_{p=0}^{P-1} s_p(m) \left[a(\varphi_p) \right] + \left[\xi(m) \right] \qquad (7.6a)$$

where the signal received at the nth element at the mth instant of time is given by

$$x_n(m) = s_d(m) \exp(j2\pi n\Delta \cos\varphi_d / \lambda)$$

$$+ \sum_{p=0}^{P-1} s_p(m) \exp(j2\pi n\Delta \cos\varphi / \lambda) + \xi_n(m) \qquad (7.6b)$$

$$= s_d(m) a_n(\varphi_d) + \sum_{p=0}^{P-1} s_p(m) a_n(\varphi_p) + \xi_n(m)$$

where $s_d(m)$ is the desired signal to be extracted. The variable m therefore represents the index for that snapshot. This SOI is periodically fluctuating with a cycle frequency η. $s_p(m)$ is the interference signal and is cyclically uncorrelated with the cyclic frequency of the desired signal. $\xi_n(m)$ is the additive noise associated with the mth snapshot at the nth element, and $[a(\varphi)]$ is the steering vector associated with the various signals and is represented by

$$\left[a(\varphi) \right]_{(N+1)\times 1} = \left\{ \left[1, \ e^{j2\pi\Delta/\cos\varphi}, \ ..., \ e^{j2\pi\Delta n/\cos\varphi} \right]^T \right\}_{(N+1)\times 1}$$

$$n = 0, 1, ..., N \qquad (7.7)$$

where the superscript T denotes the transpose of a matrix.

From (7.6) and (7.7), the received signal can be represented in matrix form as

$$[x(m)] = [A] [s(m)] + [\xi(m)]$$

$$[A] = [a(\varphi_d), a(\varphi_0), ..., a(\varphi_{P-1})]_{(N+1)\times(P+1)}$$

$$[s(m)] = [s_d(m), s_0(m), ..., s_{P-1}(m)]^T_{(P+1)\times 1}$$

$$[\xi(m)] = [\xi_0(m), \xi_1(m), ..., \xi_N(m)]^T_{(N+1)\times 1}$$

$$\qquad (7.8)$$

Now, we consider the DOA estimation of the SOI having the cyclic frequency η under the following assumptions:

1.) The desired signal exhibits the second-order cyclostationary with cycle frequency $\eta = 2f_c$, and it is cyclically uncorrelated with the other signals for the cycle frequency.
2.) The noise represented by $\xi_n(m)$ is also cyclically uncorrelated with the desired signal at the same cyclic frequency.
3.) Matrix [A] is not necessarily a full-rank matrix (i.e., there can be other interfering signals coming from the same DOA as the SOI, but it is cyclically uncorrelated with the desired signal for the cycle frequency).

In the next section we present the mathematical procedure to carry out the estimation from a computational standpoint.

7.3   DOA ESTIMATION USING CYCLOSTATIONARITY

Here, we propose a direct data domain approach exploiting the cyclostationary property of the SOI to estimate the DOA.

First, a second-order nonlinear transform is applied to the signals so as to generate a spectral line at the cyclic frequency of the SOI. From (7.8), we observe that the square of the received temporal signal at each element of the array can be written as

$$[x_0{}^2(m),\ x_1{}^2(m),\ \ldots,\ x_N{}^2(m)] \tag{7.9}$$

where each of the elements of the row vector can be written as

$$x_n^2(m) = \left[s_d(m)a_n(\varphi_d) + \sum_{p=0}^{P-1} s_p(m)a_n(\varphi_p) + \xi_n(m) \right]^2$$

$$= s_d^2(m)a_n^2(\varphi_d) + 2s_d(m)a_n(\varphi_d)\left[\sum_{p=0}^{P-1} s_p(m)a_n(\varphi_p) + \xi_n(m) \right]$$

$$+ \left[\sum_{p=0}^{P-1} s_p(m)a_n(\varphi_p) + \xi_n(m) \right]^2 \tag{7.10}$$

$$= s_d^2(m)a_n^2(\varphi_d) + \xi_n'(m)$$

where

$$\xi_n'(m) = 2s_d(m)a_n(\varphi_d)\left[\sum_{p=0}^{P-1} s_p(m)a_n(\varphi_p) + \xi_n(m) \right]$$

$$+ \left[\sum_{p=0}^{P-1} s_p(m)a_n(\varphi_p) + \xi_n(m) \right]$$

Here, squaring the signals received at each element of the antenna array produces a term that consists of squaring the desired signal, which results in the second-order cyclic frequency and another component, which consists of signals that do not contain the second-order cycle frequency η.

Using (7.9), we construct a new matrix form utilizing $M+1$ snapshots $(m = 0, 1, 2, ..., M)$. This matrix will not contain the SOI component at the cycle frequency $\eta = 2f_c$ and can be reconstructed from short signal sequences, which is given by

$$[Z][W] = [Y] \tag{7.11a}$$

where

$$[Z] = \begin{bmatrix} 1 & e^{j2\pi\eta} & \cdots & e^{j2\pi(M-1)\eta} \\ x_0^2(0) - e^{-j2\pi\eta}x_0^2(1) & x_0^2(1) - e^{-j2\pi\eta}x_0^2(2) & \cdots & x_0^2(M-1) - e^{-j2\pi\eta}x_0^2(M) \\ x_1^2(0) - e^{-j2\pi\eta}x_1^2(1) & x_1^2(1) - e^{-j2\pi\eta}x_1^2(2) & \cdots & x_1^2(M-1) - e^{-j2\pi\eta}x_1^2(M) \\ \vdots & & & \vdots \\ x_N^2(0) - e^{-j2\pi\eta}x_N^2(1) & x_N^2(1) - e^{-j2\pi\eta}x_N^2(2) & \cdots & x_N^2(M-1) - e^{-j2\pi\eta}x_N^2(M) \end{bmatrix} \tag{7.11b}$$

where the unknown weight vector is given by

$$[W]^T = [W_0, W_1, W_2 ..., W_{L-1}] \tag{7.11c}$$

and the excitation is

$$[Y]^T = [C, 0, 0, ..., 0] \tag{7.11d}$$

Here, the number of weights L is chosen to be the same as the number of snapshots minus one $(i.e., L \equiv M)$. The arbitrary constant C is specifically chosen in this case by

$$C = W_0 + W_1 e^{j2\pi\eta} + ... + W_{M-1} e^{j2\pi(M-1)\eta} \tag{7.11e}$$

However, that is not necessary as the value of the unknown constant C can be chosen arbitrarily since it gets factored out in the final computation. Therefore, the DOA of the desired signal of interest can be evaluated from

$$\text{DOA}_{\text{SOI}} = \arccos\left[\frac{\lambda}{4\pi \Delta n} \text{ phase angle of}\left\{\frac{1}{C}\sum_{m=0}^{M-1} x_n^2(m)W_m\right\}\right], n \geq 1 \tag{7.12}$$

Here n can have any value between 1 and N. For $n = 0$, the DOA term does not appear. The correctness of the expression (7.12) can be obtained through induction.

Consider the case of a signal in the presence of two interferers incident on a four-element array. Let $j_{pn}(m)$ be the contribution of the pth interferer at the antenna element n corresponding to the snapshot m. Therefore, the signals at each of the four antenna elements at the time instant m can be written as

$$
\begin{aligned}
x_0(m) &= s_d(m) + j_{00}(m) + j_{10}(m) \\
x_1(m) &= s_d(m)Z + j_{01}(m) + j_{11}(m) \\
x_2(m) &= s_d(m)Z^2 + j_{02}(m) + j_{12}(m) \\
x_3(m) &= s_d(m)Z^3 + j_{03}(m) + j_{13}(m)
\end{aligned}
\tag{7.13}
$$

where

$$
Z = e^{j\frac{2\pi \Delta}{\lambda}\cos\varphi_d}
\tag{7.14}
$$

Here s_d is the SOI coming from a direction φ_d, which is to be determined using the concept of cycle frequency. Next we square each of the signals received at the antenna elements measured for all the snapshots. Then we subtract from the squared voltages corresponding to the first snapshot and the voltages received at the second snapshot at the same antenna elements weighted by $e^{-j2\pi\eta}$. This cancels out the SOI component at the known cycle frequency at the same antenna element and leaves behind only the interference and noise components. The weight vectors should be such that they minimize the undesired signals at the output. An additional equation, representing the first row of the matrix equations, is introduced as a constraint. Its expression is such that the same weights when they operate on the SOI produce a gain factor C. Hence, (7.11) representing $[Z]$ $[W] = [Y]$ would be

$$
\begin{bmatrix}
1 & e^{j2\pi\eta} & e^{j2\pi 2\eta} & e^{j2\pi 3\eta} \\
x_0^2(0)-e^{-j2\pi\eta}x_0^2(1) & x_0^2(1)-e^{-j2\pi\eta}x_0^2(2) & x_0^2(2)-e^{-j2\pi\eta}x_0^2(3) & x_0^2(3)-e^{-j2\pi\eta}x_0^2(4) \\
x_1^2(0)-e^{-j2\pi\eta}x_1^2(1) & x_1^2(1)-e^{-j2\pi\eta}x_1^2(2) & x_1^2(2)-e^{-j2\pi\eta}x_1^2(3) & x_1^2(3)-e^{-j2\pi\eta}x_1^2(4) \\
x_2^2(0)-e^{-j2\pi\eta}x_2^2(1) & x_2^2(1)-e^{-j2\pi\eta}x_2^2(2) & x_2^2(2)-e^{-j2\pi\eta}x_2^2(3) & x_2^2(3)-e^{-j2\pi\eta}x_2^2(4)
\end{bmatrix}
$$

$$
\times
\begin{bmatrix}
W_0 \\ W_1 \\ W_2 \\ W_3
\end{bmatrix}
=
\begin{bmatrix}
C \\ 0 \\ 0 \\ 0
\end{bmatrix}
\tag{7.15}
$$

where $W_0 + W_1 e^{j2\pi\eta} + W_2 e^{j2\pi 2\eta} + W_3 e^{j2\pi 3\eta} = C$.

From (7.15), and after multiplying the second row of matrix $[Z]$ with $[W]$, we obtain

$$\left\{ x_0^2(0) - e^{-j2\pi\eta} \, x_0^2(1) \right\} W_0 + \left\{ x_0^2(1) - e^{-j2\pi\eta} \, x_0^2(2) \right\} W_1$$
$$+ \left\{ x_0^2(2) - e^{-j2\pi\eta} \, x_0^2(3) \right\} W_2 + \left\{ x_0^2(3) - e^{-j2\pi\eta} \, x_0^2(4) \right\} W_3 = 0$$

(7.16)

Then, we can rearrange equation (7.16) to yield

$$\left(s_d^2(0) + 2 s_d(0) \left\{ j_{00}(0) + j_{10}(0) \right\} + \left\{ j_{00}(0) + j_{10}(0) \right\}^2 \right.$$
$$- e^{j2\pi\eta} \left[s_d^2(1) + 2 s_d(1)\{ j_{00}(1) + j_{10}(1)\} + \{ j_{00}(1) + j_{10}(1)\}^2 \right] \left. \right) W_0$$
$$+ \left(s_d^2(1) + 2 s_d(1) \left\{ j_{00}(1) + j_{10}(1) \right\} + \left\{ j_{00}(1) + j_{10}(1) \right\}^2 \right.$$
$$- e^{j2\pi\eta} \left[s_d^2(2) + 2 s_d(2)\{ j_{00}(2) + j_{10}(2)\} + \{ j_{00}(2) + j_{10}(2)\}^2 \right] \left. \right) W_1$$
$$+ \left(s_d^2(2) + 2 s_d(2)\{ j_{00}(2) + j_{10}(2)\} + \{ j_{00}(2) + j_{10}(2) \}^2 \right.$$
$$- e^{j2\pi\eta} \left[s_d^2(3) + 2 s_d(3) \left\{ j_{00}(3) + j_{10}(3) \right\} + \left\{ j_{00}(3) + j_{10}(3) \right\}^2 \right] \left. \right) W_2$$
$$+ \left(s_d^2(3) + 2 s_d(3) \left\{ j_{00}(3) + j_{10}(3) \right\} + \left\{ j_{00}(3) + j_{10}(3) \right\}^2 \right.$$
$$- e^{j2\pi\eta} \left[s_d^2(4) + 2 s_d(4) \left\{ j_{00}(4) + j_{10}(4) \right\} + \left\{ j_{00}(4) + j_{10}(4) \right\}^2 \right] \left. \right) W_3 = 0$$

(7.17)

Next, we can again rearrange equation (7.17) to yield

$$\left(\begin{array}{l} \left[2 s_d(0)\{ j_{00}(0) + j_{10}(0) \} + \{ j_{00}(0) + j_{10}(0) \}^2 \right] W_0 \\ + \left[2 s_d(1)\{ j_{00}(1) + j_{10}(1) \} + \{ j_{00}(1) + j_{10}(1) \}^2 \right] W_1 \\ + \left[2 s_d(2)\{ j_{00}(2) + j_{10}(2) \} + \{ j_{00}(2) + j_{10}(2) \}^2 \right] W_2 \\ + \left[2 s_d(3)\{ j_{00}(3) + j_{10}(3) \} + \{ j_{00}(3) + j_{10}(3) \}^2 \right] W_3 \end{array} \right)$$
$$- e^{j2\pi\eta} \left(\begin{array}{l} \left[2 s_d(1)\{ j_{00}(1) + j_{10}(1) \} + \{ j_{00}(1) + j_{10}(1) \}^2 \right] W_0 \\ + \left[2 s_d(2)\{ j_{00}(2) + j_{10}(2) \} + \{ j_{00}(2) + j_{10}(2) \}^2 \right] W_1 \\ + \left[2 s_d(3)\{ j_{00}(3) + j_{10}(3) \} + \{ j_{00}(3) + j_{10}(3) \}^2 \right] W_2 \\ + \left[2 s_d(4)\{ j_{00}(4) + j_{10}(4) \} + \{ j_{00}(4) + j_{10}(4)\}^2 \right] W_3 \end{array} \right) = 0$$

(7.18)

As $e^{j2\pi\eta} \neq 0$ and $W_i \neq 0$, (7.18) will be true for all possible choices of W_i if and only if each summation in parentheses (•) is equal to zero. In other words,

$$
\begin{aligned}
&\left[2s_d(0)\{j_{00}(0) + j_{10}(0)\} + \{j_{00}(0) + j_{10}(0)\}^2\right]W_0 \\
+&\left[2s_d(1)\{j_{00}(1) + j_{10}(1)\} + \{j_{00}(1) + j_{10}(1)\}^2\right]W_1 \\
+&\left[2s_d(2)\{j_{00}(2) + j_{10}(2)\} + \{j_{00}(2) + j_{10}(2)\}^2\right]W_2 \\
+&\left[2s_d(3)\{j_{00}(3) + j_{10}(3)\} + \{j_{00}(3) + j_{10}(3)\}^2\right]W_3 = 0
\end{aligned}
\tag{7.19}
$$

$$
\begin{aligned}
&\left[2s_d(1)\{j_{00}(1) + j_{10}(1)\} + \{j_{00}(1) + j_{10}(1)\}^2\right]W_0 \\
+&\left[2s_d(2)\{j_{00}(2) + j_{10}(2)\} + \{j_{00}(2) + j_{10}(2)\}^2\right]W_1 \\
+&\left[2s_d(3)\{j_{00}(3) + j_{10}(3)\} + \{j_{00}(3) + j_{10}(3)\}^2\right]W_2 \\
+&\left[2s_d(4)\{j_{00}(4) + j_{10}(4)\} + \{j_{00}(4) + j_{10}(4)\}^2\right]W_3 = 0
\end{aligned}
\tag{7.20}
$$

Similarly, for the third row of $[Z]$

$$
\begin{aligned}
&\left[2s_d(0)\{j_{01}(0) + j_{11}(0)\} + \{j_{01}(0) + j_{11}(0)\}^2\right]W_0 \\
+&\left[2s_d(1)\{j_{01}(1) + j_{11}(1)\} + \{j_{01}(1) + j_{11}(1)\}^2\right]W_1 \\
+&\left[2s_d(2)\{j_{01}(2) + j_{11}(2)\} + \{j_{01}(2) + j_{11}(2)\}^2\right]W_2 \\
+&\left[2s_d(3)\{j_{01}(3) + j_{11}(3)\} + \{j_{01}(3) + j_{11}(3)\}^2\right]W_3 = 0
\end{aligned}
\tag{7.21}
$$

$$
\begin{aligned}
&\left[2s_d(1)\{j_{01}(1) + j_{11}(1)\} + \{j_{01}(1) + j_{11}(1)\}^2\right]W_0 \\
+&\left[2s_d(2)\{j_{01}(2) + j_{11}(2)\} + \{j_{01}(2) + j_{11}(2)\}^2\right]W_1 \\
+&\left[2s_d(3)\{j_{01}(3) + j_{11}(3)\} + \{j_{01}(3) + j_{11}(3)\}^2\right]W_2 \\
+&\left[2s_d(4)\{j_{01}(4) + j_{11}(4)\} + \{j_{01}(4) + j_{11}(4)\}^2\right]W_3 = 0
\end{aligned}
\tag{7.22}
$$

Similarly, for the fourth row of $[Z]$,

$$\left[2s_d(0)\{j_{02}(0) + j_{12}(0)\} + \{j_{02}(0) + j_{12}(0)\}^2\right] W_0$$
$$+ \left[2s_d(1)\{j_{02}(1) + j_{12}(1)\} + \{j_{02}(1) + j_{12}(1)\}^2\right] W_1$$
$$+ \left[2s_d(2)\{j_{02}(2) + j_{12}(2)\} + \{j_{02}(2) + j_{12}(2)\}^2\right] W_2 \qquad (7.23)$$
$$+ \left[2s_d(3)\{j_{02}(3) + j_{12}(3)\} + \{j_{02}(3) + j_{12}(3)\}^2\right] W_3 = 0$$

$$\left[2s_d(1)\{j_{02}(1) + j_{12}(1)\} + \{j_{02}(1) + j_{12}(1)\}^2\right] W_0$$
$$+ \left[2s_d(2)\{j_{02}(2) + j_{12}(2)\} + \{j_{02}(2) + j_{12}(2)\}^2\right] W_1$$
$$+ \left[2s_d(3)\{j_{02}(3) + j_{12}(3)\} + \{j_{02}(3) + j_{12}(3)\}^2\right] W_2 \qquad (7.24)$$
$$+ \left[2s_d(4)\{j_{02}(4) + j_{12}(4)\} + \{j_{02}(4) + j_{12}(4)\}^2\right] W_3 = 0$$

From (7.16) – (7.24), we observe that for a particular value for n,

$$x_n^2(0)\, W_0 + x_n^2(1)\, W_1 + x_n^2(2)\, W_2 + x_n^2(3)W_3$$
$$= e^{\,j\,\frac{2\pi\Delta}{\lambda}2n\cos\varphi_d}\, (W_0 + e^{j2\pi\eta}W_1 + e^{j2\pi2\eta}W_2 + e^{j2\pi3\eta}W_3) \qquad (7.25)$$
$$= e^{\,j\,\frac{2\pi\Delta}{\lambda}2n\cos\varphi_d} \times C$$

which yields the estimate for the direction of arrival as

$$\text{DOA} = \arccos\left[\frac{\lambda}{4\pi n\Delta}\ \text{phase angle of}\ \left\{\frac{1}{C}\sum_{m=0}^{3} x_n^2(m)W_m\right\}\right], n \ge 1 \qquad (7.26)$$

For $n = 0$, the antenna element is located at the origin. It is not possible to estimate the DOA using (7.26) as it does not contain any information.

Finally, to solve the operator equation (7.11), $[Z][W]=[Y]$, the conjugate gradient method is applied. This method may also be implemented to operate in real time utilizing, for example, a DSP32C signal processing chip [5,6]. Various versions of this iterative method have been described in Appendix B. Use of the iterative conjugate gradient method to solve the matrix equation (7.11) is terminated when the error criterion

$$\frac{\big\|\,[Z][W]_k - [Y]\,\big\|}{\|Y\|} \le 10^{-6}$$

at iteration k is satisfied

7.4 MULTIPLE CYCLE FREQUENCY APPROACH

In a direct data domain approach, the cyclic frequency of the desired signal is utilized to form an appropriate matrix that we solve for the weight vector to estimate the DOA of the desired signal. We assume that we know the cyclic frequency η with a high degree of accuracy. However, a very accurate estimate of the cyclic frequency of the desired signal in a practical situation is limited due to the length of the received data and its signal-to-noise ratio, and so on. Hence, we propose a multiple cycle frequency approach to deal with any error that results in assuming *a priori* the value of the actual cycle frequency and thereby improve the performance of the proposed method.

To minimize the error in the estimation of the DOA which may result from an inaccurate assumption of the actual cycle frequency, we propose a multipoint cycle frequency approach. For example, given the estimated cycle frequency, which is η_e, we choose three multiple cycle frequencies centered around the actual one, such as $\eta_1 = \eta_e - 0.1\eta_e$, $\eta_2 = \eta_e$, and $\eta_3 = \eta_e + 0.1\eta_e$. Then, using these three cycle frequencies, we can construct a block matrix equation corresponding to each cyclic frequency as

$$[Z_B][W] = [Y_B] \tag{7.27a}$$

where

$$[Z_B] = \begin{bmatrix}
1 & e^{j2\pi\eta_1} & \cdots & e^{j2\pi\eta_1(M-1)} \\
1 & e^{j2\pi\eta_2} & \cdots & e^{j2\pi\eta_2(M-1)} \\
1 & e^{j2\pi\eta_3} & \cdots & e^{j2\pi\eta_3(M-1)} \\
x_0^2(0)-e^{-j2\pi\eta_1}x_0^2(1) & x_0^2(1)-e^{-j2\pi\eta_1}x_0^2(2) & \cdots & x_0^2(M-1)-e^{-j2\pi\eta_1}x_0^2(M) \\
x_1^2(0)-e^{-j2\pi\eta_1}x_1^2(1) & x_1^2(1)-e^{-j2\pi\eta_1}x_1^2(2) & \cdots & x_1^2(M-1)-e^{-j2\pi\eta_1}x_1^2(M) \\
\vdots & \vdots & \vdots & \vdots \\
x_N^2(0)-e^{-j2\pi\eta_1}x_N^2(1) & x_N^2(1)-e^{-j2\pi\eta_1}x_N^2(2) & \cdots & x_N^2(M-1)-e^{-j2\pi\eta_1}x_N^2(M) \\
x_0^2(0)-e^{-j2\pi\eta_2}x_0^2(1) & x_0^2(1)-e^{-j2\pi\eta_2}x_0^2(2) & \cdots & x_0^2(M-1)-e^{-j2\pi\eta_2}x_0^2(M) \\
x_1^2(0)-e^{-j2\pi\eta_2}x_1^2(1) & x_1^2(1)-e^{-j2\pi\eta_2}x_1^2(2) & \cdots & x_1^2(M-1)-e^{-j2\pi\eta_2}x_1^2(M) \\
\vdots & \vdots & \vdots & \vdots \\
x_N^2(0)-e^{-j2\pi\eta_2}x_N^2(1) & x_N^2(1)-e^{-j2\pi\eta_2}x_N^2(2) & \cdots & x_N^2(M-1)-e^{-j2\pi\eta_2}x_N^2(M) \\
x_0^2(0)-e^{-j2\pi\eta_3}x_0^2(1) & x_0^2(1)-e^{-j2\pi\eta_3}x_0^2(2) & \cdots & x_0^2(M-1)-e^{-j2\pi\eta_3}x_0^2(M) \\
x_1^2(0)-e^{-j2\pi\eta_3}x_1^2(1) & x_1^2(1)-e^{-j2\pi\eta_3}x_1^2(2) & \cdots & x_1^2(M-1)-e^{-j2\pi\eta_3}x_1^2(M) \\
\vdots & \vdots & \vdots & \vdots \\
x_N^2(0)-e^{-j2\pi\eta_3}x_N^2(1) & x_N^2(1)-e^{-j2\pi\eta_3}x_N^2(2) & \cdots & x_N^2(M-1)-e^{-j2\pi\eta_3}x_N^2(M)
\end{bmatrix}$$

$$\tag{7.27b}$$

The unknown weight vector is also given by

$$[W]^T = [W_0, W_1, W_2 \ldots, W_{M-1}] \tag{7.27c}$$

and the excitation is

$$[Y_M]^T = [C, C, C, 0, 0, \ldots, 0] \qquad (7.27d)$$

For the sake of simplification we separate equation (7.27) into three different matrix equations, corresponding to each cyclic frequency. Then it is observed that each block of the matrix has the same structure as that of the system matrix arising for the single-constraint case of Section 7.3, so that for each cycle frequency it is equivalent to writing, for $\eta_1 = \eta_e - 0.1\eta_e$,

$$
\begin{bmatrix}
1 & e^{j2\pi\eta_1} & \cdots & e^{j2\pi\eta_1(M-1)} \\
x_0^2(0) - e^{-j2\pi\eta_1}x_0^2(1) & x_0^2(1) - e^{-j2\pi\eta_1}x_0^2(2) & \cdots & x_0^2(M-1) - e^{-j2\pi\eta_1}x_0^2(M) \\
x_1^2(0) - e^{-j2\pi\eta_1}x_1^2(1) & x_1^2(1) - e^{-j2\pi\eta_1}x_1^2(2) & \cdots & x_1^2(M-1) - e^{-j2\pi\eta_1}x_1^2(M) \\
\vdots & \vdots & \vdots & \vdots \\
x_N^2(0) - e^{-j2\pi\eta_1}x_N^2(1) & x_N^2(1) - e^{-j2\pi\eta_1}x_N^2(2) & \cdots & x_N^2(M-1) - e^{-j2\pi\eta_1}x_N^2(M)
\end{bmatrix}
$$

$$
\times
\begin{bmatrix} W_0 \\ W_1 \\ \vdots \\ W_{M-1} \end{bmatrix}
=
\begin{bmatrix} 1 \\ 0 \\ \vdots \\ 0 \end{bmatrix}
\qquad (7.28)
$$

for $\eta_2 = \eta_e$,

$$
\begin{bmatrix}
1 & e^{j2\pi\eta_2} & \cdots & e^{j2\pi\eta_2(M-1)} \\
x_0^2(0) - e^{-j2\pi\eta_2}x_0^2(1) & x_0^2(1) - e^{-j2\pi\eta_2}x_0^2(2) & \cdots & x_0^2(M-1) - e^{-j2\pi\eta_2}x_0^2(M) \\
x_1^2(0) - e^{-j2\pi\eta_2}x_1^2(1) & x_1^2(1) - e^{-j2\pi\eta_2}x_1^2(2) & \cdots & x_1^2(M-1) - e^{-j2\pi\eta_2}x_1^2(M) \\
\vdots & \vdots & \vdots & \vdots \\
x_N^2(0) - e^{-j2\pi\eta_2}x_N^2(1) & x_N^2(1) - e^{-j2\pi\eta_2}x_N^2(2) & \cdots & x_N^2(M-1) - e^{-j2\pi\eta_2}x_N^2(M)
\end{bmatrix}
$$

$$
\times
\begin{bmatrix} W_0 \\ W_1 \\ \vdots \\ W_{M-1} \end{bmatrix}
=
\begin{bmatrix} 1 \\ 0 \\ \vdots \\ 0 \end{bmatrix}
\qquad (7.29)
$$

and for $\eta_3 = \eta_e + 0.1\eta_e$,

$$
\begin{bmatrix}
1 & e^{j2\pi\eta_3} & \cdots & e^{j2\pi\eta_3(M-1)} \\
x_0^2(0)-e^{-j2\pi\eta_3}x_0^2(1) & x_0^2(1)-e^{-j2\pi\eta_3}x_0^2(2) & \cdots & x_0^2(M-1)-e^{-j2\pi\eta_3}x_0^2(M) \\
x_1^2(0)-e^{-j2\pi\eta_3}x_1^2(1) & x_1^2(1)-e^{-j2\pi\eta_3}x_1^2(2) & \cdots & x_1^2(M-1)-e^{-j2\pi\eta_3}x_1^2(M) \\
\vdots & \vdots & \vdots & \vdots \\
x_N^2(0)-e^{-j2\pi\eta_3}x_N^2(1) & x_N^2(1)-e^{-j2\pi\eta_3}x_N^2(2) & \cdots & x_N^2(M-1)-e^{-j2\pi\eta_3}x_N^2(M)
\end{bmatrix}
$$

$$
\times
\begin{bmatrix}
W_0 \\
W_1 \\
\vdots \\
W_{M-1}
\end{bmatrix}
=
\begin{bmatrix}
1 \\
0 \\
\vdots \\
0
\end{bmatrix}
\tag{7.30}
$$

From (7.28), (7.29), and (7.30), it is also observed that the weight vector [W] satisfies simultaneously the three constraints corresponding to each cyclic frequency. Then, solving those coupled matrix equations, we obtain the weight vector constrained to multiple cycle frequencies which are close to the true cyclic frequency of the desired signal. In this way we can improve the performance of the proposed method in the presence of the cycle frequency error.

Next we present some numerical examples.

7.4 SIMULATION RESULTS USING IDEAL OMNIDIRECTIONAL POINT SENSORS

Computer simulations are carried out to illustrate the performance of the direct data domain approach when available snapshots are extremely limited. We consider a 15-element linearly equispaced array whose spacing is a half wavelength. All the elements are assumed to be ideal and consist of isotropic point sources radiating in free space.

All the signals, including the desired signals and interferers, are assumed to have BPSK modulations with different carrier frequencies. Symbols are transmitted using a rectangular modulation pulse. The received signals impinging at the array are sampled at a rate 10 times faster than the symbol rate. For the following examples, we utilize 15 snapshots to form the data block.

For the first example, it is assumed that the SOI arrives at the array from 80° to the azimuth direction. The signal is corrupted only by Gaussian noise. The signal-to-noise ratio (SNR) is set at 20 dB. The spectral line that this desired signal generates at $2f_c$ when it is squared is selected to be the cycle frequency for the direct data domain approach. Figure 7.3 shows the result for the estimated DOA based on the procedure above. A total of 50 simulations have been used for this example. As a result, the DOA of the desired signal is estimated properly in a Gaussian noise environment.

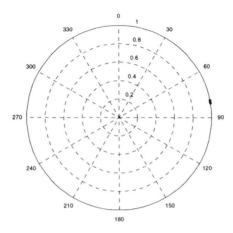

Figure 7.3. DOA estimation of a single signal in the presence of Gaussian noise.

For the second example, the SOI is perturbed by the presence of external interferences which are cyclically uncorrelated with the SOI at the same cycle frequency of the desired signal. In other words, the SOI is the only signal in the environment that generates a desired cycle frequency when squared. In this case, one interferer arrives at the array from the same direction as the SOI, and the other interferer arrives at the array from different directions. The signal-to-interference ratio is fixed at 0 dB. The parameters of the signals are represented in detail in Table 7.1. In this example, we ignore the effects of Gaussian noise. Figure 7.4 shows the result of the performance of the direct data domain approach exploiting cyclostationarity, which illustrates that the correct estimation for the DOA of the desired signal has been observed.

In the third example, the SOI is disturbed not only by the presence of external interferences but also by the presence of Gaussian noise. The desired signal arrives at the array from an angle of 10°, the external interferences are chosen as in the second example, and the signal-to-noise ratio at the antenna elements is fixed at 20 dB. The detail parameters of the signals are summarized in Table 7.1. Figure 7.5 shows the results of the direct data domain approach and illustrates the accuracy of this method.

TABLE 7.1
Input signals

	Direction	Carrier Frequency	Magnitude	Phase
SOI	80°	100 MHz	1 V/m	0
Interference 1	50°	1.2 MHz	1 V/m	0
Interference 2	20°	15 MHz	1 V/m	0
Interference 3	80°	1.3 MHz	1 V/m	0

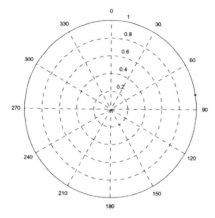

Figure 7.4. DOA estimation in the presence of external interferences.

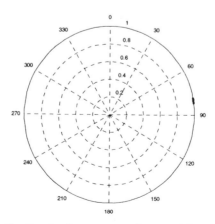

Figure 7.5. DOA estimation in the presence of external interferences plus Gaussian noise.

In the fourth example, one of the interferers has a cycle frequency which is very close to the true cycle frequency of the desired signal. The desired signal arrives at the array from 80° and signal-to-noise ratio at the antenna elements is fixed at 25 dB. The detail parameters of the desired signal and external interferences are summarized in Table 7.2. Here, the carrier frequency of one of the interferers is varied from 20 to 80 MHz in 20-MHz steps while the carrier frequency of the desired signal is fixed at 100 MHz. Figure 7.6 show the results of the proposed method for the four different carrier frequencies of the interference which is close to that of the signal. As shown in Figure 7.6, the performance of the proposed method is improved as the separation between the carrier frequency of the desired signal and that of the interferences increases. It is interesting to note that if the signals were to be separated using filtering one would need a data record of approximately 50 ns. Use of cyclostationarity can help filter signals where there is not enough data.

TABLE 7.2

Input signals

	Direction	Carrier Frequency	Magnitude	Phase
SOI	80°	100 MHz	1 V/m	0
Interference 1	50°	20–80 MHz	1 V/m	0
Interference 2	20°	190 MHz	1 V/m	0
Interference 3	80°	200 MHz	1 V/m	0

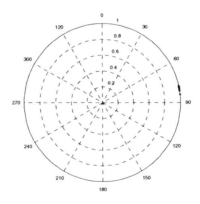

(a)

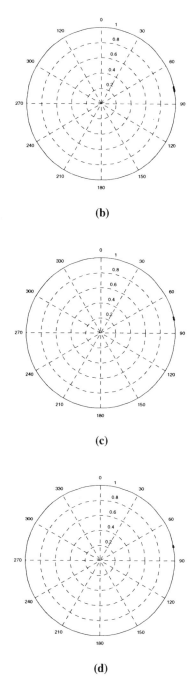

Figure 7.6. DOA estimation in the presence of interference close to the carrier frequency of 100 MHz: interference at (a) 80 MHz; (b) 60 MHz; (c) 40 MHz; (d) 20 MHz.

In the next example, we consider the root-mean-squared error (RMSE) versus SNR as the performance measure of the DOA estimation from the data measured at the antenna elements of the first example. The RMSE is defined for K different trials as

$$\text{RMSE} = \sqrt{\frac{1}{K} \sum_{k=1}^{K} \left[\text{DOA}_{\text{True}} - \text{DOA}_{\text{Estimated}}(k)\right]^2} \qquad (7.31)$$

Figure 7.7 shows the RMSE of the estimation versus SNR. The result is the average of 200 trials. As shown in Figure 7.7, the proposed method operates properly above a threshold of SNR = 20 dB.

In the next example, we evaluate the performance of the proposed method in the presence of error in the specification of the cycle frequency. We choose five multiple cycle frequencies at $\eta_e - 0.2\eta_e$, $\eta_e - 0.1\eta_e$, η_e, $\eta_e + 0.1\eta_e$ and $\eta_e + 0.2\eta_e$ to obtain a good bound for the actual cycle frequency of the desired signal. It is expected that the actual cycle frequency will be in between these five chosen frequencies. The detail values of the signal and external interferences are the same as in the second example and have been summarized in Table 7.1. The SNR at the antenna elements is fixed at 25 dB. Figure 7.8 plots the RMSE as a function of the error in the *a priori* knowledge of the cyclic frequency on the performance of the proposed method using a single constraint. However, as shown in Figure 7.9, the performance of the proposed method improves greatly when utilizing the five-cycle-frequencies approach to bound the uncertainty on the *a priori* knowledge of the cycle frequency. Use of a five-constraint approach appears to be adequate for most engineering applications.

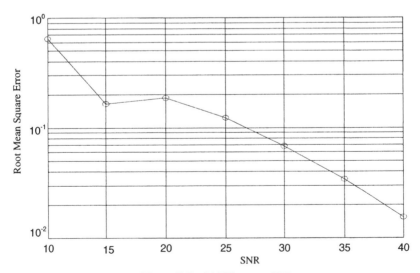

Figure 7.7. RMSE versus SNR.

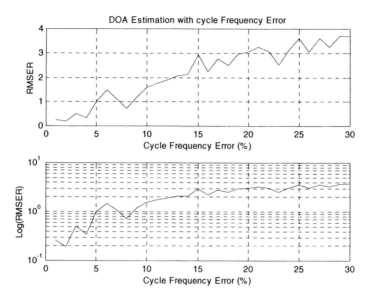

Figure 7.8. RMSE due to a single constraint when there is error in the *a priori* knowledge of the cyclic frequency.

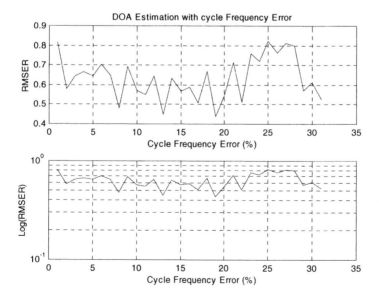

Figure 7.9. RMSE due to multiple constraints when there is error in the *a priori* knowledge of the cyclic frequency.

7.6   APPLICATION OF CYCLOSTATIONARITY USING AN ARRAY OF DIPOLES AND MICROSTRIP PATCH ANTENNAS

The direct data domain algorithms exploiting the cyclostationarity of the desired signal are based on the fact that a far-field source presents a linear phase front at the elements of the antenna array. The presence of mutual coupling between the elements of an array and near-field scatterers disturbs the capability of this algorithm, as it destroys the planar wavefront assumption. Hence, we need to preprocess the data to account for these undesired electromagnetic effects to correct for the uncertainty associated with the cyclostationarity of the desired signal, which is known *a prori*. We now use the interpolation technique described in Chapter 6 to compensate for the various mutual coupling effects, including the presence of near-field scatterers. Preprocessing to compensate the induced voltages contaminated by the effects of mutual coupling is based on transforming the real array into a uniform linear virtual array in the absence of the mutual coupling effect and other undesired electromagnetic effects using a compensation matrix, so that we can use cyclostationarity of the desired signal to obtain an accurate DOA estimation.

Our basic assumption is that the array manifold of the uniform linear virtual array corresponding to the desired signal carrying a cycle frequency can be obtained by a linear interpolation of the real array manifold, which is disturbed by various undesired electromagnetic couplings. Hence, our goal is to select the best-fit transformation matrix, $[\Im]$, between the real array manifold, $[A(\varphi)]$, and the array manifold corresponding to a uniform linear virtual array, $[A_v(\varphi)]$ such that $[\Im] [A(\varphi)] = [A_v(\varphi)]$ for all angles φ within a predefined sector, as explained in Chapter 6. Since such a transformation matrix is defined within a predefined sector, the various undesired electromagnetic effects, such as nonuniformity in spacing and mutual coupling between the elements and near-field obstacles for an array, is compensated for through the transformation above. This transformation is independent of the azimuth angle of the incident signals.

7.7   SIMULATION RESULTS USING REALISTIC ANTENNA ELEMENTS

Computer simulations are carried out to illustrate the performance of the direct data domain approach exploiting cyclostationarity in the presence of the various electromagnetic effects when available snapshots are extremely limited. As shown in Figure 7.10, we consider a 17-element linearly equispaced array whose interelement spacing is a quarter wavelength. It consists of half-wave thin-wire dipole antenna elements loaded at the center. Each element of the array is identically point loaded by 50 Ω at the center. The dipoles are z-directed, of length $L = \lambda/2$, and radius $r = \lambda/200$. The details of the chosen array are presented in Table 7.3.

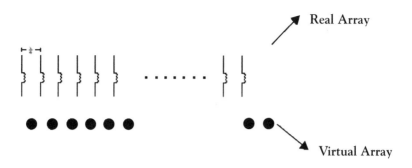

Figure 7.10. Array of dipoles with quarter-wavelength interelement spacing along with the ULVA.

TABLE 7.3

Details of the physical size of an array

Number of elements in real array	17
Length of z-directed wires	$\lambda/2$
Radius of wires	$\lambda/200$
Loading at the center	$50\,\Omega$

The desired signal and other interference signals are assumed to have BPSK modulation with different carrier frequencies. Symbols are transmitted using rectangular modulation pulses. The received signals impinging at the array are sampled at a rate 10 times faster than the symbol rate. In the following examples, we use a uniform linear virtual array (ULVA) utilizing 17-snapshots to form the data block.

As described in Chapter 6, the real array is interpolated into the virtual array consisting of $N + 1$ $(= 17)$ uniformly spaced omnidirectional point sources separated by a distance d/λ. By choosing the reference point at the first element of the real array, the steering vectors associated with the virtual array are given by

$$[\bar{a}(\varphi)] = \left[1, e^{\frac{j\,2\pi d}{\lambda}\cos\varphi}, e^{\frac{j\,2\pi 2d}{\lambda}\cos\varphi}, \ldots, e^{\frac{j\,2\pi N d}{\lambda}\cos\varphi} \right]^{T}_{(N+1)\times 1}$$

(7.32)

The incremental size Δ in the interpolation region, $[\Phi] = [\varphi_q, \varphi_{qq}] = [30°, 150°]$, is chosen to be $1°$. The sector chosen here, for example, is of $120°$ and symmetrically located. Then, a set of real steering vectors are measured/

computed for sources located at each of the angles φ_q, $\varphi_q + \Delta$, $\varphi_q + 2\Delta$, ..., φ_{qq}. The measured/ computed vector $[A(\varphi)]$ is then distorted from the ideal steering vector due to the presence of mutual coupling between the elements of the real array. Then, using (6.13), we obtain the transformation matrix to compensate for the effects of the presence of mutual coupling between the elements of the real array. Finally, using (6.14), we can obtain the corrected input voltage in which the mutual coupling effects are eliminated from the actual voltage, and then we can apply the direct data domain approach exploiting the cyclostationarity of the desired signal.

As an example, it is assumed that the SOI arrives at the array from an azimuth angle of 40°. The signal is corrupted by the Gaussian noise and external interferences. The signal-to-interference ratio is fixed at 0 dB. The parameters for the signal and the interferers are given in Table 7.4. The spectral line that this desired signal generates at $2f_c$ is 600 MHz when it is squared. This spectral line is selected in the direct data domain approach to extract the SOI. Figure 7.11 shows the results for the estimated DOA based on the procedure above. A total of 30 simulations have been carried out for this example. Figure 7.11a and c represent the results for a 10-dB and a 5-dB signal-to-noise ratio, respectively, without compensation of the mutual coupling between the elements of an array. Noise is assumed to be Gaussian and zero mean at each antenna element. Figure 7.11b and d represent the results for a 10-dB and a 5-dB signal-to-noise ratio, respectively, with compensation for mutual coupling between the 17 antenna elements for the same scenario. As shown in Figure 7.11, although the SNR is low, the DOA of the desired signal after correction of the mutual coupling between the elements of the array is estimated correctly in the presence of Gaussian noise and external interferences.

TABLE 7.4
Parameters of the signal incident on the array

	Direction	Carrier Frequency	Magnitude	Phase
SOI	40°	300 MHz	1 V/m	0
Interference 1	50°	1.2 MHz	1 V/m	0
Interference 2	35°	15 MHz	1 V/m	0

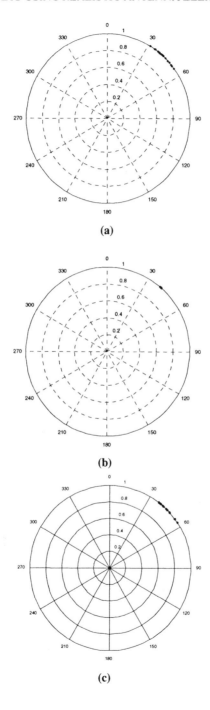

(a)

(b)

(c)

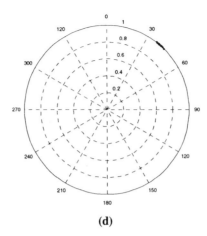

(d)

Figure 7.11. DOA estimates of the SOI in the presence of Gaussian noise and an (a, c) interference without compensation; (b, d) interference with compensation.

For the next example, the SOI is disturbed by the presence of a strong external interference, which is varied from 0 to 40 dB. The desired signal arrives at the array from 40°. The amplitudes and the DOA of the SOI and the interferers are given in Table 7.5. Figure 7.12 shows the results of root-mean-squared error [as defined by (7.31)], in which the dashed line represents the result of the procedure without compensation of the mutual coupling and the solid line represents the result of the procedure after compensation of the mutual coupling. Correct estimates for the DOA in the presence of strong external interferences is also possible. For this example, one of the interferences has a cycle frequency which is very close to the true cycle frequency of the desired signal. The desired signal arrives at the array from 40°, with a signal-to-noise ratio fixed at 15 dB. The detail parameters of the desired signal and external interferences are summarized in Table 7.6. The carrier frequency of one of the interferences is varied from 250 to 270 MHz in 20-MHz steps while the carrier frequency of the desired signal is fixed at 300 MHz. Figure 7.13a and c show results of the proposed method without compensation for the mutual coupling, and Figure 7.13b and d show the results of the proposed method with compensation for the mutual coupling. As shown in Figure 7.13, the performance of the proposed method after compensation for the mutual coupling is very good even though the carrier frequency of the desired signal is located close to that of the desired signal.

TABLE 7.5
Parameters of the signal incident on the array

	Direction	Carrier Frequency	Magnitude	Phase
SOI	40°	300 MHz	1 V/m	0
Interference 1	50°	1.2 MHz	1–100 V/m	0
Interference 2	35°	15 MHz	1 V/m	0

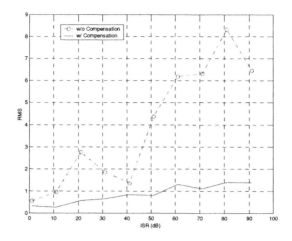

Figure 7.12. Errors in the estimate of the DOA both with and without use of the transformation matrix as a function of the strength of the interference.

TABLE 7.6
Parameters of the signal incident on the array

	Direction	Carrier Frequency	Magnitude	Phase
SOI	40°	300 MHz	1 V/m	0
Interference 1	50°	250–270 MHz	1 V/m	0
Interference 2	35°	10 MHz	1 V/m	0

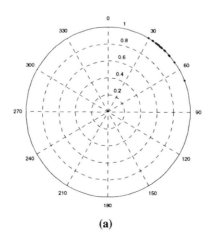

(a)

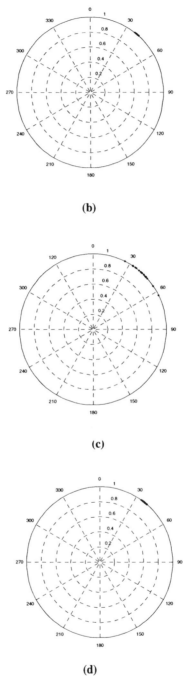

(b)

(c)

(d)

Figure 7.13. Estimates of the DOA in the presence of external interferences located close to the carrier frequency of the SOI are at 300 MHZ after compensation for mutual coupling: interference at (a, b) 250 MHz; (c, d) 270 MHz.

Next, we consider the array configuration consisting of 13 microstrip patch antenna elements as shown in Figure 7.14. Each element of the microstrip patch array is located at linearly equispaced intervals with a spacing of 0.8 wavelength. The thickness of the dielectric is 0.01λ. Then, as described in the previous example, the real array is interpolated into the virtual array, consisting of $N + 1$ ($= 13$) uniformly spaced omnidirectional point sources separated by 0.25 wavelength. The signal-to-noise ratio is fixed at 20 dB and the signal-to-interference ratio is fixed at 0 dB. The desired signal arrives at the array from $40°$, and the other external interferences are chosen similar to the ones in Table 7.7. Figure 7.15 represents the results of DOA estimation based on the procedure above. Figure 7.15a shows the result of DOA estimation without compensation of the mutual coupling between the patch elements of the array, and Figure 7.15b shows the result of DOA estimation with compensation of the mutual coupling between the patch elements of the array. As a result, we obtain the correct DOA estimation of the desired signal carrying a cycle frequency after compensating for the unwanted electromagnetic effects by the interpolation technique. If the various electromagnetic effects are not properly accounted for the results for the estimation of DOA are seriously biased.

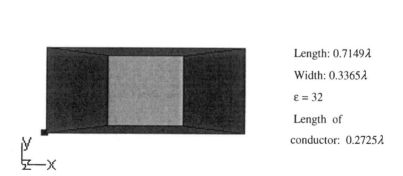

Length: 0.7149λ

Width: 0.3365λ

$\varepsilon = 32$

Length of conductor: 0.2725λ

Figure 7.14. Microstrip patch antenna array.

TABLE 7.7
Parameters of the signal incident on the array

	Direction	Carrier Frequency	Magnitude	Phase
SOI	40°	100 MHz	1 V/m	0
Interference 1	50°	1.2 MHz	1 V/m	0
Interference 2	35°	1.5 MHz	1 V/m	0

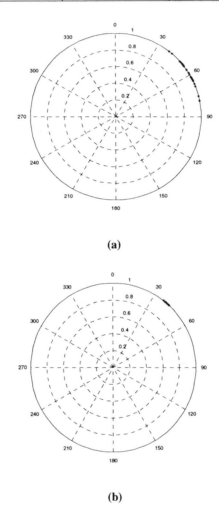

(a)

(b)

Figure 7.15. DOA estimation using a microstrip patch antenna array (a) without and (b) with compensation.

For the final example, we evaluate the performance of this algorithm using the same microstrip patch antenna array of the previous example now located on the side of an aircraft as shown in Figure 7.16. It is important to observe that some of elements of the array are spaced nonuniformly. The number of elements of the patch array is 11 and is interpolated into the virtual array consisting of $N+1 (=11)$ uniformly spaced omnidirectional point sources separated by 0.25 wavelength. The desired signal arrives at the array from 85°, with a signal-to-noise ratio fixed at 30 dB. The detail parameters of the desired signal and external interferences are summarized in Table 7.8. Figure 7.17 shows that the results of this algorithm exploiting the prespecified cycle frequency are evaluated properly after compensation of the mutual coupling between the array and the airplane.

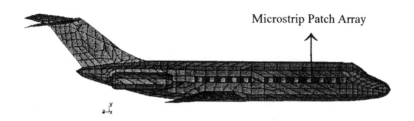

Figure 7.16. Conformal unequally spaced microstrip patch antenna array located on the body of an aircraft.

TABLE 7.8
Parameters of the signal incident on the array

	Direction	Carrier Frequency	Magnitude	Phase
SOI	85°	100 MHz	1 V/m	0
Interference 1	69°	10 MHz	0.8 V/m	0
Interference 2	50°	20 MHz	0.8 V/m	0

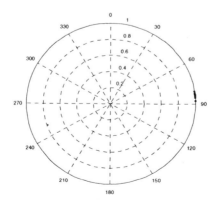

Figure 7.17. DOA estimation using the microstrip patch antenna array of Figure 7.16.

7.8   DOA ESTIMATION IN A MULTIPATH ENVIRONMENT

The purpose of this section is to estimate the DOA of the desired signal in the presence of multipaths and interferers utilizing a combined temporal and spatial processing of the data. The temporal and spatial processing is based on a direct data domain approach. The number of incoherent signals impinging at the array can be greater than the number of antenna elements in the phased array. In this section we use the concept of cyclostationarity described in Section 7.7, which is the temporal information of the desired signal to separate it from the interferences. By exploiting the principle of cyclostationarity we can extract signals with the same cycle frequency and null out the co-channel interferences and additive noise. Hence, the signal detection capability can be significantly increased. Next we use spatial processing to differentiate the various coherent/noncoherent multipaths of the desired signal which will have the same cycle frequency. The various multipaths may be coherent. We apply the Matrix Pencil [7] method to the temporally processed data to estimate the DOA of the various multipaths to separate them spatially. A survey of the Matrix Pencil method is given in Appendix C. The main contribution of this section is that by combining temporal and spatial processing based on the direct data domain approach one can handle multipath signals under the conditions that the number of signals impinging at the array at all the frequencies can be greater than the number of antenna elements. However, the number of multipaths both coherent and noncoherent has to be less than approximately half the number of antenna elements. Since we do not form a covariance matrix of the data, this method is quite suitable for short data lengths when the environment is quite dynamic and the multipaths are coherent. In this algorithm, while the estimation of the cyclic array covariance matrix is avoided, we develop a new matrix form using extremely short data samples. As a result, the computational load in the proposed approach is relatively reduced and the robustness of the estimation of the signal of interest is significantly improved when the number of available snapshots is extremely limited. Numerical results are presented to illustrate the efficiency and accuracy of this method.

7.8.1   Reformulation of the D^3 Approach Exploiting the Temporal Information of Cyclostationarity in a Multipath Environment

Consider an antenna array consisting of $N + 1$ elements that are uniformly spaced along a line. The elements in this section are assumed to be isotropic omnidirectional point radiators. The mutual coupling between the elements of an array can be taken into account as shown in Section 7.7. Then the signals received by the array at time t can be represented as

$$[x(t)] = [x_0(t), x_1(t), \ldots , x_N(t)]^T \tag{7.33}$$

$$x_n(t) = s_d(t)\, e^{j2\pi nd\cos\varphi_d/\lambda} + \sum_{q=0}^{Q-1} \beta_q\, s_d(t-\tau_q)\, e^{j2\pi nd\cos\varphi_q/\lambda}$$
$$+ \sum_{p=0}^{P-1} s_p(t)\, e^{j2\pi nd\cos\varphi_p/\lambda} + \xi_n(t) \qquad (7.34)$$

$$\text{for } n = 0, 1, \ldots, N$$

where $s_d(t)$ and $s_p(t)$ are the desired signal with a given cycle frequency and the interference signals, respectively. The sources are assumed to be located in the far field. The second term represents the multipath signals of the desired signal with the same cyclic frequency. f_c is the carrier frequency and τ_q denotes the time delay of the multipath. The basic difference between (7.6b) and (7.34) is that the effects of multipaths have been incorporated.

Without loss of generality, if the time delays are small compared with the reciprocal of the bandwidth of the signal $s_d(t)$, then $s_d(t-\tau_q) = s_d(t)e^{-j2\pi f_c\tau_q}$, which may be used in (7.34) to yield

$$x_n(t) = s_d(t)\, e^{j2\pi nd\cos\varphi_d/\lambda} + \sum_{q=0}^{Q-1} \beta_q\, s_d(t)\, e^{-j2\pi f_c\tau_q}\, e^{j2\pi nd\cos\varphi_q/\lambda}$$
$$+ \sum_{p=0}^{P-1} s_p(t)\, e^{j2\pi nd\cos\varphi_p/\lambda} + \xi_n(t) \qquad (7.35)$$

If there is no time delay, that is, τ_q is 0, it simply becomes the coherent case.

From (7.34) and (7.35), the received signal can be represented in matrix form as

$$[x(t)] = [A][s(t)] + [\xi(t)]$$
$$[A] = \left[a(\varphi_d),\, a(\varphi_0),\, \ldots,\, a(\varphi_{P+Q-1})\right]_{(N+1)\times(P+Q+1)}$$
$$[s(t)] = \left[s_d(t),\, s_0(t),\, \ldots,\, s_{P+Q-1}(t)\right]^T_{(P+Q+1)\times 1} \qquad (7.36)$$
$$[\xi(t)] = \left[\xi_0(t),\, \xi_1(t),\, \ldots,\, \xi_N(t)\right]^T_{(N+1)\times 1}$$

We assume in this section that the desired signal and their multipath component have the same cycle frequency ($\eta = 2f_c$). The interferers and noise are cyclically uncorrelated at the cycle frequency of the desired signal.

To derive the direct data domain algorithm exploiting the principles of cyclostationarity, we first square the input signals. Let us chose the first antenna element as the reference element. Next, we have evaluate the products containing the terms $x_0(t)x_{1+n}(t)$ for $n = 0, 1, \ldots, N-1$. A typical term is of the form

$$x_0(t)\,x_{1+n}(t) \;=$$

$$\left\{ s_d(t) + \sum_{p=0}^{P-1} s_p(t) + \sum_{q=0}^{Q-1} \beta_q\, s_d(t) e^{-j2\pi f_c \tau_q} + \xi_1(t) \right\} \times \left\{ s_d(t)\, e^{j2\pi d(1+n)\cos\varphi_d/\lambda} \right.$$

$$+ \sum_{q=0}^{Q-1} \beta_q\, s_d(t) e^{-j2\pi f_c \tau_q} e^{j2\pi d(1+n)\cos\varphi_d/\lambda} + \sum_{p=0}^{P-1} s_p(t) e^{j2\pi d(1+n)\cos\varphi_d/\lambda} + \left. \xi_{1+n}(t) \right\}$$

$$= s_d^2(t)\, e^{j2\pi d(1+n)\cos\varphi_d/\lambda} + \sum_{q=0}^{Q-1} \beta_q\, s_d^2(t) e^{-j2\pi f_c \tau_q} e^{j2\pi d(1+n)\cos\varphi_d/\lambda}$$

$$+ \sum_{q=0}^{Q-1} \beta_q\, s_d^2(t) e^{-j2\pi f_c \tau_q} e^{j2\pi d(1+n)\cos\varphi_d/\lambda} + s_d(t) \sum_{p=0}^{P-1} s_p(t) e^{j2\pi d(1+n)\cos\varphi_d/\lambda}$$

$$+ \sum_{q=0}^{Q-1} \beta_q\, s_d(t) e^{-j2\pi f_c \tau_q} \sum_{q=0}^{Q-1} \beta_q\, s_d(t) e^{-j2\pi f_c \tau} e^{j2\pi d(1+n)\cos\varphi_d/\lambda} \qquad (7.37)$$

$$+ \sum_{q=0}^{Q-1} \beta_q\, s_d(t) e^{-j2\pi f_c \tau_q} \left\{ \sum_{p=0}^{P-1} s_p(t) e^{j2\pi d(1+n)\cos\varphi_d/\lambda} + \xi_{1+n}(t) \right\}$$

$$+ \sum_{p=0}^{P-1} s_p(t)\,\xi_{1+n}(t) + s_d(t)\,\xi_{1+n}(t) + \sum_{p=0}^{P-1} s_p(t) \left\{ s_d(t) e^{j2\pi d(1+n)\cos\varphi_d/\lambda} \right.$$

$$+ \sum_{q=0}^{Q-1} \beta_q\, s_d(t) e^{-j2\pi f_c \tau_q} e^{j2\pi d(1+n)\cos\varphi_d/\lambda} + \left. \sum_{p=0}^{P-1} s_p(t) e^{j2\pi d(1+n)\cos\varphi_d/\lambda} \right\}$$

$$+ \xi_1(t) \left\{ s_d(t) e^{j2\pi d(1+n)\cos\varphi_d/\lambda} + \sum_{q=0}^{Q-1} \beta_q\, s_d(t) e^{-j2\pi f_c \tau_q} e^{j2\pi d(1+n)\cos\varphi_d/\lambda} \right.$$

$$+ \left. \sum_{p=0}^{P-1} s_p(t) e^{j2\pi d(1+n)\cos\varphi_d/\lambda} + \xi_{1+n}(t) \right\}$$

Here, squaring the signals received at the antenna elements produces a term in the temporal domain which consists of the signals that produce the second-order cycle frequency and other components that do not contain the second-order cycle frequency.

Next we sample the received signal and form the difference function using the cycle frequency of the SOI, which is known *a priori* as

$$x_0(m)\,x_{1+n}(m) - e^{-j2\pi\eta}\,x_0(m+1)\,x_{1+n}(m+1) \qquad (7.38)$$

Equation (7.38) will contain no components of the signals at the cycle frequency $\eta = 2f_c$. Hence, our goal is to construct a least squares estimate of the signal which exhibits the same cyclostationarity property as the SOI, by solving for a set of weights. Then by forming a weighted sum of these terms of (7.37), we obtain a signal that has the signal components having the same cycle frequency as the SOI. We then apply the Matrix Pencil technique described in Appendix C to this estimate to separate signals spatially which have the same cycle frequency and resolve their DOA.

First, using (7.37), we construct a new matrix form based on (7.38), utilizing $M+1$ snapshots containing short signal sequences as

$$[Z][W] = [Y] \tag{7.39a}$$

where

$$[Z] = \begin{bmatrix}
1 & e^{j2\pi\eta} & \cdots \\
x_0(0)x_1(0) - e^{-j2\pi\eta}x_0(1)x_1(1) & x_0(1)x_1(1) - e^{-j2\pi\eta}x_0(2)x_1(2) & \cdots \\
x_0(0)x_2(0) - e^{-j2\pi\eta}x_0(1)x_2(1) & x_0(1)x_2(1) - e^{-j2\pi\eta}x_0(2)x_2(2) & \cdots \\
\vdots & \vdots & \cdots \\
x_0(0)x_N(0) - e^{-j2\pi\eta}x_0(1)x_N(1) & x_0(1)x_N(1) - e^{-j2\pi\eta}x_0(2)x_N(2) & \cdots
\end{bmatrix}$$

$$\begin{matrix}
\cdots & e^{j2\pi(M-1)\eta} \\
\cdots & x_0(M-1)x_1(M-1) - e^{-j2\pi\eta}x_0(M)x_1(M) \\
\cdots & x_0(M-1)x_2(M-1) - e^{-j2\pi\eta}x_0(M)x_2(M) \\
& \vdots \\
\cdots & x_0(M-1)x_N(M-1) - e^{-j2\pi\eta}x_0(M)x_N(M)
\end{matrix}$$

$$\tag{7.39b}$$

where the unknown weight vector is given by

$$[W]^T = [W_0, W_1, W_2, \ldots, W_{M-1}] \tag{7.39c}$$

and the excitation is assumed to be

$$[Y]^T = [C, 0, 0, \ldots, 0] \tag{7.39d}$$

Here, the arbitrary constant C is related to the weights and is given by

$$C = W_0 + W_1 e^{j2\pi\eta} + \cdots + W_{M-1} e^{j2\pi(M-1)\eta} \tag{7.39e}$$

Since $[Y]$ is an excitation function, the value of the unknown constant can be chosen arbitrarily. The actual value is not important as it gets factored in the final computation. Setting the weighted sum of the elements of the matrix $[Z]$ starting from the second row to zero, nulls interferences which have different cycle frequencies other than the SOI in a least square sense. The equation represented by the first row of (7.39 a) constraints the array gain in the direction of the desired signal for the prespecified cycle frequency. Therefore, only the signals (desired signal plus multipath signals of the desired signal) with the same cyclic frequency can be extracted while nulling out the other co-channel interferences and additive thermal noise by using the following weighted sum:

$$
\begin{aligned}
G(n) = \frac{1}{C} \sum_{m=0}^{M-1} x_0(m)\, x_{1+n}(m) W_m \\
= \left\{ 1 + \sum_{q=0}^{Q-1} \beta_q\, e^{-j2\pi f_c \tau_q} \right\} e^{j2\pi n d \cos \varphi_d / \lambda} \\
+ \sum_{q=0}^{Q-1} \left\{ \sum_{u=0}^{Q-1} \beta_q \beta_u\, e^{-j2\pi f_c(\tau_q + \tau_u)} \right\} e^{j2\pi n d \cos \varphi_q / \lambda} \\
= \gamma_d\, e^{j2\pi n d \cos \varphi_d / \lambda} + \sum_{q=0}^{Q-1} \gamma_q\, e^{j2\pi n d \cos \varphi_q / \lambda} \qquad \text{for} \quad n = 0,1,\ldots, N-1
\end{aligned}
$$

$$ \tag{7.40a} $$

$$
\begin{aligned}
\gamma_d &= \left\{ 1 + \sum_{q=0}^{Q-1} \beta_q\, e^{-j2\pi f_c \tau_q} \right\} \\
\gamma_q &= \left\{ \sum_{u=0}^{Q-1} \beta_q \beta_u\, e^{-j2\pi f_c(\tau_q + \tau_u)} \right\}
\end{aligned}
$$

$$ \tag{7.40b} $$

Hence, the signal $G(n)$ at the nth antenna element provides a least squares estimate of the sum of all the signals induced at that antenna element having the same cycle frequency of interest while eliminating the other co-channel interferences and noise.

Now we can estimate the DOA of the various signals having the same cyclic frequency by processing them spatially using the Matrix Pencil method described in Appendix C, which is one of the direct data domain methods to resolve a sum of complex exponentials.

7.8.2   Simulation Result in a Multipath Environment

We consider a 12-element linearly equispaced array whose interelement spacing is a half wavelength. All the elements are assumed to be ideal and to consist of omnidirectional point sources radiating in free space. The desired signal and other interference signals are assumed to have BPSK modulation with different carrier frequencies. Also, there are multipath signals and they have the same cyclic frequency as that of the desired signal. Symbols are transmitted using rectangular modulation pulses. The received signals impinging at the array are sampled at a rate 10 times faster than the symbol rate. For all the examples, we utilize 12 snapshots (i.e., the voltages at each of the antenna elements are sampled 12 times).

For the first example, it is assumed that the SOI arrives at the array from an angle of 55° with respect to the broadside direction. There are three multipaths of the desired signal arriving at the array from different directions. Each multipaths have different delays, corresponding to 10, 20, and 30 μs. Also, there are 13 interferences with different carrier frequencies, as described in Table 7.9. The signal-to-noise ratio (SNR) at each antenna element is 25 dB. Here interference 3 and the desired signal impinge at the array from the same angular direction but with a different carrier frequency. In this example, the number of the signals incident at the array is greater than the number of antenna elements of the array. Figure 7.18 represents the results of the performance of the proposed two-step processing with an assumed cycle frequency of 200 MHz. The results of Figure 7.18 illustrate that the proposed method estimates the various DOAs correctly even when the SOI and all its multipaths have the same cycle frequency. The method is also capable of nulling out the other interferences and noise. A total of 30 simulations have been carried out for this example and the results superposed on Figure 7.18.

TABLE 7.9

Parameters of all the signals incident on the array

	Direction	Carrier Frequency	Magnitude	Phase
SOI	55°	100 MHZ	1 V/m	0
Multipath 1	85°	100 MHz	1 V/m	0
Multipath 2	110°	100 MHz	1 V/m	0
Multipath 3	15°	100 MHz	1 V/m	0
Interference 1	5°	1.2 MHz	1 V/m	0
Interference 2	20°	1.5 MHz	1 V/m	0
Interference 3	55°	10 MHz	1 V/m	0
Interference 4	40°	1.8 MHz	1 V/m	0
Interference 5	60°	3.5 MHz	1 V/m	0
Interference 6	110°	0.1 MHz	1 V/m	0
Interference 7	60°	0.35 MHz	1 V/m	0
Interference 8	30°	50 MHz	1 V/m	0
Interference 9	43°	70 MHz	1 V/m	0
Interference 10	12°	65 MHz	1 V/m	0
Interference 11	18°	80 MHz	1 V/m	0
Interference 12	10°	75 MHz	1 V/m	0
Interference 13	52°	0.12 MHz	1 V/m	0

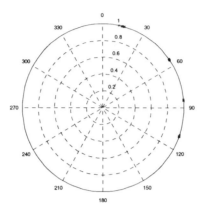

Figure 7.18. Estimates of the DOA of the SOI (its amplitude is 1.0) along with its various multipaths.

For the second example we consider the performance of the proposed approach as we vary the magnitude of the SOI. The magnitude of the desired signal has been varied from 0.8 V/m to 0.2 V/m. The other interferences and multipath signals are the same as in the previous example. Figure 7.19 shows the results of the performance of the proposed approach in terms of different magnitudes of the desired signal. As shown in Figure 7.19, the DOA of the SOI is estimated correctly even though its amplitude is small in comparison to the other multipath components. Figure 7.20 represents the RMSE associated with the estimation of the DOAs of the various signal components for Figure 7.19. As expected, the RMSE associated with the SOI arriving from 55° is slightly higher than that of the signals arriving from 15°, 85°, and 110°. The RMSE associated with the signal arriving from 15° is also slightly higher than that of the other signals (110°, 85°).

In the third example we want to present the performance of the algorithm in detecting signals that are spaced very close to each other in space but have the same cyclic frequency. In this case the SOI of Table 7.9 is changed to 65° and the DOA of the remaining signals, along with their multipaths, remains the same. Figures 7.21 show that closely spaced sources are properly identified even when they are located within the beamwidth of the main lobe. The spatial separation between the two signals with the same cyclic frequency has been varied from 20° to 5° in steps of 5°. Figure 7.22 represents the RMSE associated with estimation of the DOA of the various signal components for Figure 7.21.

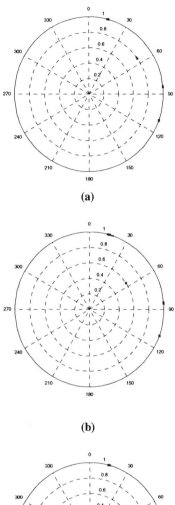

(a)

(b)

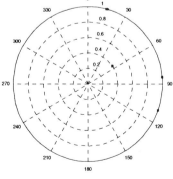

(c)

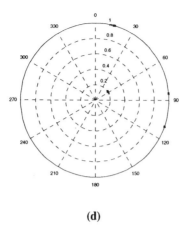

(d)

Figure 7.19. Estimates of the DOA of the SOI along with its various multipaths: amplitude is (a) 0.8, (b) 0.6, (c) 0.4, (d) 0.2.

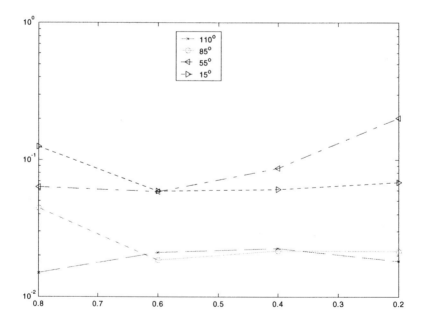

Figure 7.20. RMSE associated with estimation of the DOA of the various signal components.

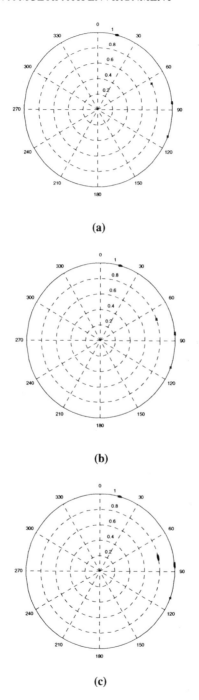

Figure 7.21. Estimation of the DOA of the various signal components and their multipaths: angular separation is (a) 20°, (b) 15°, (c) 5°.

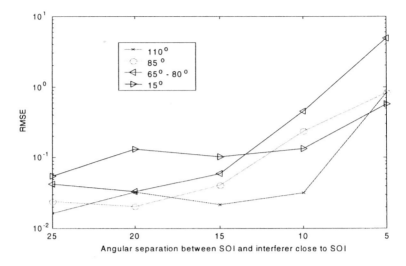

Figure 7.22. RMSE in estimates of the DOA of the various signals along with their multipaths.

In the fourth example we observe the performance of the proposed approach, specially, in the case of coherent multipaths. The DOA of the SOI is still 65° and the remainder of the signals is given by the entries of Table 7.9. For this special situation, we assume that there is no time delay between the desired signal and their multipath components. Figure 7.23 shows the performance of the proposed method as a function of the SNR varying from 30 to 5 dB.

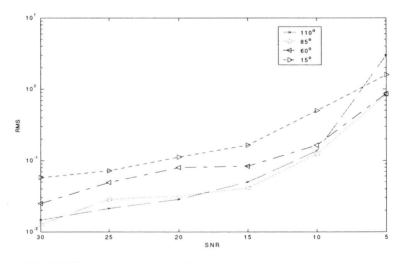

Figure 7.23. RMSE in estimates of the DOA of the various signals along with their multipaths.

The next example presents the performance of the proposed approach associated with the estimation of the amplitudes of the SOI in case of coherent multipaths (i.e., we assume no phase difference between the SOI and the different multipaths). The DOA of the SOI is 65° and the amplitudes and DOA of the remaining signals are given in Table 7.9. Figures 7.24 illustrates that it is possible to identify the SOI along with its multipaths even though the amplitude of the SOI may be smaller in amplitude than the multipath component. Figure 7.25 represents the RMSE associated with estimation of the DOA of the SOI and its various multipaths corresponding to Figure 7.24. As expected, the error associated with estimation of the DOA of the SOI at an angle of 65° is slightly higher than that of the other signals (85°, 110°, 15°), as its amplitude is smaller than the multipath components.

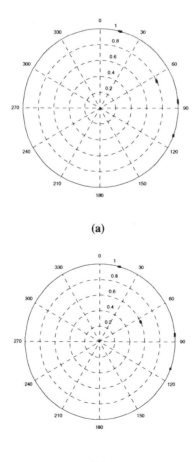

(a)

(b)

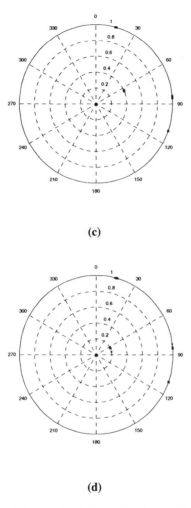

(c)

(d)

Figure 7.24. Estimates of the DOA of the SOI along with its various multipaths: amplitude is (a) 0.8, (b) 0.6, (c) 0.4, (d) 0.2.

For the final example we want to compare the performance of this algorithm in its ability to detect closely spaced signals in space having the same cyclic frequency in a coherent multipath environment. The DOA of the SOI is still 65° and the DOA and amplitude of the other signals are given in Table 7.9. Figure 7.26 shows that closely spaced sources are properly separated even when the SOI along with their multipaths are located within the beam width of the main lobe. The spatial separation between the two signals, namely the SOI and its multipath at the same cyclic frequency, has been varied from 20° to 5° in steps of 5°. Figure 7.27 represents the error associated with the estimation of the DOA of the SOI along with their multipaths at the same cycle frequency in Figure

7.26. As shown, RMSE is slightly higher when the SOI and its multipath are closely located in angle.

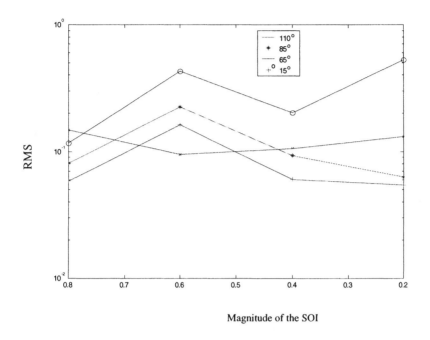

Figure 7.25. RMSE associated with the DOA of the various signal components.

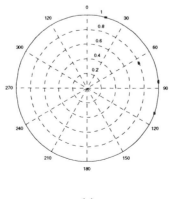

(a)

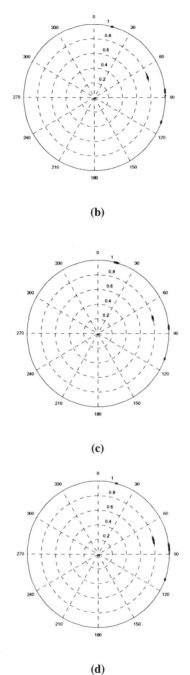

(b)

(c)

(d)

Figure 7.26. Estimation of the DOA of the desired signal and their multipaths: separation between the SOI and its multipath is (a) 20°, (b) 15°, (c) 10°, (d) 5°.

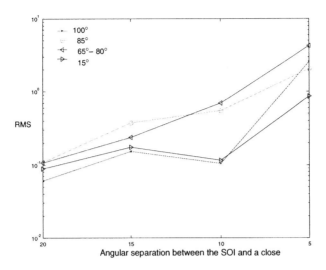

Figure 7.27. RMSE for the DOA of the SOI and the various multipaths.

7.9   CONCLUSION

This chapter described a direct data domain approach to estimate the direction of arrival of the desired signal in the presence of noise and interferences based on exploiting the cyclostationary property of the signals. A two-step approach, first temporal followed by spatial processing, has also been presented, to identify the SOI along with its various multipath components impinging on an antenna array along with other signals which may have a different carrier frequency.

In this chapter we have observed that the number of signals impinging at the array can be greater that the number of antenna elements which is generally the situation in a typical mobile communication environment. The numerical examples show that this algorithm is efficient and robust in the case when the available snapshots are extremely limited.

REFERENCES

[1] W. A. Gardner, *Cyclostationarity in Communication and Signal Processing,* IEEE Press, Piscataway, NJ.

[2] W. A. Gardner, "Simplification of MUSIC and ESPRIT by Exploitation of Cyclostationarity," *Proceedings of the IEEE*, Vol. 76, No. 7, July 1988.

[3] L. Castedo and A. R. Figueiras-Vidal, "An Adaptive Beamforming Technique Based on Cyclostationary Signal Properties," *IEEE Transactions on Signal Processing*, Vol. 43, No. 7, July 1995.

[4] B. G. Agee, S. V. Schell, and W. A. Gardner, "Spectral Self-Coherence Restoral: A New Approach to Blind Adaptive Signal Extraction Using Antenna Arrays," *Proceedings of the IEEE*, Vol. 78, Apr. 1990.

[5] T. K. Sarkar and N. Sangruji, "An Adaptive Nulling System for a Narrow-Band Signal with a Look-Direction Constraint Utilizing the Conjugate Gradient Method," *IEEE Transactions on Antennas and Propagation*, Vol. 37, No. 7, 1989.

[6] T. K. Sarkar, S. Park, J. Koh, and R. A. Schneible, "A Deterministic Least Square Approach to Adaptive Antennas," *Digital Signal Processing: A Review Journal*, Vol. 6, p.p. 185–194, 1996.

[7] T. K. Sarkar and O. Pereira, "Using the Matrix Pencil Method to Estimate the Parameters of a Sum of Complex Exponentials," *IEEE Antennas and Propagation Magazine*, Vol. 37, No. 1, Feb. 1995.

[8] Q. Wu and K. M. Wong, "Blind Adaptive Beamforming for Cyclostationary Signals," *IEEE Transactions on Signal Processing*, Vol. 44, No. 11, Nov. 1996.

[9] G. Xu and T. Kailath, "Direction-of-Arrival Estimation via Exploitation of Cyclostationarity: A Combination of Temporal and Spatial Processing," *IEEE Transactions on Signal Processing*, Vol. 40, No. 7, July 1992.

8

A SURVEY OF VARIOUS PROPAGATION MODELS FOR MOBILE COMMUNICATION

SUMMARY

For mobile systems, in order to estimate the signal parameters accurately, it is necessary to estimate its propagation characteristics through a medium. The propagation analysis provides a good initial estimate of the signal characteristics. The ability to accurately predict radio propagation behavior for wireless personal communication systems such as cellular mobile radio is becoming crucial to system design. Since site measurements are costly, propagation models have been developed as a suitable low-cost, convenient alternative. Channel modeling is required to predict path loss and to characterize the impulse response of the propagating channel. The path loss is associated with the design of base stations, as this tells us how much a transmitter should need to radiate to service a given region. Channel characterization, on the other hand, deals with the fidelity of the received signals and has to do with the nature of the waveform received at a receiver. The objective here is to design a suitable receiver that will receive the distorted transmitted signal due to the multipath and dispersion effects of the channel and decode the transmitted signal. An understanding of the various propagation models can actually address both problems. This chapter begins with a review of the available information provided by the various propagation models for both indoor and outdoor environments. The existing models can be classified into two major classes: statistical models and site-specific models. The main characteristics of a radio channel, such as path loss, fading, and time delay spread, are discussed. Currently, a third alternative, which includes many new numerical methods, is being introduced into propagation prediction. The advantages and disadvantages of some of these methods are summarized. In addition, an impulse response characterization for the propagation path is presented, including models for small-scale fading.

8.1 INTRODUCTION

The commercial success of cellular communications since its initial implementation in the early 1980s has led to an intense interest among wireless engineers in understanding and predicting radio propagation characteristics in various urban and suburban areas, even within buildings. As the explosive growth of mobile communications continues, it is very valuable to have the capability of determining optimum base station locations, obtaining a suitable data rate, and estimating their coverage without conducting a series of propagation measurements, which are very expensive and time consuming. In order to provide design guidelines and installations of the mobile systems, it is therefore important to develop effective propagation models for mobile communication.

8.2 DEFINITIONS AND TERMINOLOGIES USED FOR CHARACTERIZING VARIOUS PARAMETERS OF A PROPAGATION CHANNEL

In order to understand the nature of the models that are going to be presented, several definitions and terminologies for both narrowband and wideband wave propagation over a radio channel are first described to make the reader familiar with the terminologies and parameters of the problem.

8.2.1 Path Loss

Path loss (PL) is a measure of the average RF attenuation suffered by a transmitted signal when it arrives at the receiver after traversing several wavelengths. It is defined by [1]

$$PL(dB) = 10 \log \frac{P_t}{P_r} \tag{8.1}$$

where P_t and P_r are the transmitted and received power, respectively. In free space, power reaching the receiving antenna, which is separated from the transmitting antenna by a distance d, is given by the Friis free-space equation:

$$P_r(d) = \frac{P_t G_t G_r \lambda^2}{(4\pi)^2 d^2 L} \tag{8.2}$$

where G_t and G_r are the gain of the transmitting and receiving antenna, respectively, L is the system loss factor not related to propagation, and λ is the wavelength in meters. It is clear that equation (8.2) does not hold for $d = 0$.

Hence, many propagation models use a different representation for a close-in distance, d_0, known as the received power reference point. It is typically chosen to be 1 m. In realistic mobile radio channels, free space is not the appropriate propagation medium. A general PL model uses a parameter, γ, to denote the power law relationship between the separation distance and the received power. So path loss (in decibels) can be expressed as [2]

$$PL(d) = PL(d_0) + 10\gamma \log(d/d_0) + X_\sigma \qquad (8.3)$$

where $\gamma = 2$ characterizes free space. However, it is generally higher for wireless channels. X_σ denotes a zero mean Gaussian random variable of standard deviation σ that reflects the variation on the average of the received power that naturally occurs when a PL model of this type is used. Path loss is the main ingredient of a propagation model. It is related to the area of coverage of mobile systems.

8.2.2   Power Delay Profile

The random and complicated radio propagation channels can be characterized using the impulse response approach. For each point in the three-dimensional environment, the channel is a linear filter with the impulse response $h(t)$. The impulse response provides a wideband characterization of the propagating channel and contains all the information necessary to simulate or analyze any type of radio transmission through that channel.

Multipath propagation causes severe dispersion of the transmitted signal. The expected degree of dispersion is determined through measurement of the power delay profile of the channel. The power delay profile provides an indication of the dispersion or distribution of transmitted power over various paths in a multipath model for propagation. The power delay profile of the channel is calculated by taking the spatial average of $|h(t)|^2$ over a local area. By making several local area measurements of $|h(t)|^2$ for different locations, it is possible to build an ensemble of power delay profiles, each representing a possible small-scale multipath channel state [3, 10]. A typical plot of the power delay profile is shown in Figure 8.1.

Many multipath channel parameters are derived from the power delay profile. Power delay profiles are measured using wideband channel sounding techniques and are presented in the form of plots of the received power as a function of an additional or excess delay with respect to a fixed time delay reference. There is a delay between when the signal is transmitted to the time when it is received, due to the finite velocity of propagation of the electromagnetic signal. However, additional delay may be introduced by the propagation medium as well. A mobile channel exhibits a continuous multipath structure, hence the power delay profile can be thought of as a density function of the form

$$P(\tau) \;=\; \frac{\left|h(t)\right|^2}{\int_{-\infty}^{\infty}\left|h(t)\right|^2 dt} \tag{8.4}$$

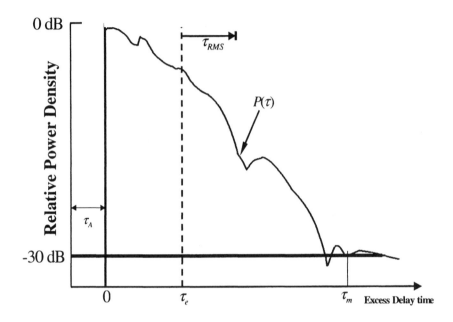

Figure 8.1. Typical power delay profile and definition of the delay parameters.

8.2.3   Time Delay Spread

Time dispersion varies widely in a mobile radio channel, due to the fact that reflections and scattering occur at seemingly random locations, and the resulting multipath channel response appears random as well. Because time dispersion is dependent on the geometric positional relationships between the transmitter, receiver, and the surrounding physical environment, some parameters, which can grossly quantify the multipath channel, are used. They are as follows.

8.2.3.1 First Arrival Delay (τ_A). This time delay corresponds to the arrival of the first transmitted signal at the receiver. It is usually measured at the receiver. This delay is set by the minimum possible propagation path delay from the transmitter to the receiver. It serves as a reference, and all delay measurements are taken relative to it. How the origin is defined is shown in Figure 8.1. Any delay measured longer than this reference delay is called an *excess delay*.

8.2.3.2 Mean Excess Delay (τ_e). This is the first moment of the power delay profile, as shown in Figure 8.1 with respect to the first delay and is expressed as

$$\tau_e = \int (\tau - \tau_A) P(\tau) d\tau \tag{8.5}$$

8.2.3.3 RMS Delay (τ_{RMS}). This is the square root of the second central moment of a power delay profile, as seen in Figure 8.1. It is the standard deviation about the mean excess delay and is expressed as

$$\tau_{RMS} = [\int (\tau - \tau_e - \tau_A)^2 P(\tau) d\tau]^{1/2} \tag{8.6}$$

RMS delay is a good measure of the multipath spread. It gives an indication of the nature of the intersymbol interference (ISI). Strong echoes (relative to the shortest path) with long delays contribute significantly to τ_{RMS}. The effects of dispersion on the performance of a digital receiver can be related reliably only to the RMS delay, independent of the shape of the power-delay profile, so long as it is small compared to the symbol period (T) of the digital modulation. It is also used to give an estimate of the maximum data rate of transmission.

8.2.3.4 Maximum Excess Delay (τ_m). This is measured with respect to a specific power level which is characterized as the threshold of the signal. When the signal level is lower than the threshold, it is processed as noise. For example, in Figure 8.1 the maximum excess delay spread can be specified as the excess delay (τ_m) for which $P(\tau)$ falls below −30 dB of its peak value, as shown in Figure 8.1.

8.2.4 Coherence Bandwidth

Whereas the delay spread is a natural phenomenon caused by reflection and scattering of the transmitted signal in a radio channel, the coherence bandwidth, B_C, is defined in terms of the RMS delay spread. It is a statistical measure of the range of frequencies over which the channel can be considered "flat ." It is defined as the bandwidth over which the variation of the signal is about 10% and is approximated by [4, 10]

$$B_C \approx \frac{1}{50 \, \tau_{RMS}} \tag{8.7}$$

It is important to note that an exact relationship between the coherence bandwidth and the RMS delay spread does not exist. The real coherence bandwidth depends on the actual impulse response of the channel.

8.2.5   Types of Fading

The type of fading experienced by a signal propagating through a mobile radio channel depends on the nature of the transmitted signal as well as on the characteristics of the channel. Different transmitted signals will undergo different types of fading according to the relation between the signal parameters, [such as path loss, bandwidth (BW), symbol period, etc.] and the channel parameters (such as RMS delay spread and Doppler spread). Figure 8.2 describes the different types of fading and the different relationships that exist between them [5].

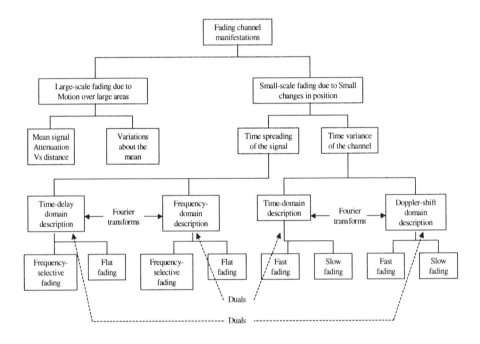

Figure 8.2. Types of fading.

The phenomenon of large-scale fading is affected primarily by the presence of hills, forests, and buildings between the transmitter and the receiver. The statistics of large-scale fading provide a way of computing an estimate of the path loss as a function of distance and other factors.

A channel is said to exhibit frequency-selective fading when the delay spread is greater than the symbol period. This condition occurs whenever the received multipath components of a symbol extend beyond the time duration of the symbols. Such multipath dispersion of the signal yields a kind of inter-symbol interference (ISI) called *channel-induced ISI*. When the delay spread is less than the symbol period, a channel is said to exhibit *flat fading* and there is no channel-induced ISI distortion. But there can still be performance degradation due to the irresolvable phasor components that add up destructively to yield a substantial reduction in signal-to-noise ratio (SNR) at the receiver.

Fast fading and slow fading are classified on the basis of how rapidly transmitted baseband signal changes as compared to the rate of electrical parameter changes of the channel. If the channel impulse response changes at a rate much faster than the transmitted signal, the channel may be assumed to be a fast-fading channel. Otherwise, it is assumed to be a slow-fading channel. It is important to note that the velocity of the mobile or velocity of objects using the channel through a baseband signal determines whether a signal undergoes fast or slow fading.

8.2.6 Adaptive Antenna

An application of antenna arrays has been suggested in recent years for mobile communications systems to overcome the problems of single-antenna systems. The use of adaptive antenna arrays helps in improving the system performance by increasing channel capacity and spectrum efficiency, extending range coverage, tailoring beam shape, steering multiple beams to track many mobiles, and compensating for the aperture distortion electronically. It also reduces multipath fading, co-channel interferences, system complexity and cost, bit error rate (BER), and outage probability [6].

A phased array antenna uses an array of simple antennas and combines the signal induced on the elements to form the output. The term *adaptive antenna* is used for a phased array when the gain and the phase of the signals induced in the various elements are weighted before combining to adjust the gain of the array in a dynamic fashion along a particular look direction while placing nulls along the undesired directions. A block diagram of a typical adaptive antenna system is shown in Figure 3.1.

The propagation models used for an adaptive antenna are different from those for a single antenna. Details of a phased array antenna and joint estimation of channel parameters can be obtained in [6–8].

8.3   MULTIPATH PROPAGATION

In a typical mobile radio application, the base station is fixed in position while the mobile unit is moving, usually subject to such a condition that the propagation between them is largely through scattering, either by reflection or diffraction from buildings and terrain or objects within buildings, because of obstruction of the line-of-sight (LOS) path. Radio waves therefore arrive at the mobile receiver from different directions with different amplitude, phase, and time delays, resulting in a phenomenon known as *multipath propagation*. The radio channel is then obtained as the sum of contributions from all the paths.

If the input signal is a unit impulse, $\delta(\tau)$, the output will be the channel impulse response that can be written as [9]

$$h(t) = \sum_{n=1}^{N} A_n \, \delta(t - \tau_n) \exp(-j\varphi_n) \tag{8.8}$$

It can thus be characterized by N time-delayed impulses, each represented by an attenuation and phase-shifted version of the original transmitted impulse. Here, A_n, τ_n, and φ_n are the attenuation, delay in time of arrival, and phase corresponding to path n, respectively. This model is only valid for a very narrowband signal. For a wideband signal the phase term has to be appropriately dealt with.

Although multipath interference seriously degrades the performance of communication systems, little can be done to eliminate it. However, if we characterize the multipath medium well and have sound knowledge of the propagation mechanisms and their influence on the system, the best design for the system can be selected to achieve good propagation performance and hence to achieve better quality of service.

8.3.1   Three Basic Propagation Mechanisms

Reflection, diffraction, and scattering are the three basic propagation mechanisms [10], that affect propagation in mobile communication systems. They are briefly explained below.

8.3.1.1 Reflection. Reflection occurs when a propagating electromagnetic wave impinges upon an object that has very large dimensions compared to those of the wavelength of the propagating wave. Reflection occurs from the surface of the ground, walls, and furniture. When reflection occurs, the wave may also be partially refracted. The coefficient of reflection and refraction is a function of the material properties of the medium, and generally depends on the wave polarization, angle of incidence, and the frequency of the propagating wave.

8.3.1.2 Diffraction. Diffraction occurs when the radio path between a transmitter and receiver is obstructed by a surface that has sharp edges. The waves produced by the obstructing surface are present throughout the space and even behind the obstacle, giving rise to bending of waves around the obstacle, even when a line-of-sight (LOS) path does not exist between the transmitter and receiver. At high frequencies, diffraction, like reflection, depends on the geometry of the object as well as the amplitude, phase, and polarization of the incident wave at the point of diffraction.

8.3.1.3 Scattering. Scattering occurs when the medium through which the wave propagates consists of objects with dimensions that are small compared to the wavelength, and where the number of obstacles per unit volume is large. Scattered waves are produced by rough surfaces, small objects, or by other irregularities in the channel. In practice, foliage, street signs, lampposts and stairs within buildings can induce scattering in mobile communication systems. Sound knowledge of the physical details of the objects can be used to accurately predict scattered signal strength.

In most cases, the scattering can be neglected [11] and the complex received field from the various incident paths is given by [12]

$$\vec{E}_{\text{receive}} = \sum_i \vec{E}_i \tag{8.9}$$

where

$$E_i = E_0 \wp_{ti} \wp_{ri} L_i(d) \prod_j \Gamma(\phi_{ji}) \prod_p T(\phi_{pi}) e^{-jkd} \tag{8.10}$$

and

E_0 amplitude of the reference incident field (V/m)

$\wp_{ti}$ and $\wp_{ri}$ transmitting and receiving antenna field radiation patterns along the direction of interest for the ith multipath component

$L_i(d)$ path loss characterization for the ith multipath component

$\Gamma(\phi_{ji})$ reflection coefficient for the jth reflection of the jth multipath component;

$T(\phi_{pi})$ transmission coefficient for the pth transmission of the ith multipath component

e^{-jkd} propagation phase factor due to the path length $d (k = 2\pi/\lambda$ with λ representing the wavelength)

d path length (meters)

E_i field strength of the ith multipath component

For diffraction, the product of the complex reflection and transmission coefficients is replaced by the complex diffraction coefficient.

8.3.2 Propagation in Outdoor and Indoor Environments

With the growth in the capacity of mobile communications, the size of a cell is becoming smaller and smaller from macrocell to microcell and then to picocell. The service environments include both outdoor and indoor areas.

When propagation is considered in an outdoor environment, one is interested primarily in three types of areas: urban, suburban, and rural. The terrain profile of a particular area also needs to be taken into account. The terrain profile may vary from a simple curved earth to a highly mountainous region. The presence of trees, buildings, moving cars, and other obstacles must also be considered. Direct path, reflections from the ground and buildings, and diffractions from the corners, and roofs of buildings are the main contributors to the total field generated at a receiver due to radio-wave propagation.

There is also a great deal of interest in characterizing radio propagation inside buildings with the advent of personal communication systems (PCSs). The indoor radio channel differs from the traditional outdoor mobile radio channel in two aspects: the distance covered is much smaller, and the variability of the environment is much greater for a much smaller range of transmitter and receiver separation distance [10]. Propagation into and inside buildings has, to some extent, a more complex multipath structure than for an outdoor propagation environment. This is mainly because of the nature of the structures used for the buildings, the layout of rooms, and most important, the type of construction materials used. Table 8.1 represents the categories of buildings where propagation measurements have been made [13].

TABLE 8.1
Classification of buildings

Category	Description
1	Residential houses in suburban areas
2	Residential houses in urban areas
3	Office buildings in suburban areas
4	Office buildings in urban areas
5	Factory buildings with heavy machinery
6	Other factory buildings, sports halls, exhibition centers
7	Open environment (e.g., railway stations, airports, etc.)
8	Underground (e.g., subways, underground streets, etc.)

8.3.3   Summary of Propagation Models

In mobile communications, signals from the mobile arrive at a base station with multipaths, each with its own angle of arrival (AOA), path delay, and attenuation. When the communication system uses an adaptive processing methodology, it is also important to estimate joint angle and delay for various signals.

There are two main models for characterizing path loss, which are empirical (or statistical) models and site-specific (or deterministic) models. The former is based on the statistical characterization of the received signal. They are easier to implement, require less computational effort, and are less sensitive to the environment geometry. The latter has a certain physical basis and requires a vast amount of data regarding geometry, terrain profile, locations of building and furniture in buildings, and so on. These deterministic models require more computations and are more accurate.

Most models regarding fading apply stochastic processes to describe the distribution of the received signal. It is useful to use these models to simulate propagation channels and estimate the performance of the system in a homogeneous environment. Models of time delay spread both for outdoor and indoor environments are generally derived from a lot of measurements. In [142] some of the propagation models have been discussed and here we have included additional new models.

8.4   EMPIRICAL OR STATISTICAL MODELS FOR PATH LOSS

8.4.1   Outdoor Case

A number of empirical or statistical models are suitable for both macrocell and microcell scenarios for outdoor environment. Some of them are described below.

8.4.1.1   *Okumura et al. Model.* This is one of the most widely used models for propagation in urban areas [14]. The model can be expressed as

$$L_{50}(\text{dB}) = L_F + A_{mu}(f, d) - G(h_{te}) - G(h_{re}) - G_{\text{AREA}} \qquad (8.11)$$

where L_{50} is the median value of the propagation path loss, L_F is the free-space propagation loss, A_{mu} is the median attenuation in the medium relative to free space at frequency f, and d corresponds to the distance between the base and the mobile, $G(h_{te})$ and $G(h_{re})$ are the gain factors for the base station antenna and mobile antenna, respectively, h_{te} and h_{re} are the effective height of the base station and the mobile antennas (in meters), respectively, and G_{AREA} is the gain generated by the environment in which the system is operating. Both $A_{mu}(f,d)$ and G_{AREA} can be found from empirical curves. Okumura et al.'s model is considered to be among the simplest and best in terms of accuracy in predicting path loss for early cellular systems. It is very practical and has become a standard for system planning in Japan. The major disadvantage of this model is its slow response to rapid changes in terrain profile.

8.4.1.2   *Hata Model.* It is an empirical formulation [15] of the graphical path loss data provided by Okumura's model. The formula for the median path loss in urban areas is given by

$$L_{50}(\text{urban})(\text{dB}) = 69.55 + 26.16 \log f_c - 13.82 \log h_{te}$$
$$- a(h_{re}) + (44.9 - 6.55 \log h_{te}) \log d \qquad (8.12)$$

where f_c is the frequency and varies from 150 to 1500 MHz, h_{te} and h_{re} are the effective height of the base station and the mobile antennas (in meters), respectively, d is the distance from the base station to the mobile antenna, and $a(h_{re})$ is the correction factor for the effective antenna height of the mobile which is a function of the size of the area of coverage. For small to medium-sized cities, the mobile antenna correction factor is given by

$$a(h_{re}) = (1.1\log f_c - 0.7)h_{re} - (1.56\log f_c - 0.8)\ \text{dB} \qquad (8.13)$$

For a large city, it is given by

$$a(h_{re}) = 8.29(\log 1.54 h_{re})^2 - 1.1\ \text{dB} \quad \text{for} \quad f_c \le 300\,\text{MHz} \qquad (8.14a)$$

$$a(h_{re}) = 3.2(\log 11.75 h_{re})^2 - 4.97\ \text{dB} \quad \text{for} \quad f_c \ge 300\,\text{MHz} \qquad (8.14b)$$

To obtain the path loss in a suburban area, the standard Hata formula is modified as

$$L_{50}(\text{dB}) = L_{50}(\text{urban}) - 2[\log(f_c/28)]^2 - 5.4 \qquad (8.15)$$

The path loss in open rural areas is expressed through

$$L_{50}(\text{dB}) = L_{50}(\text{urban}) - 4.78(\log f_c)^2 - 18.33\log f_c - 40.98 \qquad (8.16)$$

This model is quite suitable for large cell mobile systems, but not for personal communications systems, which cover a circular area of approximately 1km in radius.

8.4.1.3 COST-231–Walfisch–Ikegami Model.
This utilizes the theoretical Walfisch–Bertoni model [16], and is composed of three terms [17]:

$$L_b = \begin{cases} L_0 + L_{rts} + L_{msd} & \text{for} \quad L_{rts} + L_{msd} > 0 \\ L_0 & \text{for} \quad L_{rts} + L_{msd} \le 0 \end{cases} \qquad (8.17)$$

where L_0 represents the free-space loss, L_{rts} is the rooftop-to-street diffraction and scatter loss, and L_{msd} is the multiscreen diffraction loss. The free-space loss is given by

$$L_0 = 32.4 + 20\log d + 20\log f \qquad (8.18)$$

where d is the radio-path length (in km), f is the radio frequency (in MHz), and

$$L_{rts} = -16.9 - 10 \log w + 10 \log f + 20 \log \Delta h_{\text{Mobile}} + L_{ori} \qquad (8.19)$$

Here w is the street width (in m) and

$$\Delta h_{\text{Mobile}} = h_{\text{Roof}} - h_{\text{Mobile}} \qquad (8.20)$$

is the difference between the height of the building on which the base station antenna is located, h_{Roof}, and the height of the mobile antenna, h_{Mobile}. L_{ori} is given by

$$L_{ori} = \begin{cases} -10 + 0.354\,\phi & 0° \le \phi < 35° \\ 2.5 + 0.075\,(\phi - 35) & \text{for} \quad 35° \le \phi < 55° \\ 4.0 - 0.114\,(\phi - 55) & 55° \le \phi \le 90° \end{cases} \qquad (8.21)$$

where ϕ is the angle of incidence relative to the direction of street. L_{msd} is given by

$$L_{msd} = L_{bsh} + k_a + k_d \log d + k_f \log f - 9 \log b \qquad (8.22)$$

where b is the distance between the buildings along the signal path and L_{bsh} and k_a represent the increase of path loss due to a reduced base station antenna height. Using the abbreviation

$$\Delta h_{\text{Base}} = h_{\text{Base}} - h_{\text{Roof}} \qquad (8.23)$$

where h_{Base} is the base station antenna height, we observe that L_{bsh} and k_a are given through

$$L_{bsh} = \begin{cases} -18 \log(1 + \Delta h_{\text{Base}}) & h_{\text{Base}} > h_{\text{Roof}} \\ 0 & h_{\text{Base}} \le h_{\text{Roof}} \end{cases} \qquad (8.24)$$

$$k_a = \begin{cases} 54 & & h_{\text{Base}} > h_{\text{Roof}} \\ 54 - 0.8\,\Delta h_{\text{Base}} & d \ge 0.5\,km & \text{and} & h_{\text{Base}} \le h_{\text{Roof}} \\ 54 - 1.6\,\Delta h_{\text{Base}}\,d & d < 0.5\,km & \text{and} & h_{\text{Base}} \le h_{\text{Roof}} \end{cases} \qquad (8.25)$$

The terms k_d and k_f control the dependence of the multiscreen diffraction loss versus distance and the radio frequency of operation, respectively. They are given by

$$k_d = \begin{cases} 18 & h_{\text{Base}} > h_{\text{Roof}} \\ 18 - 15 \dfrac{\Delta h_{\text{Base}}}{h_{\text{Roof}}} & h_{\text{Base}} \le h_{\text{Roof}} \end{cases} \qquad (8.26a)$$

and

$$k_f = -4 + 0.7\left(\frac{f}{925} - 1\right) \qquad (8.26b)$$

for medium-sized cities and suburban centers with moderate tree densities and

$$k_f = -4 + 1.5\left(\frac{f}{925} - 1\right) \qquad (8.26c)$$

for metropolitan centers.

This model is being considered for use by International Telecommunication Union-Radiocommunication Sector (ITU-R) in the International Mobile Telecommunications-2000 (IMT-2000) standards activities. Some improved solutions for diffraction by multiple absorbing half-planes have also been developed [18, 19]. Other solutions based on the uniform theory of diffraction (UTD) are also available [20, 21]. The performance of the various methods in estimating the multiple diffraction loss term and the final diffraction loss term is given in [22]. Recently, a correction to the COST-231–Walfisch–Ikegami Model has been reported [23].

8.4.1.4 Dual-Slope Model. This is based on a two-ray model [24, 25] which is used commonly when the transmitting antenna is several wavelengths or more above the horizontal ground plane and suitable for the line-of-sight (LOS) propagation regions. The propagation loss, $L(d)$, in that case is described by a dual-slope model. This can be represented as a function d, the distance between the base station and the receiver. It is given by [26]

$$L(d) = L_b + \begin{cases} 10n_1 \log d + P_1 & 1 < d < d_{brk} \\ 10(n_1 - n_2)\log d_{brk} + 10n_2 \log d + P_1 & d \ge d_{brk} \end{cases} \qquad (8.27)$$

where $P_1 = \text{PL}(d_0)$ and is the path loss in dB at the reference point d_0 and d_{brk} represents the breakpoint or the turning point distance. The "point" where this transition occurs is often called the *Fresnel breakpoint*. L_b is a basic transmission loss parameter which depends on frequency, and antenna heights n_1 and n_2 represent the slopes of the best-fit line before and after the breakpoint. If the transmitter and receiver antenna heights are known, along with the distance between them, the path loss can be computed based on the two parameters n_1 and

n_2. It is very reasonable to let $n_1 = 2$ for the region prior to the Fresnel breakpoint. There is much more variability in the path loss and the exponent for the region beyond the Fresnel breakpoint, with values of n_2 ranging from 2 to 7.

8.4.1.5 Other Models. Other models, including use of wideband measurements for different situations, have been discussed in recent times [27, 28, 141]. These models have been developed from measurements and use different parameters for different situations.

8.4.2 Indoor Case

Indoor radio propagation is not influenced by terrain profile as is outdoor propagation, but it can be affected by the layout in a building, especially if there exists various building materials. Owing to reflection, refraction, and diffraction of the radio wave by objects such as walls, windows, and doors inside a building, the transmitted signal often reaches the receiver through more than one path.

Distance/power model is the main propagation model for path loss. As shown in equation (8.3), many researchers estimate the rate of decay of a transmitter signal through this relation [29–34]. In an enclosed environment, the value of γ in equation (8.3) may be 1.5 to 1.8 when the transmitter and receiver are placed in the same hallway and are in sight of each other. When the receiver is located within a room off the hallway, γ ranges from 3 to 4 [30–33]. γ also varies with frequency [33] and is dependent on building materials used in a particular environment [34]. References [5] and [32] provide reviews on early works of these models. Table 8.2 [29] shows the parameters of (8.3) determined by measurement for different buildings [35–40].

TABLE 8.2
Path loss measured in different buildings

Building	γ	PL (dB)	Frequency (MHz)	Ref.
Grocery store	1.8	5.2	914	[35]
Retail store	2.2	8.7	914	[35]
Open-plan factories	2.2	7.9	1300	[36]
	1.4–3.3	—	910	[37]
Open-plan factories B	2.0	3.7	1300	[38]
	2.1	4.0	1300	[38]
Open-plan factories C	2.4	9.2	1300	[38]
	2.1	9.7	1300	[38]
Suburban office building open plan	2.4	9.6	915	[39]
	2.6	14.1	1900	[39]
Suburban office building soft partition	2.8	14.2	915	[39]
	3.8	12.7	1900	[39]
	3.0	—	850	[40]

In order to take into account the attenuation due to walls and floors, two additional terms are added to (8.3) to result in [35, 41]

$$PL(d) = PL(d_0) + 10n\log(d/d_0) + \sum_{q=1}^{Q} FAF(q) + \sum_{p=1}^{P} WAF(p) \quad (8.28)$$

where $FAF(q)$ and $WAF(p)$ are the floor and wall attenuation factors, respectively. Table 8.3 lists the FAFs for two buildings [29, 30].

It is also observed that the propagation path loss as a function of the distance also has two distinct regions for indoor environments [42], as described in (8.28). When electromagnetic radiation is incident on a wall or a floor in an oblique fashion, less power will be transmitted through the wall than would occur at normal incidence. Reference [41] modifies the term of $WAF(p)$ as $WAF(p)/\cos \phi_p$ and $FAF(q)$ as $WAF(q)/\cos \phi_q$, where $WAF(p)$ and $FAF(q)$ are the values of the attenuation factors at normal incidence. ϕ_p and ϕ_q are the angles of incidence of the signal on the walls and floors, respectively. A diffraction term has also been added to the formula in [41]. When the base station is out of the building, the path loss in the building has been given in [43].

These empirical or statistical models described in this section are simple to implement and are used widely when the accuracy of the data is not a critical requirement.

TABLE 8.3
Average Attenuation Factor (FAF)

Location	FAF (dB)	PL (dB)
Office building 1		
Through 1 floor	12.9	7.0
Through 2 floors	18.7	2.8
Through 3 floors	24.4	1.7
Through 4 floors	27.0	1.5
Office building 2		
Through 1 floor	16.2	2.9
Through 2 floors	27.5	5.4
Through 3 floors	31.6	7.2

8.5 SITE-SPECIFIC MODELS FOR PATH LOSS

Site-specific propagation models, also called *deterministic models*, are based on the theory of electromagnetic-wave propagation. Unlike statistical models, site-specific propagation models do not rely on extensive measurements but on greater detail of the environment and provides an accurate prediction of the signal propagation.

In theory, propagation characteristics of electromagnetic waves could exactly be computed by solving Maxwell's equations. Unfortunately, this approach requires very complex mathematical operations and requires considerable computing power. In [44], this method has been applied to simplified environments.

8.5.1   Ray-Tracing Technique

Ray tracing is a technique based on geometrical optics (GO), which can easily be applied as an approximate method for estimating levels of high-frequency electromagnetic fields. GO assumes that energy can be considered to be radiating through infinitesimally small tubes, often called *rays*. These rays are normal to the surface of equal signal power, and lie along the direction of propagation and travel in straight lines, provided that the relative refractive index of the medium is constant. Therefore, signal propagation can be modeled via ray propagations. By using the concept of ray tracing, rays can be launched from a transmitter location and the interaction of the rays can be described using the well-known theory of refraction and reflection and interactions with the neighboring environment. In GO, only direct, reflected, and refracted rays are considered, and consequently, abrupt transition areas may occur, corresponding to the boundaries of the regions where these rays exist. The geometrical theory of diffraction (GTD) and its uniform extension, the uniform GTD (UTD) [45], complement the GO theory by introducing a new type of rays, known as *diffracted rays*. The purpose of these rays is to remove the field discontinuities and to introduce proper field corrections, especially in the zero field regions predicted by GO.

The Fermat principle and the principle of local field are two basic concepts used extensively by ray models. The Fermat principle states that a ray follows the shortest path from a source point to a field point, while the principle of the local field states that when hitting a surface, the high-frequency rays produce reflection, refraction, and diffraction. This depends only on the electrical and geometrical properties of the scatterer in the immediate neighborhood of the point of interaction.

The ray-tracing method is widely used in propagation model and system design [12, 42, 46–72, 145]. It is most accurate when the point of observation is many wavelengths away from the nearest scatterer. All scatterers are assumed to be large, compared to a wavelength. Two types of ray-tracing methods, namely the image method [12, 48, 55, 58] and the brute-force ray-tracing method, are generally used. They are now explained.

8.5.1.1 Image Method. This method generates the images of a source at all planes. These images then serve as secondary sources for subsequent points of reflections. If there are N reflecting planes, there are N first-order (i.e., one-reflection) images of a source, $N(N - 1)$ two-reflection images, $N(N - 1)(N - 1)$ three-reflection images, and so on [48]. To determine whether an image of the source is visible at the destination is to trace the intersection of the reflected ray at all necessary planes of interest. Thus, the energy reaches the destination through multiple reflections and contributes to the received power. Once a ray has been traced through all its reflections to the source, the attenuations associated with all the reflection terms are calculated.

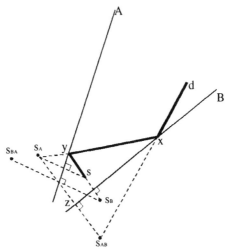

Figure 8.3. Images due to a source placed between two mirrors, A and B.

Image method is efficient but can handle simple environments only. Many environments with which we are concerned in our daily life are complicated, and the conventional image method is not adequate. Figure 8.3 shows a source and its images corresponding to two reflections [48]. The concept of a lit region has been introduced in [58] and is illustrated in Figure 8.4. Behind the plate representing a building where the image is formed, the region is termed the *unlit domain*. For the 2D case, only reflections from walls and diffraction from corners in buildings are taken into account. Ground reflection and rays over rooftops are neglected [58]. In [53–55] a modified shooting-and-bouncing-ray technique, combined with the image method, has been used to deal with the radio wave propagation in furnished rooms. The effects of diffraction have also been considered. A threshold must be set with respect to the number and order of reflection and diffraction rays that can be considered.

8.5.1.2 Brute-Force Ray-Tracing Method. This method considers a bundle of transmitted rays that may or may not reach the receiver. The number of rays considered and the distance from the transmitter to the receiver location determines the available spatial resolution and hence the accuracy of the model. This method requires more computing power than the image method.

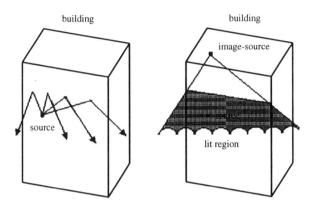

Figure 8.4. The reflection by a wall in a building is modeled by an image source placed behind it and a lit region in front of it.

A finite sample of the possible directions of the propagation from the transmitter is chosen. For each such direction, a ray is launched. If a ray hits a wall, a reflecting ray and refracting ray are generated. If a ray hits a wedge, a family of diffracting rays is generated. A reception sphere with the correct radius can describe a region that will receive exactly one ray. If the radius is too large, two rays could be received and the same specular ray may be counted twice. If the radius is too small, it is possible that none of the rays will reach the reception sphere and the specular ray will be excluded [42, 59]. Figure 8.5 shows the proper size of the reception sphere which may receive a ray. For each receiver location, the perpendicular distance d from the receiver to the ray is computed together with the total (unfolded) ray path length L from the source to the perpendicular projection point. If d is greater than or equal to $(\phi L)/2$ for the 2D case or $(\phi L)/\sqrt{3}$ for the 3D case, the ray is treated as not having reached the receiver location. Here ϕ is the angle between two rays. Otherwise, the ray is considered to be contributing to the received signal. There is no reception sphere associated with use of the image method.

The key part of the ray-tracing method is the generation and description of the rays. There are two kinds of methods to obtain the rays at the source point. One is a 2D approach, the other is a 3D method.

• **2D ray-tracing model** [42]. In two dimensions, all the rays or ray tubes are treated as ray sectors, as shown in Figure 8.6. At the source, rays are launched along different directions with the same sector angle ϕ in a plane. How to choose the angle ϕ depends on the accuracy required and the computation time. If the angle is small, it will provide high accuracy and will take much time to compute. For example, if the angle $\phi = 1°$, then there will be 360 rays to be traced.

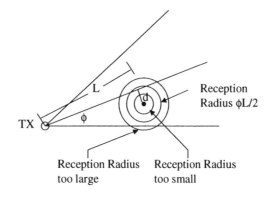

Figure 8.5. Reception sphere for a 2D ray tracing.

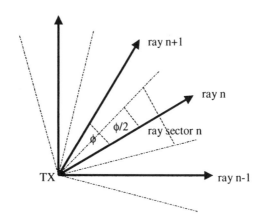

Figure 8.6. Rays generated from a source in two dimensions.

Each ray is launched from the source and can be traced through a binary tree. An intersection with a surface of an object is represented by a node in the tree. The incident ray is decomposed into an object-reflected ray and an object-penetrated ray. It is assumed that the reflected ray propagates along the specular direction (incidence angle equals the aspect angle) and the ray that penetrates the object keeps the original direction of the incident ray. Both rays then propagate to the next intersection. An intersection with a wedge is also represented by a node, and the diffraction point is processed as a source and a large number of rays must be launched. The decomposition process is repeated as a recursion process. This procedure is continued until the rays are weaker than an assumed threshold, leaves a predefined propagation area, or is received. The strength of the field at the receiver is then calculated according to (8.9).

A 2D diffraction model has been introduced in [62, 64]. The various buildings are considered as vertical knife-edges, neglecting the over-rooftop diffraction and ground reflection. Because the buildings are much higher than the base station (BS) and mobile station (MS) antennas in an urban microcellular environment, the weak contribution from the signals from the over-rooftop rays can be neglected. No rays due to a single ground reflection from the transmitter to the receiver exist in the shadow regions. For ranges less than 1 km from a transmitter (primary source), the received power may have a range R dependence according to the power law of $1/R^2$. However, for LOS regions, the ground reflections appear to be less important.

As reported in [64], the 2D ray-tracing algorithm is quite accurate when the transmitting and receiving antenna heights are much below the rooftops of the surrounding buildings. This propagation model between a transmitter and receiver located close to the ground is usually called the *canyon model*.

When using a 2D model, the inputs are: (1) 2D geometry described by means of vectors specifying the location of the building walls; (2) estimated electrical characteristics of the building walls (permittivity and conductivity or the scalar reflection coefficient); (3) base-station location; (4) antenna pattern, and (5) frequency of operation.

• *3D ray-tracing model* [12, 64–72]. The transmitter and receiver are modeled as point sources when using this ray-tracing technique. In order to determine all possible rays that may leave the transmitter and arrive at the receiver in three dimensions, it is necessary to consider all possible angles of departure and arrival at the transmitter and receiver. Rays are launched from the transmitter at an elevation angle θ and with an azimuth angle φ as defined in the usual coordinate system. Antenna patterns are incorporated to include the effects of antenna beamwidth in both azimuth and elevation.

To maintain all the ray manipulation routines as general, it is desirable that each ray tube occupies the same solid angle $d\Omega$, and each wavefront has an identical shape and size at a distance r from the transmitter. Additionally, these wavefronts must be such that they can be subdivided so that an increased ray resolution can be handled easily. For example, let $r = 1$ and the total wavefront be the surface of a unit sphere. The problem then becomes one of subdividing the sphere surface into areas of equal "patches" so that all are of the same size and shape and collectively covers the surface of interest without gaps. Hexagonal [12] and triangular [53, 54] ray wavefronts have also been used.

The procedure of ray tracing in 3D is similar to a 2D model, but more computational time is needed.

Some sectors of the walls in a corridor thus can be made of different materials, for example, wood, metal, concrete, or even glass, which may have different reflectivity for the incident wave. Neglecting the differences between the reflectivities of the various materials will degrade, predicting the accuracy of the propagation model. Therefore, in [12] it is proposed to have the concept of an *effective building material* to represent the physical and complicated constitutive materials used in the walls of a building. However, permittivity of this effective material is not easy to determine since it depends on the experimental data as well as on the propagation model. To simplify this problem, patches of different dielectric constants and physical sizes have been introduced in [65]. It is noted that the size and dielectric constant of each patch are chosen according to their physical dimensions and the material it is made of.

The key to a propagation model based on ray tracing is to find a computationally fast way to determine the dominant ray paths so as to provide accurate pathloss predictions. It is well known that for outdoor propagation prediction, in addition to specular reflections, diffraction from edges must be accounted for, especially in non-LOS regions. Unfortunately, diffractions are very time consuming to model since a single incident ray encountering an edge will generate an entire family of new rays. The generation of a large number of diffracted rays limits the number of diffractions that can be considered. For any given path we choose at most two, unless an approximation can be made to find the important contributing rays. In order to find the contributing rays in an urban environment where the building walls are nearly always vertical planar polygons, a vertical-plane-launch (VPL) method has been developed [66]. The VPL approach accounts for specular reflections from vertical surfaces and diffraction at the vertical edges and approximates diffraction along horizontal edges by restricting the diffracted rays to lie in the plane of incidence or in the plane of reflection. The VPL approach can treat many multiple forward diffractions at horizontal edges. It can also be used for rooftop antennas and areas where there exist buildings of different heights.

To improve the efficiency of ray-tracing models, many researchers have developed a large number of methods [68, 72]. In [68], a hybrid technique has been presented where the object database is held in two dimensions, but a ray-tracing engine operates in three dimensions. The 3D rays are produced by combining the results of two 2D ray tracers, one on a horizontal and the other on a vertical plane. Moreover, by significantly enhancing the concept of illumination zones the performance of the algorithm can be dramatically improved.

Comparison of 2D and 3D models has been made in [64].

8.5.2 Finite-Difference Time Domain Models

Based on geometrical optics (GO) and usually supplemented by UTD, a ray-tracing algorithm provides a relatively simple solution to radio propagation. However, it is well known that GO provides good results for electrically large objects and UTD is rigorous only for perfectly conducting wedges. For complex lossy structures with finite dimensions, ray-tracing fails to predict correctly the scattered fields. In a complicated communication environment, transmitting and receiving antennas are often inevitably installed close to a structure with complex material properties, for which no asymptotic solutions are available. Such problems can be solved by numerical solution of Maxwell's equations. In particular, the finite-difference time domain (FDTD) method is an alternative. The advantages of the FDTD method are its accuracy and that it simultaneously provides a complete solution for all the points in the map, which can give signal coverage information throughout a given area.

In a simple outdoor environment, a two-dimensional FDTD is generally applied [73]. A simple approach for introducing the correct spherical wave spreading has been developed. A comparison with the FDTD predictions could be used to evaluate and refine the GTD-based methods.

A reduced formulation for the standard FDTD technique [74], requiring four scalar field components instead of the usual six, is used to predict channel statistics inside a residential building. Measurements have also been conducted and results compared with simulated data. In order to introduce arbitrary-shaped antennas into the simulations, a two-and-a-half dimension (2.5 D) or a multimode FDTD method has been established for indoor radio propagation calculations [75].

A hybrid technique [76] based on combining a ray-tracing method with FDTD method for more accurate modeling of radio wave propagation has also been suggested. The basic idea is to use ray tracing to analyze wide areas and FDTD to study areas close to structure with complex material properties, where ray-based solutions are not sufficiently accurate.

As a numerical method, FDTD requires large amounts of computer memory to keep track of the solution at all locations, and extensive calculations to update the solution at successive instants of time. Application of an accurate numerical analysis method to model an entire area is neither practical because of the computational resource required, nor is it necessary for open areas without many objects.

8.5.3   Moment Method Models

Ray-tracing models can be used with sufficient precision to predict radio coverage for large buildings having a large number of walls between the transmitter and the receiver, while the moment method (MM) model is better when higher precision is required and the size of the buildings is smaller. A combination of these two models is also possible using the advantages of each. Where a lot of small but dominant obstacles are present or paths that cannot be taken into account by a ray-tracing model, the MM model can be used [77, 78].

The solutions determined by MM are numerically exact as long as the spatial segmentation used for the objects is small enough. Due to the limitations of the computer memory and CPU time, the MM is generally used to analyze objects which are tens of wavelengths in size. However, by choosing structures with dimensions around a few wavelengths, the MM can be used to check and verify the ray-tracing program. A 2D problem, which includes stair-shaped walls above a lossy ground, was simulated by using both the MM and ray-tracing methods [77].

The transmission of an UHF wave through a window in a wall is critical when integrating systems that include both indoor and outdoor areas. A novel simulation approach based on the moment method is presented in [79]. In the simulations, the walls have been modeled as two long dielectric slabs, long enough so that any diffraction or reflection at the outer edges would not influence the results. No distinction has been made between the concrete and brick parts. The glass plates have also been assumed to be homogeneous and the aluminum frame was modeled as a perfectly conducting material. The resulting simulations are then compared to a set of measurements, where good agreement has been achieved.

A hybrid approach combining the ray-tracing method and the periodic moment method (PMM) for material objects has been developed to study the indoor wave propagation, penetrations, and scattering due to periodic structures in buildings [80]. The PMM is applied to evaluate the specular and grating transmission and reflection coefficients of the periodic structures. Those data are then used in a ray-tracing program to find the specular and grating rays for each ray tube illuminating one of the periodic structures. Those excited ray tubes are continuously traced to determine their contributions to the receiving antennas.

8.5.4   Artificial Neural Network Models

The main problem with the statistical models is usually the accuracy, while the site-specific models lack computational efficiency. Artificial neural networks (ANNs) have shown very good performance in solving problems with mild nonlinearity on a set of noisy data. This case corresponds to a problem at a field-level prediction, as the data obtained from measurements is always noisy. Another key feature of the neural networks is the intrinsic parallelism, allowing for fast evaluation of the solutions.

The ANN model [81], which has the form of a multilayer perception, is generally used with 12 inputs and one output. It has been developed to predict the propagation in an indoor environment. In this case, a 2D floor plan is used for a database with a resolution of 10×10 cm. All particular locations are classified into 11 distinct categories, such as wall, corridor, outdoor area, laboratory, and so on. One input of the network represents the normalized distance from the transmitter to the receiver. In addition, there is an input for each defined environment category. Other inputs represent either a normalized number of occurrences (doors and windows) or an appropriate percentage (wall, corridor, and so on) of that category along the straight line drawn from the transmitter to the receiver. The process of learning may last for a couple of hours, but the process for field-level prediction is fast. The accuracy of a prediction model significantly depends on the accuracy of the environment databases.

In [82], theoretical investigations into the suitability of a neural network simulator for the prediction of field strength based on topographical and morphographical data are presented. Effective input and output data processing is developed using a deterministic and heuristic formula for the training of a neural network simulator. The network used is similar to that described in [81]. The inputs are frequency, heights of the antenna for a base and mobile stations, respectively, and the distance between them. The output is the field strength.

Although the multilayer neural network [82] is a useful method for approximating the propagation loss, however, it suffers from drawbacks of slow convergence and unpredictable solutions during learning. To overcome this difficulty, radial basis function (RBF) neural networks that have a "linear in the parameters" representation are proposed to enhance the real-time learning capability and achieve a rapid convergence [83]. The RBF neural network is a two-layer localized receptive field network whose output nodes form a combination of radial activation functions computed by the hidden layer nodes. Appropriate centers and connection weights in the RBF network lead to a network that is capable of forming the best approximation to any continuous nonlinear mapping up to an arbitrary resolution. Such an approximation introduces best nonlinear approximation capability into the prediction model in order to accurately estimate the propagation loss over an arbitrary environment based on adaptive learning from measurement data. Okumura's data are often included to demonstrate the effectiveness of the RBF neural network approach.

8.5.5   Other Models

Recently, many new methods have been introduced to predict propagation for mobile communications. Some of them are described here.

8.5.5.1 Parabolic Equation Model. This is applied to the modeling of radio-wave propagation in an urban environment. As a parabolic version of Maxwell's equations it allows full treatment [84, 138, 139] of 3D electromagnetic scattering which is not possible with scalar versions of the algorithm. It is particularly useful for accurate modeling of scattering by a single building or a group of buildings at microwave frequencies. Examples include scattering by a building with a hemispherical roof and scattering by a group of buildings with sloping roofs.

8.5.5.2   Fast Far-Field Approximation Model. This is substantially faster than conventional integral-equation (IE)-based techniques. The technique is improved by incorporating the Green's function perturbation method. The method has been applied to gently undulating terrains and compared to published experimental results in the 900-MHz band. It has also been successfully applied to more hilly terrain and to surfaces with added buildings [85]. An improved version of the "shifting function" has been introduced, which can improve the performance of the technique for more challenging problems such as scattering from a wedge.

The issues of profile truncation and small-scale roughness effects have been addressed, and numerical results presented show excellent agreement with published measured data.

8.5.5.3 Waveguide Model. In large metropolitan areas which have tall buildings, the transmitting and receiving antennas are both located below the rooftops and the city streets act as a type of waveguiding structure for the propagating signal [86, 87, 140]. In this case, there is a need to develop efficient algorithms for the computation and mapping of the field distribution in such structures. Theoretical analysis of propagation in a city street modeled as a 3D multislit waveguide is proposed. Assuming that the screens and slits are distributed by a Poisson law, the statistical propagation characteristics in such a waveguide are expressed in terms of multiple ray fields approaching the observer. Algorithms for path loss prediction have been presented and compared with experimental data in the two references cited above.

8.5.5.4 Boltzmann Model. This was initially developed for simulated fluid flows. It describes a physical system in terms of the motion of fictitious microscopic particles on a lattice [88, 143]. This technique can take complicated boundary conditions into account. Two-dimensional simulations have been performed starting from a city map and a renormalization scheme has been proposed. The method, which is simple and easy to implement, provides good path loss predictions compared with on-site measurements.

8.6   SUMMARY OF MODELS FOR PATH LOSS

A summary of the various propagation models dealing with path loss has been reviewed. A brief comparison of some of the main models is presented in Table 8.4. Propagation models dealing with path loss for mobile communication have been emphasized using two very different approaches. First, a simple empirical or statistical model of the path loss has been considered where some of the parameters used are determined empirically from measurements. The second approach used is site-specific methods. Ray tracing is the main method. Some other numerical methods used in electromagnetic fields computation have also been applied.

Each of these two kinds of approaches makes a very different trade-off of accuracy versus complexity. The empirical (statistical) models are extremely simple (no environmental information is used other than of a very type in the choice of the parameters), but the predictions are not very accurate. On the other hand, site-specific models are considerably more accurate than the empirical models, but require a great deal of specific information about the area of interest (the locations of all the objects at a minimum and possibly the locations of large objects).

TABLE 8.4
Comparison of models for path loss

Model Name	Suitable Environment	Complexity	Experimental Data	Details of Environment Request	Accuracy	Time	Other
Okumura model	Macrocell	Simple	Based on experiments	No	Good	Little	Graphical path loss data
Hata model	Macrocell (early cellular)	Simple	No	No	Good	Little	
Cost-231	Microcell (outdoor)	Simple	No	No	Good	Little	
Dual-slope	Microcell and picocell (LOS region)	Simple	No	No	Good	Little	
Ray-tracing	Outdoor and indoor	Complex	No	Yes	Very good	Very much	
FDTD	Indoor (small)	Complex	No	Every detail	Best	Very much	Often combined with ray-tracing
MOM	Indoor (small)	Complex	No	Every detail	Best	Very much	
ANN	Outdoor and indoor	Complex	Yes	Detail	Very good	Little	Take time to learn from experimental data

8.7 EFFICIENT COMPUTATIONAL METHODS FOR PROPAGATION PREDICTION FOR INDOOR WIRELESS COMMUNICATION

This section presents two different deterministic methods for efficient characterization of an indoor channel. First an improved version of a ray-tracing method is presented. Next the finite-difference-time-domain (FDTD) method is used to calculate the effects of walls in an indoor wave propagation environment.

8.7.1   Efficient Ray-Tracing Methods

The application of several ray-tracing techniques in combination with the uniform theory of diffraction is an efficient method for prediction of propagation in the UHF (communication) band in an indoor environment. This is discussed in detail [136] as we improve the computational efficiency of the two-dimensional (2D) ray-tracing method by reorganizing the objects in an indoor environment into irregular cells. In addition, by making use of the 2D ray-tracing results, a new three-dimensional (3D) propagation prediction model is developed, which can save 99% of the computation time of traditional 3D model. This new hybrid model is more accurate than 2D models and more efficient than traditional 3D models in computing the path loss to any point in the building. In this model, reflection and refraction by layered materials and diffraction from the corners of the wall are considered.

As shown in Figure 8.7 there are four main types of rays from the transmitter to the receiver. There are two types of ray-tracing methods, one is called the *image method* [53, 54, 57] and the other is the *brute force ray-tracing method* [12, 42]. The image method is well suited to analysis of radio propagation associated with geometry of low complexity and with small number of reflections. The brute-force method launches a bundle of rays that may or may not reach the receiver. It requires numerous ray object intersection tests and extensive data arrays for ray tracing. In the brute-force method, both refraction and diffraction can be considered. In these two methods, both the 2D and 3D ray-tracing models are used widely. In the 2D model, only those rays in a plane are traced, so it needs less computation time. In the 3D model, all rays must be traced, so it needs much more computation time.

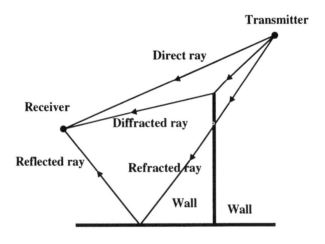

Figure 8.7.   Four main types of rays in an indoor environment.

2D ray-tracing technique is widely used for indoor propagation prediction [58, 128, 129]. When the indoor environment is large and complex, it will take much CPU time to calculate the propagation characteristics. It is therefore important to improve the computational efficiency. In [128, 129], only those rays that have certain contribution to the receiver are traced. In [130], effective-propagation-area method and dominant-corner extraction method are used. However, by making use of the geometrical characteristics of an indoor environment and reorganizing the objects, a new 2D ray-tracing model termed an irregular cell model is presented. Almost 60% of the CPU time can be saved using this technique and without any loss in accuracy.

Based on that, we develop a new efficient 3D ray-tracing model that integrates several improved ray-tracing techniques. In this model, all objects are projected onto the floor. By using a 2D ray-tracing algorithm and a suitable selection of an elevation angle, all effective rays (which can reach the receiver) can be determined. Because fewer effective rays need to be traced, much computation time can be saved. In addition, the patched-wall technique and the ray-fixed coordinate system, respectively, are used in the analysis in order to improve the computation accuracy and to reduce the dimension (from three to two) for the dyadic characterization of reflection, refraction, and diffraction coefficients. The computation results are validated by measurements carried out in a building at Shanghai Jiao Tong University at 1.7 GHz.

8.7.1.1 Rays in an Indoor Environment. As shown in Figure 8.7, the direct ray makes the main contribution to the received signal, if it exists. When the receiver is out of sight from the transmitter, reflected, refracted, and diffracted rays carry energy to the receiver. They are expanded as follows.

• *Reflected and refracted rays.* In an indoor environment, objects always have a certain thickness, as shown in Figure 8.8. In addition, they also introduce losses. Generally, when a ray in air illuminates an object, a reflected ray and a refracted ray are produced in the upper and lower areas of the space, respectively. The reflected ray can be considered as the rays coming from the mirror image of the object. The refracted ray in the lower area of the space is parallel to the incident ray, but it has a deviation Δd, as shown in Figure 8.8. The deviation is given by

$$\Delta d = d \sin(\phi_i - \phi_t) / \cos\phi_t \qquad (8.29)$$

where ϕ_i and ϕ_t are the angles of incidence and refraction, respectively. In an indoor environment, the thickness of a typical wall is 20 to 30 cm, so the general distance between the refracted ray and the incident ray may be less than 20 cm. If the wall is thicker, the refracted ray may be too weak to be considered in the calculations. When using a ray-tracing technique, the ray is considered as a tube and it diffuses as it propagates. For example, assume the angle of the tube to be 1°; the radius of its wavefront will be 17.46 cm after it propagates a distance of 20 m. At most, it is reasonable to ignore the offset between the refracted ray, and the incident ray as has been done in most papers. In [131], an equivalent source is introduced to consider this offset. On modern interior walls which are construced from two layers of gypsumboard nailed onto the studs, the offset is also very small. Sometimes, there are multiple reflections within the wall, and the offset between the first reflected ray and other higher-order reflected rays can be treated in a fashion similar to that used for the refracted ray.

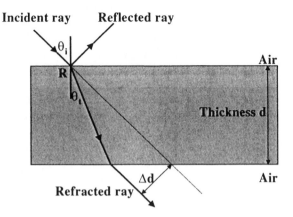

Figure 8.8. Typical reflected and refracted rays in an indoor environment.

The coefficients of reflection (R) and refraction (T) for layered lossy materials are given by (8.30) [132]. It is assumed that there are n layers of homogeneous and nonmagnetic lossy materials. Layer 1 and layer n are the air regions. One can therefore write

$$\begin{bmatrix} 1 \\ R^{\perp//} \end{bmatrix} = [U_{1n}] \begin{bmatrix} T^{\perp//} \\ 0 \end{bmatrix} \tag{8.30}$$

where $[U_{1n}] = [U_{12}][U_{23}] \cdots [U_{(n-1)n}]$ with

$$[U_{i(i+1)}] = \frac{1}{2}\left[1 + \frac{k_{(i+1)x}}{k_{ix}}\right]\begin{bmatrix} e^{jk_{(i+1)x}d_i} & R_{i(i+1)}e^{-jk_{(i+1)x}d_i} \\ R_{i(i+1)}e^{jk_{(i+1)x}d_i} & e^{-jk_{(i+1)x}d_i} \end{bmatrix}$$

k_i = wavenumber of medium $i = \omega\sqrt{\varepsilon_i\mu_0}$

$k_{ix}^2 = k_i^2 - (k_1\sin\theta)^2$, for $i = 1, 2, ..., n-1$

d_i = thickness of layer i with ($d_1 = d_n = 0$)

ε_i = complex dielectric constant of layer i

ϕ = angle of incidence

ω = angular frequency of the incident ray

Superscripts $\perp$ and $//$ represent the perpendicular polarization and parallel polarization, respectively. The values for the reflection coefficients for different polarizations are given by

$$R_{i(i+1)}^{\perp} = \frac{1 - \dfrac{k_{(i+1)x}}{k_{ix}}}{1 + \dfrac{k_{(i+1)x}}{k_{ix}}} \quad \text{for perpendicular polarization,}$$

and

$$R_{i(i+1)}^{//} = \frac{1 - \dfrac{\varepsilon_i k_{(i+1)x}}{\varepsilon_{i+1} k_{ix}}}{1 + \dfrac{\varepsilon_i k_{(i+1)x}}{\varepsilon_{i+1} k_{ix}}} \quad \text{for parallel polarization for } i = 1, 2, \ldots, n-1.$$

When a ray-fixed coordinate system is used, reflected field $\vec{E}_r(R)$ at a position R is determined from the incident field $\vec{E}_i(R)$ at position R by using the following equation:

$$\begin{bmatrix} E_r^{\perp} \\ E_r^{//} \end{bmatrix} = \begin{bmatrix} R^{\perp} & 0 \\ 0 & R^{//} \end{bmatrix} \begin{bmatrix} E_i^{\perp} \\ E_i^{//} \end{bmatrix} \tag{8.31}$$

We have a similar relation to obtain the refracted field at the position R, by substituting the reflection coefficients $R^{\perp}$ and $R^{//}$ with the respective transmission coefficients by $T^{\perp}$ and $T^{//}$, respectively, with $T^{\perp}$ and $T^{//}$ given by (8.30).

The magnitude of the reflection coefficients $|R|$ at 1.8 GHz for various materials is shown in Figure 8.9. Some of the materials described, for example, are wall (thickness $d = 20$ cm, $\varepsilon_r = 3.25$), a wooden door ($d = 4$ cm, $\varepsilon_r = 1.94$), and a glass window ($d = 0.4$ cm, $\varepsilon_r = 3.95$). From this figure it can be seen that the magnitude of the reflection coefficient for all the materials are close to 1 when the incident angle is close to $90°$.

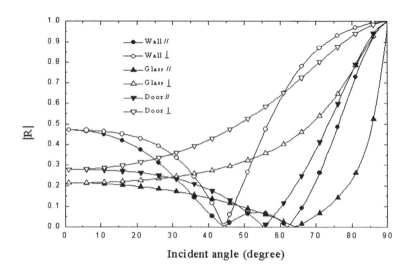

Figure 8.9. Magnitude of the reflection coefficients as a function of polarization for different materials.

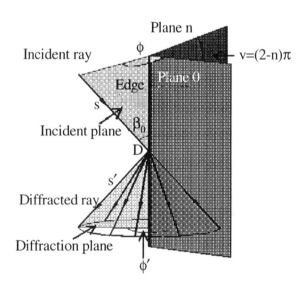

Figure 8.10. Diffraction due to a corner.

• **Diffracted ray.** The uniform theory of diffraction (UTD) is applied to calculate the diffracted field from the corners in an indoor environment. In Figure 8.10 a wedge with an angle $(2 - n)\pi$ is used for illustration purposes. The field is incident on it at an oblique angle. The coefficients of diffraction $D^{\perp //}$ are given by [54]

$$
\begin{aligned}
D^{\perp //} = \frac{-e^{-j\pi/4}}{2n\sqrt{2\pi k}\,\sin\beta_0} &\left\{ \cot\frac{\pi + (\phi' - \phi)}{2n} F[kLa^+(\phi' - \phi)] \right. \\
&+ \cot\frac{\pi - (\phi' - \phi)}{2n} F[kLa^-(\phi' - \phi)] \\
&+ R_0^{\perp //} \cot\frac{\pi - (\phi + \phi')}{2n} F[kLa^-(\phi + \phi')] \\
&\left. + R_n^{\perp //} \cot\frac{\pi + (\phi + \phi')}{2n} F[kLa^+(\phi + \phi')] \right\}
\end{aligned}
$$

(8.32)

where

$F(x) \quad = 2j\sqrt{x}\,e^{jx}\int_{\sqrt{x}}^{\infty} e^{-j\tau^2}\,d\tau \quad$ is a Fresnel integral

$a^{\pm}(\phi' \pm \phi) \quad = 2\cos^2\dfrac{2n\pi N^{\pm} - (\phi' \pm \phi)}{2}$

$N^{\pm} \qquad\quad =$ integer that approximately satisfies equation $2n\pi N^{\pm} - (\phi' + \phi) = \pm\pi$ as close as possible

$\beta_0 \qquad\quad =$ angle between the incident ray and the edge of the wedge

$\phi \qquad\quad =$ angle between the plane of diffraction and plane 0, as shown in Figure 8.16

$\phi' \qquad\quad =$ angle between the plane of diffraction and plane 0, as shown in Figure 8.16

$k \qquad\quad =$ wavenumber

$L \qquad\quad =$ distance parameter dependent on the form of the incident wave for an incident spherical wave $L = \dfrac{ss'}{s+s'}\sin^2\beta_0$, and s and s' are distances from the diffracted point D to the source and observation point, respectively

$R_0^{\perp //}$ and $R_n^{\perp //} =$ reflection coefficients related to plane 0 with an incident angle ϕ and form a plane n with a reflection angle $n\pi - \phi'$, respectively.

In a ray-fixed coordinate system, we have a relation similar to (8.31) for determining the diffracted fields. For an incident angle $\phi = 45°$ and $\beta = 90°$, the magnitude of the diffraction coefficients for a 90° concrete edge as a function of the diffraction angle at 1.8 GHz is shown in Figure 8.11. In our model, only the diffractions from the vertical corners of the walls that are recorded in the database are considered because most receivers and transmitters are generally vertically polarized. When a ray hits the edge, the edge is treated as a secondary source and all rays emanating from it are traced as a normal ray. Because the strength of the diffracted ray is weak, the second- and higher-order diffraction effects are neglected.

8.7.1.2 Improvement of the Computational Efficiency for 2D Ray Tracing.

A typical indoor environment is shown in Figure 8.12, where some furniture is present in the room. When applying a 2D ray-tracing algorithm, a database must be built. In order to deal with the differences in the dielectric properties among materials, each different object, such as a wooden door, a glass window, or a concrete wall, is defined as different regions. A boundary between any two regions is denoted by a line segment. All the data related to that segment, such as the point of origin, direction, length, thickness, and dielectric properties are specified. For example, there are 17 different segments for each sidewall of the corridor, and there are more than 80 segments in the environment of Figure 8.12.

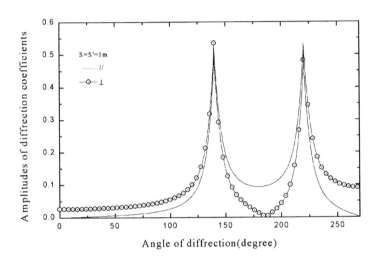

Figure 8.11. Magnitudes of the diffraction coefficients.

When a ray is traced through a usual 2D ray-tracing model, one needs to locate the ray intersections for every segment to determine whether there is an intersection. The equation for the ray is given by

$$\begin{cases} x = d_1 t + x_0 \\ y = d_2 t + y_0 \end{cases} \quad (t \geq 0) \tag{8.33}$$

where (x_0, y_0) is the origin and (d_1, d_2) is the direction vector of the ray. The equation for a segment is given by

$$\begin{cases} x = a_1 m + x_1 \\ y = a_2 m + y_1 \end{cases} \qquad (0 \le m \le l) \qquad (8.34)$$

where (x_1, y_1) is the origin and (a_1, a_2) are the direction cosines for each of the segments. The length of each segment is L. From (8.33) and (8.34), it is easy to obtain t and m and determine the point of intersection of the segments. From a geometrical point of view, one ray may intersect with many segments. It is important to get the point of intersection that is nearest the origin of the ray, where the reflection and refraction really occur.

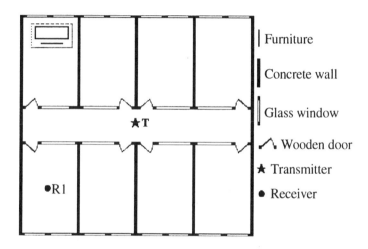

Figure 8.12.   Typical environment for modeling propagation.

In fact, a ray usually intersects with only one segment before it changes its direction. This means that all computations of intersections with the remainder of the segments are not necessary. If all these computations are omitted, the computational efficiency of ray tracing will be greatly improved. In [133] the propagation area is divided into equal cells. But there are several disadvantages associated with this technique. It is difficult to determine the best size for the cell. Sometimes, there is no segment in a cell, and sometimes one segment crosses more than one cell. All these make the computations complicated. To overcome these difficulties, we have developed an irregular cell technique.

It is possible that physically disjoint segments of an indoor environment may lie on a straight line but are not continuous (e.g., open windows). We reorganize these segments into a line that can be described as a special segment without having any dielectric properties. According to this linear grouping method, the indoor environment as shown in Figure 8.12 has only nine lines. When there is furniture within the room, we can first introduce a rectangular box (indicated by a dashed line in Figure 8.12) to encompass it, and then consider the two diagonals of the box as two virtual segments to see if the ray of interest intersects them. If the answer is yes, we take the real segments of the furniture into account; otherwise, we omit the entire box. That can result in considerable saving of the CPU time.

In our model we first build the database of segments of the indoor environment for prediction. All the segments in the database will be reorganized automatically into lines or boxes that form another database. When a 2D ray-tracing method is applied, only the intersections between rays and lines or boxes need to be determined. If a ray has no intersection with a line or two diagonals of a box, it has no intersection with the segment members of that line or box, and therefore tracing of the ray is terminated. For all the lines or boxes intersecting

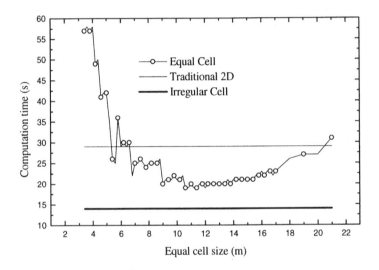

Figure 8.13. Computation time for predicting path loss at location R1 of Figure 8.12.

with the ray, the nearest line or box to the point of origin of the ray can be found and then the corresponding segments intersecting with the ray can also be determined. When the real intersecting segment is found, the electric field is calculated and both the reflected and refracted rays can be traced further.

In order to verify our model, numerical simulations have been carried out. The path losses in an indoor environment shown in Figure 8.12 have been predicted. The environment is composed of an area 17.9×20.8 m^2 and consists of concrete walls, wooden doors, glass windows, and furniture. The transmitter is located at the central part of the floor. The transmitting frequency is 1800 MHz. Figure 8.13 plots the computation time on a PC for calculating the path loss at the receiver location $R1$ as shown in Figure 8.12. It indicates that the proposed irregular cell method can save almost 60% of CPU time on the average over other traditional methods. The conclusion is also true for other locations of transmitter and receivers in this environment. Because no ray intersections with objects have been omitted, the accuracy of the prediction does not degrade.

8.7.1.3 New Improved Model.

• *Patched-wall model.* The layout for the measurement system shown in Figure 8.14 is on the 3rd floor of a building with many classrooms at Shanghai Jiao Tong University, Shanghai. In Figure 8.14, T denotes the location of the transmitter and R_n indicates the location of the receiver. The field is measured at 39 receiver locations. The width, length, and height of the floor are 18.26 m, 76.39 m, and 3.74 m, respectively. There are 20 rooms on this floor. The ceiling and the floor are made of concrete.   The partition board between classrooms is

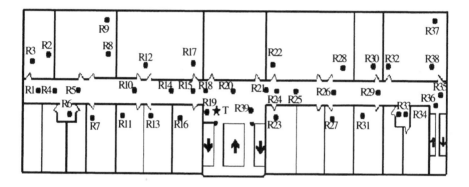

Figure 8.14. Layout of a building in the University (arrows mean stairs).

made of wood. The sidewalls of the corridor are mainly made of concrete. There are wooden doors and glass windows on these walls. Figure 8.15 shows a part of the sidewall. The front and back walls of the building are composed of concrete with inserts for glass windows.

Because of the differences in the dielectric characteristics between different materials, it is necessary to introduce patches of different dielectric constants and physical sizes to represent actual objects. Four different patches have been used. They are concrete walls, wooden doors, wooden partition boards, and glass windows. A database is built to record the location, thickness, and dielectric constants of all the patches. In order to simplify the calculation of the intersections, every patch is divided into several rectangles and is given an integer number for identification (ID). As shown in Figure 8.15, the patches are separated by dashed lines. There are 10 patches in Figure 8.15. The IDs of the patches having the same vertical projection on the floor are different only in the last digit. The ceiling and the floor are considered to be special patches. The measurement environment is divided into more than 400 patches that are stored in the database.

- *Model description.* The process of ray tracing is a complicated program of recursion. For the 3D model, there are many rays that need to be traced. But in fact, there are only a few rays that will reach the receiver. It is advantageous to search for the rays that will probably reach the receiver for tracing. In some simplified models, only singly reflected rays from the floor and ceiling are considered. Even though there may be no direct rays, however, multireflected rays from the floor and ceiling cannot be neglected.

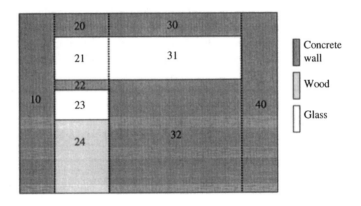

Figure 8.15. Part of an actual wall and how to divide it into patches.

Indoor environments are composed of floors, ceilings, and walls vertical to the floor. On a single floor (considering the floor to be the x–y plane and the direction vertical to the floor the z-axis), one can find the following properties for propagation when using a ray-tracing technique. In this coordinate system, transmitter and receiver are regarded as points in a 3D space. Therefore:

1.) When the floor or the ceiling reflects a ray, the incident and reflected rays have the same azimuth angle. Namely, perpendicular projections of the two rays on the floor are in a line.
2.) When a wall vertical to the floor reflects a ray, the angle of incidence is equal to the angle of reflection along the perpendicular projection plane on the floor.
3.) The angle between the ray reflected by a wall and the z-axis is equal to that between the incident ray and the z-axis.
4.) Rays that have the same azimuth angle in a 3D space have the same perpendicular projection on the floor.

In our model, first we project all walls, objects, the transmitter, and the receiver (except the ceiling and the floor) vertically onto the floor and use a 2D ray-tracing technique on the projection plane and then apply a 3D ray-tracing technique based on the results of a 2D ray tracing. This is explained through the following figures. Figure 8.16 characterizes a simple indoor environment with a wall, a ceiling, and a floor. The height of the transmitter and the receiver from the floor are h_1 and h_2, respectively. The height of the ceiling is h above the floor. In Figure 8.17, the wall, the transmitter, and the receiver are projected vertically onto the floor. By performing a 2D ray tracing on the floor, one sees that there are only two paths from the projections of the transmitter and the receiver. If no reflection from the ceiling and the floor is considered, there are only two actual paths in the 3D space when using a 3D ray-tracing model. One is the direct path from the transmitter to the receiver; and the other includes one reflection from the wall.

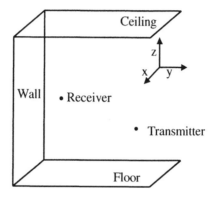

Figure 8.16. Simplified room structure.

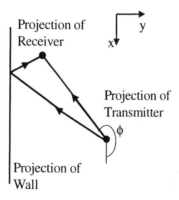

Figure 8.17. Projection onto the floor of various objects described in Figure 8.16.

When multiple reflections from the ceiling and floor are considered, the situation is different. We define the unfolded length of a path to be d and the azimuth angle of the projection for the corresponding launching ray to be ϕ. Using the properties of propagation, one can observe that rays launched in the 3D space that may reach the receiver is given by

$$\theta = \frac{\pi}{2} + \alpha \arctan \frac{2nh + \alpha h_0}{d} \qquad (8.35)$$

where $h_0 = h_1 \pm h_2$, the positive sign is used when the total number of reflections from the ceiling and the floor is odd, and the negative sign is used when it is even, respectively. Possible values are $\alpha = \pm 1$. The positive value is used when the first reflection (does not include reflections from walls) occurs on the floor, and the negative value is used when it is from the ceiling.

These rays are all significant and should be launched in a 3D environment. The paths from the transmitter to the receiver for all these rays launched in 3D have the same perpendicular projection on the floor. The launching of the rays are carried out for $n = 0, 1, 2, \ldots, M$ ($n \neq 0$ when $\alpha = -1$), where n is an index related to the order of reflection from the ceiling and the floor and M is an index related to the maximum order of reflection from the ceiling and the floor. When $n = 0$, two 3D paths are included; one has no reflection from the ceiling or the floor, and the other has only one reflection from the floor. When $n = 1$, there are four 3D paths that are considered. The first path has one reflection from the ceiling. The second path has one reflection from the ceiling and then one reflection from the floor. The third one has one reflection from the floor and then one reflection from the ceiling, and the last one had one reflection from the floor, one reflection from the ceiling, and then one reflection from the floor again. So, the total number of useful 3D rays included in each 2D path is $2 + 4M$. The total 2D paths should include all reflections from all the walls.

During the 2D tracing on the projection plane, all 2D paths from the projection of the transmitter to the projection of the receiver are recorded and the unfolded length for each path is also calculated. After an integer M is given, a 3D ray tracing is performed for all useful 3D launched rays corresponding to each 2D path. Because only parts of all 3D launching rays need to be traced, much time of computation can be saved. Based on Figure 8.17, Figure 8.18 shows the various paths in 3D space according to our model. If $M = 1$, then $n = 0$ and there are six paths in the 3D space corresponding to each 2D path as shown in Figure 8.17. Therefore, a total of 12 [$= 2 \times (2 + 4 \times 1)$] 3D rays are considered.

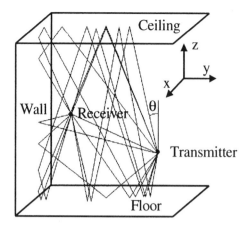

Figure 8.18. Tracing of rays in a three-dimensional space.

The field intensity of each ray is calculated using a 3D ray tracing. Finally, the total field intensity can be obtained by summing up all the individual field intensities. Our model includes direct, reflected, refracted, and diffracted fields that are represented by the rays. Each propagation mechanism is treated separately.

8.7.1.4 Results of Simulation and Measurement. The simulation environment and the measurement site are shown in Figure 8.14. In the prediction simulation, all the advantages of a 3D ray-tracing model have been taken into account in the new model. Because of the high computational efficiency of the 2D tracing technique, fewer useful 3D rays need to be traced. This 3D model just took 1% of the computation time over traditional 3D models. And because no useful rays have been ignored, the prediction accuracy is quite high.

To carry out the measurement, a 1.7-GHz narrowband CW signal generator is used as the transmitter. A 0- to 15-dBm CW signal is transmitted by a half-wavelength dipole antenna at a height of 1.6m above the floor. The signal is received by a half-wavelength dipole antenna located at a height of 1.5m from the floor. To assure that the propagation channel is stationary in time, the measured data has been averaged over 10 instantaneous sampled values.

Figure 8.19 shows both the predicted and measured results for the path loss at each location in the building. Predicted results from both the new model and the traditional 3D model have been presented along with measurements. However, the latter takes about 100 times more computation time than the new model. Because the position of receivers R18 and R19 are close to the transmitter, it is difficult to measure the path loss accurately. In the corridor, there is a LOS path or one reflected path so that the signal received at these locations is stronger than those inside the rooms.

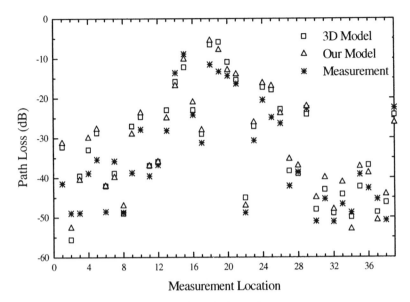

Figure 8.19. Results of simulation and measurement.

The method used in [61] has been extended to an indoor environment. In [61], the authors used the image method and projected objects on the ground to find an exact 3D path for an outdoor street scene. In this new model using multiple reflections from the floor and the ceiling and from one 2D path on a projection plane, many 3D paths can be determined. A formula to determine the 3D path has also been provided.

8.7.1.5 Conclusion. A computationally efficient method is described which transforms the results of a 2D ray-tracing method to a 3D model which has correlations with experimental results.

8.7.2. Analysis of the Effects of Walls on Indoor Wave Propagation Using FDTD

The ray-tracing technique does not take the effects of the inner structure of walls into account for indoor propagation prediction. A numerical approach to treat this problem using the FDTD method is now described. Numerical results for path loss calculated by a FDTD method are compared with those obtained by the ray-tracing technique. Use of a ray-tracing technique in propagation prediction has been under development for more than a decade. It is powerful and easy to use.

There have been a number of recent investigations on indoor radio propagation modeling using a ray-tracing method [57, 65, 126], but few pay attention to the inner structure of the walls [134, 135]. When using a ray-tracing method, it is assumed that reflection from walls has a substantial specular component. Typical concrete blocks used in wall construction are shown in Figure.8.20. The air hole in the block forms a periodicity of blocks. It is found that nonspecular reflection occurs when the operating frequency is above 1.2 GHz [134].

In this approach, the reflection and transmission characteristics of the walls can be derived by solving for higher-order Floquet modes. Also, the method of homogenization can be used to determine the effective material properties of walls, but they did not consider nonspecular reflection.

In this section, the FDTD method is used to predict propagation properties for indoor environments [137]. The periodic structure of walls is also considered. Numerical results for the path loss are calculated by a FDTD method and are compared to those obtained by the ray-tracing method. It is proved that the inner structure of walls has a considerable influence on the path loss when predicting propagation, and the FDTD method can give more accurate results.

In this approach, the reflection and transmission characteristics of the walls can be derived by solving for higher-order Floquet modes. Also, the method of homogenization can be used to determine the effective material properties of walls, but they did not consider nonspecular reflection.

In this section, the FDTD method is used to predict propagation properties for indoor environments [137]. The periodic structure of walls is also considered. Numerical results for the path loss is calculated by a FDTD method and are compared to those obtained by the ray-tracing method. It is proved that the inner structure of walls has a considerable influence on the path loss when predicting propagation, and the FDTD method can give more accurate results.

8.7.2.1 *Description of the Procedure.*

For the analysis of a wall at a frequency of 1.8 GHz, the parameters for the model of Figure 8.20 are chosen as $d = \lambda_0/20$ and the relative dielectric constants of a concrete wall as $\varepsilon_r = 3$. When a plane wave in the x–z plane is incident on the layered periodic structure at an angle θ to the z-axis with an electric field polarized along y, the inner periodic layer of the structure will support an infinite set of modes with different wavenumbers, km $(m = 0, \pm1, \pm2, \ldots)$. The field produced by these modes can in turn be decomposed into a series of space harmonics $(n = 0, \pm1, \pm2, \ldots)$, each having a wavenumber $\beta_n = k_0 \sin\theta + 2\pi n/T$ along x, where k_0 is the wavenumber of the incident wave in free space and T is the period of the inner structure. The space harmonics in the periodic region will couple to the air region. As a result, there will exist reflected and transmitted space harmonics whose directions of propagation in the air are given by $\sin\phi_n = \beta_n/k_0$, where ϕ_n is the outbound angle relative to the normal [134]. At higher frequencies, more modes have real ϕ_n and will carry power along nonspecular directions. Figure 8.21 shows the relation of the incident angle and the outbound angle for the structure of Figure 8.20 at 1.8 GHz when $n = 0, \pm1$.

In order to observe the effects of the inner structure of the walls on wave propagation, consider a room with a wooden door, glass windows, and concrete walls which are placed in the computation area as shown in Figure 8.22. A two-dimensional FDTD method is applied. An artificial outer boundary encloses the room defining the computational area. A second-order radiation boundary condition has been used at the outer boundary. The incremental spatial step is d and the time step is $dl/(2C)$, where C is the wave velocity in free space. The thickness of the wooden door, which has a permittivity of $\varepsilon_r = 1.9$, is $6d$. The thickness of the glass windows is d and it has a permittivity of $\varepsilon_r = 3.9$. A line source (T) to simulate the antenna is placed outside the room. A CW signal with an electric field polarized along the y-axis is transmitted at 1.8 GHz. The path loss at 250 points from left to right along a line is evaluated and compared with the results of ray tracing. When using the ray-tracing method, the transmitter is treated as a point source.

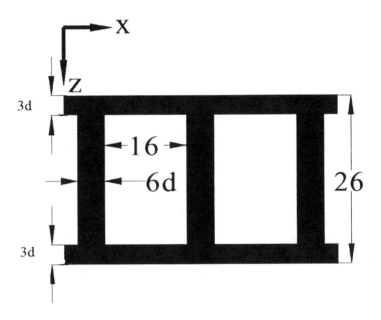

Figure 8.20. Typical periodic inner structure of the walls.

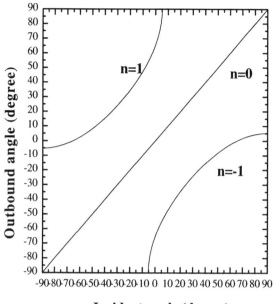

Incident angle (degree)

Figure 8.21. Outbound angle (in degrees) of the $n = 0$, ± 1 space harmonic modes as a function of the incident angle (in degrees).

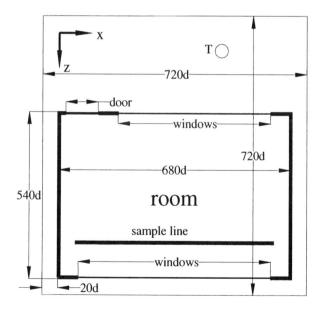

Figure 8.22. Two-dimensional geometry of the grid for the FDTD method.

8.7.2.2   Numerical Results. The fields of the environment are calculated using a 2D FDTD method when the wall is (1) assumed to be a periodic structure, and (2) uniform in nature with an effective dielectric constant. For the FDTD method, 5000 time steps are calculated and it takes more than 6 hours of CPU time on a PC. A 2D ray-tracing method is used to analyze the uniform structure, and it takes only 10 minutes to compute the path loss for 250 sample points.

Figure 8.23 shows the simulation results of path loss for the sample points. It indicates that the effects of the inner structure of the walls cannot be neglected for propagation prediction. The maximum difference in field strength when the wave is assumed to propagate through a periodic structure and a uniform structure is more than 10 dB when using the FDTD method. As a full-wave analysis tool, FDTD is more accurate than the ray-tracing method. All the propagation phenomena, such as reflection, transmission, and diffraction, are included. A ray-tracing method considers only reflection and transmission. The maximum difference in the field strength when propagating through a periodic structure using the FDTD method and the ray-tracing method is around 18 dB.

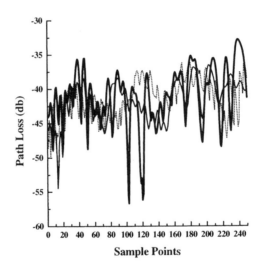

Figure 8.23. Simulation results using three different FDTD methods.
━━━━━ , FDTD (periodic); ────, FDTD (uniform); -------, ray tracing (uniform).

8.8   MODELS FOR SMALL-SCALE FADING

Small-scale fading refers to dramatic changes in signal amplitude and phase that can be experienced as a result of small changes (as small as a half-wavelength) in the spatial separation between a receiver and a transmitter. These changes in the envelope of the received signal is statistically described by a stochastic process.

In order to get a sound understanding of the channel, it is important to study the distribution of the envelope of the received signal. A few possible choices of the statistical distributions to model the envelope are explained below. As explained and emphasized in Appendix A, we have to be extremely careful when using a statistical model, as a complete lack of knowledge of the system cannot be supplemented by a stochastical distribution.

8.8.1 Ricean Distribution

When there is a dominant stationary (nonfading) signal component present, such as a line-of-sight propagation path, the fading distribution is Ricean. The Ricean distribution (also called *Rice distribution* or *Rician distribution*) is given by

$$p(r) = \begin{cases} \dfrac{r}{\sigma} \exp\left[\dfrac{r^2 + Ar}{2\sigma^2} \right] I_0\left(\dfrac{Ar}{\sigma^2} \right) & (A \ge 0,\, r \ge 0) \\[2ex] 0 & r < 0 \end{cases} \tag{8.36}$$

where r is the amplitude of the envelope of the received signal, $2\sigma^2$ is the predicted mean power of the multipath signal, A denotes the peak amplitude of the dominant signal, and $I_0(\bullet)$ is the modified Bessel function of the first kind and zero order. The Ricean distribution is often described in terms of a parameter K, which is defined as the ratio between the deterministic signal power and the variance of the multipath. It is given by [10]

$$K(dB) = 10 \log \frac{A^2}{2\sigma^2} \tag{8.37}$$

K is known as the *Ricean factor* and completely specifies the distribution.

It is possible to estimate the Ricean K-factor of a signal, from measurements of received power versus time. One approach is to compute the distributions of the measured data, then compare the result to a set of hypothesis distributions using a suitable goodness-of-fit test. Another is to compute a maximum likelihood estimate using an expectation/maximization (EM) algorithm. However, both of these approaches are relatively cumbersome and time consuming. A simple and rapid method has been developed based on calculating the first and second moments of the time series data. When perfect moment estimates of the Ricean envelopes are available, the method is exact. In that case, the factor K can only be obtained implicitly, by equating a ratio of the measured moments to a complicated function of K. By contrast, the method described in [89] yields an explicit and quite simple expression for K in terms of the measured moments.

8.8.2 Rayleigh Distribution

As the dominant signal in Ricean distribution becomes weaker, the composite signal resembles a noise signal which has the envelope of a Rayleigh distribution. For mobile radio channels, the Rayleigh distribution is widely used to describe the statistical time-varying nature of the received envelope of a flat fading signal, or an individual multipath component. The Rayleigh distribution has a probability density function (pdf) given by

$$p(r) = \begin{cases} \dfrac{r}{\sigma^2} \exp\left(-\dfrac{r^2}{2\sigma^2} \right) & (0 \le r \le \infty \\[2mm] 0 & (r < 0) \end{cases} \tag{8.38}$$

The probability that the envelope of the received signal does not exceed a specified value R is given by the corresponding cumulative distribution function (cdf):

$$P(R) = P_r(r < R) = \int_0^R p(r)\,dr = 1 - \exp\left(-\frac{R^2}{2\sigma^2} \right) \tag{8.39}$$

where r is the envelope amplitude of the received signal and $2\sigma^2$ is the predicted mean power of the multipath signal.

Since the fading data are usually measured in terms of the fields, quantities for a particular distribution cannot be assumed. The median value is often used other than the mean values and it is easy to compare different fading distributions which may have widely varying means. A typical Rayleigh fading envelope for a moving mobile at 900 MHz is shown in Figure 8.24 [90].

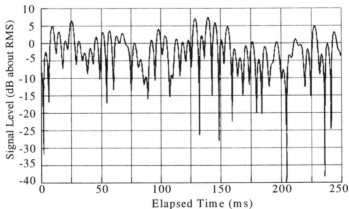

Figure 8.24. Typical Rayleigh fading envelope at 900 MHz received by a mobile traveling at 120 km/hr.

8.8.3   Lognormal Fading Model

The lognormal fading model is quantified to represent the distribution of rays which experience multiple reflections and diffractions between a transmitter and a receiver. The lognormal pdf can be expressed as

$$p(r) = \frac{1}{r\sqrt{2\pi\sigma^2}} \exp\left\{ -\frac{[\ln(r) - m]^2}{2\sigma^2} \right\}$$ (8.40)

where m is the median value and σ is the standard deviation of the corresponding normal distribution, obtained by using the transformation $y = \ln(r)$[91]. Techniques such as the Monte Carlo method and Schwartz and Yeh method have been developed to simulate the power sum for the lognormal components [92].

8.8.4   Suzuki Model

The Suzuki model combines the lognormal and Rayleigh distributions and provides a more accurate approximation of the sum of correlated complex lognormals for a wider variety of channel behaviors. Usually, the Rayleigh distribution is obtained from two statistical independent normal processes, $\mu_1(t)$ and $\mu_2(t)$, with zero means and identical autocorrelation functions according to the relation

$$\xi(t) = \sqrt{\mu_1^2(t) + \mu_2^2(t)}$$ (8.41)

where $\xi(t)$ can be regarded as the envelope of one complex-valued normal random process $\lambda(t)$. The requirement of statistical independence between $\mu_1(t)$ and $\mu_2(t)$ is identical with the demand for a symmetrical power spectrum for $\lambda(t)$. The received power averaged over a period of a few seconds can vary considerably due to various shadowing effects. In order to adapt the model to this behavior, the process $\xi(t)$ is substituted by the product $\eta(t) = \zeta(t) \times \xi(r)$, where the lognormal process $\zeta(t) = \exp[\mu_3(t)]$ is defined by a normal process $\mu_3(t)$ with variance s^2 and mean m. The product process with this particular amplitude density distribution is called the *Suzuki process* and is given by [93]

$$p_\eta(r) = \int_0^\infty \frac{r}{\sigma^2} \exp\left(\frac{r^2}{2\sigma}\right) \frac{1}{\sqrt{2\pi} s\sigma} \exp\left[-\frac{(\ln\sigma - m)^2}{2s^2} \right] d\sigma$$ (8.42)

where σ is the standard deviation and r is the amplitude. The assumption of statistical independence does not always meet the real conditions in a multipath wave propagation, so it has been modified [94] and simulated in [95].

It is a widely accepted statistical model for the received signal envelope in macrocellular mobile radio channels, where there is no direct LOS path.

8.8.5   Nakagami Model

The Nakagami Model was developed in the early 1940s. The corresponding probability density function is written as [96]

$$p(r) = \frac{2m^m r^{2m-1} \exp\left(-\frac{m}{\Omega}r^2\right)}{\Gamma(m)\,\Omega^m} \tag{8.43}$$

Here, r is the envelope amplitude of the received signal, $\Omega = <r^2>$ is the time-averaged power of the received signal, and $m = <r^2>^2 / <(r^2 - <r>^2)^2>$ is the inverse of the normalized variance of r^2. $\Gamma(\bullet)$ is the Gamma function. The Nakagami pdf may be shown to be a more general expression of other well-known density functions. For $m = 1$, the Rayleigh probability density function is obtained. It can also be approximated by both Rice and lognormal distributions over certain domains given the appropriate bounds on the parameters.

A Nakagami model parameterized by the fading severity parameter m has been shown to fit well to some urban multipath propagation data. A simulation of the Nakagami model has been presented in [97].

8.8.6   Weibull Model

The Weibull model arises when results from mobile radio propagation measurements are plotted on graph paper that is scaled such that a Rayleigh distribution appears as a straight line with a slope of -1. The Weibull pdf can be written as [91]

$$p(r) = \frac{\alpha b}{r_0}\left(\frac{br}{r_0}\right)^{\alpha-1} \exp\left[-\left(\frac{br}{r_0}\right)^{\alpha}\right] \tag{8.44}$$

where α is a shape parameter, which is chosen so as to yield a best fit to the measurement results. r_0 is the RMS value for r. $b = \sqrt{(2/\alpha)\Gamma(2/\alpha)}$ is a normalization factor. For the special case in which $\alpha = 1/2$, it becomes a Rayleigh pdf. The Weibull distribution provides flexibility to model any shape offered by a Nakagami distribution, but it lacks a theoretical basis. In [146] it is shown that the Weibull distribution characterizes indoor propagation path losses quite well.

8.8.7   Other Fading Models

Many other models for fading have been developed. They are the *Rice-lognormal model* [98], *Nakagam–Rice model* [99], *Nakagami-lognormal model* [100], and the *K*-distribution, which is a substitute for the *Rayleigh-lognormal distribution* [101]. They are mixtures of two kinds of distributions and are now widely used. In [102], propagation models that include both the effects of shadowing and multipath fading have been developed and have been used in studying terrestrial and satellite channels. In [103, 104] a new theoretical model for the prediction of fast fading in an indoor environment has been developed. This model makes the assumption that the number of dominant propagation paths that contribute to the signal in the receiver in a multipath environment is not infinite but rather small (e.g., 15). This assumption leads to the development of a new theoretical model called POCA that is more general and more efficient than the Rayleigh one, especially for indoor environments, and it fits generally for environments in which a small number of dominant propagation paths exist.

In this section, some of the statistical models for fading have been introduced. Only their pdf distribution has been described. Details of the implementation of these models can be obtained from the various references cited.

8.9 IMPULSE RESPONSE MODELS

Narrowband or continuous wave (CW) path loss is a parameter that can predict the power level of the system and the space coverage of a base station. In modern mobile communication systems with high data rates and small cell size it is necessary to model the effects of multipath delay as well as fading. The impulse response is a useful characterization of the system since the output of the system can be computed through convolution of the input with the impulse response if the system is linear. A multipath propagation channel is modeled as a linear filter and has a complex baseband impulse response [105–125, 144, 147–151].

In digital wireless communications, one of the main reasons of occurrence of bit errors is intersymbol interference (ISI) caused by multipath propagation. If the symbol rate is much lower than the coherent bandwidth, time delay spread can be neglected. In this case, multipath propagation only causes fading of the signal level, and Gaussian noise is the dominant factor that causes bit errors. If the symbol rate is relatively high, time delay spread cannot be neglected.

A large number of measurements have been made for the impulse response for both outdoor and indoor environments. Based on the measured data, some models have been derived. The results of measurements along with some typical models are described here followed by some deterministic models.

8.9.1   Models Based on Measurement Results

Because of the importance of determining the time delay spread for wireless communications, a number of wideband channel sounding techniques have been developed. These techniques may be classified as: (1) direct pulse measurements, (2) spread spectrum sliding correlator measurement, and (3) swept frequency measurements.

The block diagram for each of these techniques is shown in Figure 8.25 [10]. Almost the same result can be obtained from the last two methods [105]. Some measurements, both for indoor and outdoor environments and the calculated RMS time delay, have been summarized in Table 8.5, where Med indicates median values, T-R indicates the distance between transmitter and receiver, and Ave. implies the mean average values. In Table 8.5 it is found that the RMS delay spread varies from several nanoseconds to several microseconds, corresponding to different environments and frequencies.

There are many factors affecting the impulse response, such as frequency, height of transmitting and receiving antennas, fixed objects (mountains, buildings, or indoor furniture), and moving humans or objects close to the transmitter and/or the receiver. It is shown in Table 8.5 that the RMS time delay varies little for different frequencies for indoor environments [106], but a different conclusion has been obtained in [107]. In [108] it is found that at 11.5 GHz, RMS time delay in most cases is significantly smaller than at 2.4 GHz and 4.75 GHz, which are in general about the same. The reason that two different conclusions have been reached is that all these various conjectures have been derived from specific measurements. OLOS (obstructed line-of-sight) channels are subjected to higher variations in RMS delay spread caused by movement of the terminal around a small area because of the presence of more scatters in OLOS than in an LOS environment. On the other hand, the short time variations in the RMS delay spread in OLOS environments are found to be less than those in LOS environments [117]. It is concluded that in an empty indoor environment the RMS delay spread is a function of transmit-receive separation, and when there are furniture and mazes of semipermanent partitions, RMS delay spread becomes constant for all ranges [105]. It is shown in [121] that in many cities where terrain is flat, RMS delay spreads do not exceed 7 or 8 µs, and in urban areas with surrounding hills, RMS delay spreads do not exceed 13.5 µs. In [26, 123], RMS delay spread has been analyzed as a function of path loss and the height of antenna. The predicted results may be bounded by an exponential model of the form $\tau_{RMS}(ns) = \exp(0.065PL)$, where PL is the path loss in decibels. Table 8.6 shows the measurements of τ_{RMS} with relation to the height of the antennas. It indicates that the delay spread increases as the antenna height increases because as the antenna is raised it becomes visible to more objects scattering the electromagnetic radiation at greater distances.

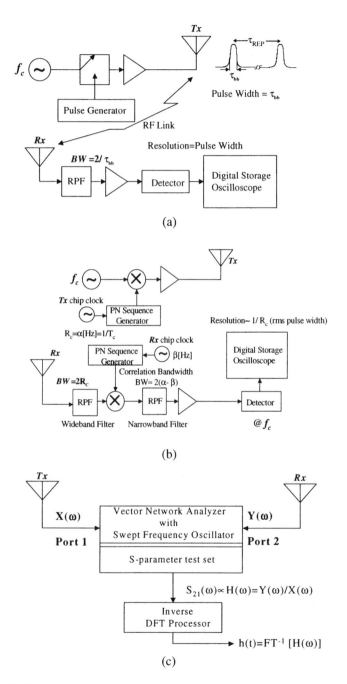

Figure 8.25. Block diagrams of the impulse response measurement system: (a) direct RF channel impulse response measurement system; (b) spread spectrum channel impulse measurement system; (c) frequency domain channel impulse measurement.

TABLE 8.5
Measurements and RMS delay spreads

Technique	Location	τ_{RMS}(ns)	Frequency (MHz)	Reference
B	Office building	30–100	850, 1,700	[106]
A	Manufacturing floors	15–29(Area A) 31–62(Area B) 48–90(Area C) 52–57(Area D) 19–37(Area E)	910	[109]
A	Indoor sports arena Open-plan factory Textile plant Office building	7–120 40–30 15–125 5–40	1,300 4,000	[110]
A	Within a room (4 rooms)	7–16 10 ns(Med.)	37,200	[111]
B	Urban area I Urban area II	98–1270 590(Ave.) 61–2940 480(Ave.)	910	[111][112]
C	Building 1 Building 2	About 12–72 About 4–25	40,000	[113]
B	Sidewalk of LOS street	0.5d + 40 (d is T-R)	2,600	[113][114]
A	Engineering bldg. Retail store	12.85–84.60 20.74–102.44	2,400	[115]
–	Laboratories at same floor	8.3 LOS (Med.) 14.1 OLOS1 (Med.) 22.3 OLOS2 (Med.)	910	[117]
B	Outdoor Site1 Site2 Site 3	60–250, 130(Med.) 40–130, 70(Med.) 60–250, 120(Med.)	–	[118]
C	In two buildings	10–50 with mean 20–30	900–1300	[119]
C	Four types of indoor locations	10–20 (Med.) 10–20 (Med.) 5–15 (Med.)	2,400 4,750 1,150	[108]
A	Hamburg Dusseldorf Frankfurt (bank) Frankfurt (apart.)	1300 (typical) 3100 (typical) 8100 (largest) 19,600 (largest)	942.225	[120]
A	Washington Greenbelt Oakland San Francisco	2500–7500 2000–7000 2500–13,500 1000–25,500	892	[121]
–	LOS (100–400m)	140.6–325.4	2,197.5	[122]

TABLE 8.6
RMS delay spread as a function of the height of an antenna

Antenna Height	Mean RMS Delay (ns)	Standard Deviation of RMS Delay (ns)	Location	Reference
3.7 m	136.8	138.0	San Francisco	[26]
8.5 m	176.8	147.1		
13.3 m	275.9	352.0		
3.7 m	134.6	127.4	Ottawa	[123]
8.5 m	173.1	156.8		

8.9.2   Statistical Models of Time Delay Spread

8.9.2.1 Two-ray Rayleigh Fading Model.
A commonly used multipath model is an independent Rayleigh fading two-ray model [10]. The impulse response of the model is represented by

$$h(t) = \alpha_1 \exp[j\theta_1(t)]\delta(t) + \alpha_2 \exp[j\theta_2(t)]\,\delta(t-\tau) \tag{8.45}$$

where α_1 and α_2 are independent random variables and have a Rayleigh pdf. θ_1 and θ_2 are also two independent random variables and their density functions are uniformly distributed over $[0, 2\pi]$. τ is the time delay between the two rays.

8.9.2.2 Saleh and Valenzuela Model.
Saleh and Valenzuela [30] reported the results of indoor measurements between two vertically polarized omni-directional antennas located on the same floor of a medium-sized office building. The measurements indicate that the statistics of the channel impulse response are independent of the polarizations of the transmitting and receiving antennas if there is no line-of-sight path between them. The model assumes that the multipath components arrive in clusters. The clusters and components within a cluster form a Poisson arrival process with different rates.

8.9.2.3 Lognormal At Any distance.
In [124], Cardoso and Mouliness present three conjectures about delay spread. The first is that τ_{RMS} is lognormal at any distance. This is derived from measurement data which are classified into urban, suburban, rural, and mountainous area. The second conjecture is that the median τ_{RMS} increases with distance, and the third conjecture is that τ_{RMS} tends to increase with shadow fading. τ_{RMS} can be given by

$$\tau_{RMS} = T_1 d^{\varepsilon} y \tag{8.46}$$

where T_1 is the median value of τ_{RMS} at $d = 1$ km, ε, is an exponent that lies between 0.5–1.0, and y is the lognormal variate.

8.9.2.4 SIRCIM Model. Based on measurements at 1300 MHz in five factory and other types of buildings, the piecewise functions of excess delay for the probabilities of multipath arrivals are derived as [3]:

$$P_R(T_K, S_1) = \begin{bmatrix} 1 - \dfrac{T_K}{367} & T_K < 110\,\text{ns} \\[2mm] 0.65 - \dfrac{T_K - 110}{360} & 110\,\text{ns} < T_K < 200\,\text{ns} \\[2mm] 0.22 - \dfrac{T_K - 200}{1360} & 200\,\text{ns} < T_K < 500\,\text{ns} \end{bmatrix} \qquad (8.47)$$

$$P_R(T_K, S_2) = \begin{bmatrix} 0.55 + \dfrac{T_K}{667} & T_K < 100\,\text{ns} \\[2mm] 0.08 + 0.62e^{-\frac{T_K - 100}{75}} & 100\,\text{ns} < T_K < 500\,\text{ns} \end{bmatrix} \qquad (8.48)$$

where S_1 and S_2 correspond to LOS and OLOS environments, respectively. T_K in units of nanoseconds is the excess delay at which a multipath component will arrive at the receiver and takes on values which are integer multiples of 7.8 ns.

8.9.2.5 Δ–K Model. This model [9, 119] takes into account the clustering property of paths caused by the grouping property of scatterers. The process is described by transitions between two states representing different mean arrival rates. Initially, the process starts in state 1 with a mean arrival rate $\lambda_0(t)$. If a path arrives at time t, transition is made to state 2 with a mean arrival rate $K\lambda_0(t)$. If no additional paths arrive in the interval $[t, t + \Delta]$, a transition is made back to State 1 at the end of the interval.

8.9.2.6 Discrete-Time Model. In this model [119, 125], the time axis is divided into small time intervals called bins. Each bin is assumed to contain either one multipath component or no multipath component. The possibility of more than one path in a bin is excluded. A reasonable bin size is the resolution of the specific measurement.

8.9.3   Deterministic Models of Time Delay Spread

The statistical models above are based on measurements made in a specific environment. It may not be suitable for prediction in other environments.

8.9.3.1 Ray Tracing. In theory, ray-tracing technique [12, 105, 118, 123, 127] can determine almost all multipath components, including their amplitude, time delay, and phase, and it is effective in predicting the time delay spread. When applying a ray tracing method, detailed knowledge of the environment is required. Another advantage of ray-tracing models over other propagation models is its ability to incorporate antenna radiation patterns and particularly to consider the effect of the radiation pattern on each ray individually.

Since the phase of each ray arriving at a receiver varies significantly with distance and cannot be accurately predicted, it will be impossible to achieve agreement with either the instantaneous measured shape or the value of the RMS delay spread for a single individual measurement. The comparison between measured and modeled results should be for average values in a small area around the actual mobile position [127]. A seven-ray and a 25-ray model consider only single or double reflections, respectively, and are used to simulate propagation in a room [105]. The RMS delay spread is predicted and agrees well with that calculated from measurements.

8.9.3.2 VRP Model. This model is similar to a seven-ray model but in an outdoor environment. It assumes some virtual reflection points (VRPs) located at the intersection points along the LOS on streets and at building walls [122]. It does not consider the effects of traffic and moving human. The predicted results of the *RMS* delay spreads are verified by measurements.

8.10 CONCLUSION

Propagation models are not only needed for installation guidelines, but also play a key part of any analysis or design that strives to mitigate interference. In this chapter we have surveyed some of the typical propagation models which provide good estimates for both large- and small-scale fading channels. Despite the enormous efforts to date, much work remains in the understanding and predicting the characters of mobile communications channels. In addition, an efficient ray-tracing method has been presented for tracing rays in an indoor propagation system. A FDTD method has been described to analyze wave propagation through the walls in a building.

REFERENCES

[1] H. L. Bertoni, *Radio Propagation for Modern Wireless Systems*, Prentice Hall, Upper Saddle River, NJ, pp. 90–92, 2000.

[2] J. B. Andersen, T. S. Rappaport, and S. Youshida, "Propagation Measurement and Models for Wireless Communications Channels," *IEEE Communications Magazine*, Vol. 33, No. 1, pp. 42–49, Jan. 1995.

[3]   T. S. Rappaport, S. Y. Seidel, and K. Takamizawa, "Statistical Channel Impulse Response Models for Factory and Open Plan Building Radio Communicate System Design," *IEEE Transactions on Communications*, Vol. 39, No. 5, pp. 794–807, May 1991.

[4]   W. C. Y. Lee, *Mobile Cellular Telecommunications Systems*, McGraw-Hill, New York, 1989.

[5]   B. Sklar, "Rayleigh Fading Channels in Mobile Digital Communication Systems, Part I," *IEEE Communications Magazine*, Vol. 35, No. 9, pp. 136–146, 1997.

[6]   L. C. Godara, "Applications of Antenna Arrays to Mobile Communications, Part I, Performance Improvement, Feasibility, and System Considerations," *Proceedings of the IEEE*, Vol. 85, No. 7, pp. 1031–1060, July 1997.

[7]   M. Chryssomallis, "Smart Antennas," *IEEE Antennas and Propagation Magazine*, Vol. 42, No. 3, pp. 129–136, June 2000.

[8]   G. G. Raleigh and T. Boros, "Joint Space–time Parameter Estimation for Wireless Communication Channels," *IEEE Transactions on Signal Processing*, Vol. 46, No. 5, pp. 1333–1343, May 1998.

[9]   G. L. Turin, "A Statistical Model for Urban Multipath Propagation," *IEEE Transactions on Vehicular Technology*, Vol. 21, No. 1, pp. 1–9, 1972.

[10] T. S. Rappaport, *Wireless Communications: Principles and Practice*, Prentice Hall, Upper Saddle River, NJ, 1996.

[11] M. F. Catedra, J. Perez, F. Saez de Adana, and O. Gutierrez, "Efficient Ray-Tracing Techniques for Three Dimensional Analyses of Propagation in Mobile Communications: Application to Picocell and Microcell Scenarios," *IEEE Antennas and Propagation Magazine*, Vol. 40, No. 2, pp. 15–27, Apr. 1998.

[12] S. Y. Seidel and T. S. Rappaport, "Site-Specific Propagation Prediction for Wireless In-Building Personal Communication System Design," *IEEE Transactions on Vehicular Technology*, Vol. 43, No. 4, pp. 879–891, 1994.

[13] D. Molkdar, "Review on Radio Propagation into and within Buildings," *IEE Proceedings H*, Vol. 138, No. 1, pp. 61–73, Feb. 1991.

[14] T. Okumura, E. Ohmori, and K. Fukuda, "Field Strength and Its Variability in VHF and UHF Land Mobile Service," *Review Electrical Communication Laboratory*, Vol. 16, No. 9–10, pp. 825–873, 1968.

[15] M. Hata, "Empirical Formula for Propagation Loss in Land Mobile Radio Service," *IEEE Transactions on Vehicular Technology*, Vol. 29, No. 3, pp. 317–325, 1980.

[16] J. Walfisch and H. L. Bertoni, "A Theoretical Model of UHF Propagation in Urban Environments," *IEEE Transactions on Antenas and Propagation*, Vol. 36, No. 12, pp. 1788–1796, 1988.

[17] K. Low, "Comparison of Urban Propagation Models with CW-Measurements," *IEEE Vehicular Technology Society 42nd VTS Conferenc:. Frontiers of Technology—From Pioneers to the 21st Century*, Vol. 2, pp. 936–942, 1992.

[18] S. R. Saunders and F. R. Bonar, "Explicit Multiple Building Diffraction Attenuation Function for Mobile Radio Wave Propagation," *Electronic Letters*, Vol. 27, No. 14, pp. 1276–1277, July 1991.

[19] S. R. Saunders and F. R. Bonar, "Prediction of Mobile Radio Wave Propagation over Buildings of Irregular Heights and Spacings," *IEEE Transactions on Antennas and Propagation*, Vol. 42, No. 2, pp. 137–144, 1994.

[20] M. J. Neve and G. B. Rowe, "Assessment of GTD for Mobile Radio Propagation Prediction," *Electronics Letters*, Vol. 29, No. 7, pp. 618–620, Apr. 1993

[21] Juan-Llacer and N. Cardona, "UTD Solution for the Multiple Building Diffraction Attenuation Function for Mobile Radiowave Propagation," *Electronics Letters*, Vol. 33, No. 1, pp. 92–93, Apr. 1997.

[22] L. Juan-Llacer, L. Ramos, and N. Cardona, "Application of Some Theoretical Models for Coverage Prediction in Macrocell Urban Environments," *IEEE Transactions on Vehicular Technology*, Vol. 48, No. 5, pp. 1463–1468, 1999.

[23] D. Har, A. M. Watson, and A. G. Chadney, "Comment on Diffraction Loss of Rooftop-to-Street in Cost 231–Walfich–Ikegami Model," *IEEE Transactions on Vehicular Technology*, Vol. 48, No. 5, pp. 1451–1452, 1999.

[24] T. S. Rappaport and L. B. Milstein, "Effect of Radio Propagation Path Loss on DS-CDMA Cellular Frequency Reuse Efficiency for the Reverse Channel," *IEEE Transactions on Vehicular Technology*, Vol. 41, No. 3, pp. 231–242, Aug. 1992.

[25] H. Xia, H. L. Bertoni, L. R Maciel, A. Lindsay-Stewart, and R. Rowe, "Radio Propagation Characteristics for Line-of-Sight Microcellular and Personal Communications," *IEEE Transactions on Antennas and Propagation*, Vol. 41, No. 10, pp. 1439–1447, Oct. 1993.

[26] M. J. Feuerstein, K. L Blackard, T. S. Rappaport, S. Y. Seidel, and H. H. Xia, "Path Loss, Delay Spread, and Outage Models as Functions of Antenna Height for Macrocellular System Design," *IEEE Transactions on Vehicular Technology*, Vol. 43, No. 3, pp. 487–498, Aug. 1994.

[27] N. Beaunstein, "Prediction of Cellular Characteristics for Various Urban Environments," *IEEE Antennas and Propagation Magazine*, Vol. 41, No. 6, pp. 135–145, 1999.

[28] S. Ichitsubo, T. Furuno, T. Taga, and R. Kawasaki, "Multipath Propagation Model for Line-of-Sight Street Microcells in Urban Area," *IEEE Transactions on Vehicular Technology*, Vol. 49, No. 2, pp. 422–427, 2000.

[29] T. S. Rappaport and S. Sandhu, "Radio-Wave Propagation for Emerging Wireless Personal-Communication Systems," *IEEE Antennas and Propagation Magazine*, Vol. 36, No. 5, pp. 14–24, Oct. 1994.

[30] A. A. M. Saleh and R. L. Valenzuela, "A Statistical Model for Indoor Multipath Propagation," *IEEE Journal on Selected Areas in Communications*, Vol. 5, No. 2, pp. 128–137, 1987.

[31]   R. J. C. Bultitude, "Measurement Characterization and Modeling of Indoor 800/900 MHz Radio Channels for Digital Communications," *IEEE Communications Magazine*, Vol. 5, No. 6, pp. 5–12, 1987.

[32]   F. C. Owen and C. D. Pundey, "In-Building Propagation at 900 MHz and 1650 MHz for Digital Cordless Telephone," *Proceedings of the 6th International Conference on Antennas and Propagation, ICAP'89*, Part 2, *Propagation*, pp. 276–281, 1989.

[33]   P. Valch, B. Segal, J. Lebel, and T. Pavlasek, "Cross-Floor Signal Propagation inside a Contemporary Ferro-Concrete Building at 434, 864, and 1705 MHz," *IEEE Transactions on Antennas and Propagation*, Vol. 47, No. 7, pp. 1230–1232, 1999.

[34]   S. E. Alexander, "Characterising Buildings for Propagation at 900 MHz," *Electronics Letters*, Vol. 19, No. 20, p. 860, Sept. 1983.

[35]   S. Y. Seidel and T. S. Rappaport, "914 MHz Path Loss Prediction Models for Indoor Wireless Communications in Multifloored Buildings," *IEEE Transactions on Antennas and Propagation*, Vol. 40, pp. 207–217, 1992.

[36]   T. S. Rappaport, "Characterization of UHF Multipath Radio Channels in Factory Buildings," *IEEE Transactions on Antennas and Propagation*, Vol. 37, No. 8, pp. 1058–1069, Aug. 1989.

[37]   K. Pahlavan, R. Ganesh, and T. Hotaling, "Multipath Propagation Measurements on Manufacturing Floors at 910 MHz," *Electronics Letters,* Vol. 25, No. 3, pp. 225–227, Feb. 1989.

[38]   D. A. Haand, "Indoor Wide Band Radio Wave Propagation and Models at 1.3 GHz and 4.0 GHz," *Electronics Letters,* Vol. 26, No. 21, pp. 1800–1802, Oct. 1990.

[39]   Telesis Technologies Laboratory, *Experimental License Report to FCC*, Aug. 1991.

[40]   D. M. J. Davasirvatham, "A Comparison of Time Delay Spread and Signal Level Measurements within Two Dissimilar Office Buildings," *IEEE Transactions on Antennas and Propagation*, Vol. 35, No. 3, pp. 319–324, Mar. 1987.

[41]   K. W. Cheung, J. H. M. Sau, and R. D. Murch, "A New Empirical Model for Indoor Propagation Prediction," *IEEE Transactions on Vehicular Technology*, Vol. 47, No. 3, pp. 996–1001, Aug. 1998.

[42]   W. Honcharenko, H. L. Bertoni, and J. Dailing, "Mechanisms Governing UHF Propagation on Single Floors in Modern Office Buildings," *IEEE Transactions on Antennas and Propagation*, Vol. 41, No. 4, pp. 496–504, 1992.

[43]   S. Ruiz, Y. Samper, J. Perez, R. Agusti, and J. Olmos, "Software Tool for Optimising Indoor/Outdoor Coverage in a Construction Site," *Electronics Letters*, Vol. 34, No. 22, pp. 2100–2101, Oct. 1998.

[44]   G. M. Whitman, K. S. Kim, and E. Niver, "A Theoretical Model for Radio Signal Attenuation inside Buildings," *IEEE Transactions on Vehicular Technology*, Vol. 44, No. 3, pp. 621–629, Aug. 1995.

[45] R. J. Luebbers, "Finite Conductivity Uniform GTD versus Knife Edge Diffraction in Prediction of Propagation Path Loss," *IEEE Transactions on Antennas and Propagation*, Vol. 32, No. 1, pp. 70–76, Jan. 1984.

[46] F. Ikegami, T. Takeuchi, and S. Yoshida, "Theoretical Prediction of Mean Field Strength for Urban Mobile Radio," *IEEE Transactions on Antennas and Propagation*, Vol. 39, No. 3, pp. 299–302, Mar. 1991.

[47] W. K. Tam and V. N. Tran, "Propagation Modeling for Indoor Wireless Communication," *Electronics and Communication Engineering Journal*, pp. 221–228, Oct. 1995.

[48] J. W. McKown and R. L. Hamilton, Jr., "Ray Tracing as a Design Tool for Radio Networks," *IEEE Network Magazine*, Vol. 5, pp. 27–30, Nov. 1991.

[49] S. Y. Tan and H. S. Tan, "Improved Three-Dimension Ray Tracing Technique for Microcellular Propagation," *Electronics Letters*, Vol. 31, No. 17, pp. 1503–1505, Aug. 1995.

[50] S. Y. Tan and H. S. Tan, "Propagation Model for Microcellular Communications Applied to Path Loss Measurements in Ottawa City Streets," *IEEE Transactions on Vehicular Technology*, Vol. 44, No. 2, pp. 313–317, May 1995.

[51] S. Y. Tan and H. S. Tan, "A Theory for Propagation Path-Loss Characteristics in a City-Street Grid," *IEEE Transactions on Electromagnetic Compatibility*, Vol. 37, No. 3, pp. 333 –342, Aug. 1995.

[52] M. G. Sanchez, L. de Haro, A. G. Pino, and M. Calvo, "Exhaustive Ray Tracing Algorithm for Microcellular Propagation Prediction Models," *Electronics Letters*, Vol. 32, No. 7, pp. 624–625, Mar. 1996.

[53] S. H. Chen and S. K. Jeng, "SBR Image Approach for Radio Wave Propagation in Tunnels with and without Traffic," *IEEE Transactions on Vehicular Technology*, Vol. 45, No. 3, pp. 570–578, Aug. 1996.

[54] S. H. Chen and S. K. Jeng, "An SBR/Image Approach for Radio Wave Propagation in Indoor Environments with Metallic Furniture," *IEEE Transactions on Antennas and Propagation*, Vol. 45, No. 1, pp. 98–106, Jan. 1997.

[55] F. Villanese, W. G. Scanlon, N. E. Evans, and E. Gambi, "Hybrid Image/Ray-Shooting UHF Radio Propagation Predictor for Populated Indoor Environments," *Electronics Letters*, Vol. 35, No. 21, pp. 1804–1805, Oct. 1999.

[56] M. C. Lawton and J. P. McGeehan, "The Application of a Deterministic Ray Launching Algorithm for the Prediction of Radio Channel Characteristics in Small-Cell Environments," *IEEE Transactions on Vehicular Technology*, Vol. 43, No. 4, pp. 955–969, Nov. 1994.

[57] U. Dersch and E. Zollinger, "Propagation Mechanisms in Microcell and Indoor Environments," *IEEE Transactions on Vehicular Technology*, Vol. 43, No. 4, pp. 1058–1066, Nov. 1994.

[58] K. Rizk, J. F. Wagen, and F. Gardiol, "Two-Dimensional Ray-Tracing Modeling for Propagation Prediction in Microcellular Environments,"

IEEE Transactions on Vehicular Technology, Vol. 46, No. 2, pp. 508–518, May 1997.

[59]   G. M. Whitman, K. S. Kim, and E. A. Niver, "Theoretical Model for Radio Signal Attenuation Inside Buildings," *IEEE Transactions on Vehicular Technology*, Vol. 44, No. 3, pp. 621–629, Aug. 1995.

[60]   J. H. Tarng and T. R. Liu, "Effective Models in Evaluating Radio Coverage on Single Floors of Multifloor Buildings," *IEEE Transactions on Vehicular Technology*, Vol. 48, No. 3, pp. 782–789, May 1999.

[61]   S. C. Jan and S. K. Jeng, "A Novel Propagation Modeling for Microcellular Communications in Urban Environments," *IEEE Transactions on Vehicular Technology*, Vol. 46, No. 4, pp. 1021–1026, Nov. 1997.

[62]   W. Zhang, "Fast Two-Dimensional Diffraction Modeling for Site-Specific Propagation Prediction in Urban Microcellular Environments," *IEEE Transactions on Vehicular Technology*, Vol. 49, No. 2, pp. 428–436, Mar. 2000.

[63]   H. Mokhtari and P. Lazaridis, "Comparative Study of Lateral Profile Knife-Edge Diffraction and Ray Tracing Technique Using GTD in Urban Environment," *IEEE Transactions on Vehicular Technology*, Vol. 48, No. 1, pp. 255–261, Jan. 1999.

[64]   S. C. Kim, B. J. Guarino, Jr., T. M. Willis III, V. Erceg, S. J. Fortune, R. A.Valenzuela, L. W. Thomas, J. Ling, and J. D. Moore, "Radio Propagation Measurements and Prediction Using Three-Dimensional Ray Tracing in Urban Environments at 908 MHz and 1.9 GHz," *IEEE Transactions on Vehicular Technology*, Vol. 48, No. 3, pp. 931–944, 1999.

[65]   J. H. Tarng, W. R. Chang, and B. J. Hsu, "Three-Dimensional Modeling of 900-MHz and 2.44-GHz Radio Propagation in Corridors," *IEEE Transactions on Vehicular Technology*, Vol. 46, No. 2, pp. 519–527, May 1997.

[66]   G. Liang and H. L. Bertoni, "A New Approach to 3-D Ray Tracing for Propagation Prediction in Cities," *IEEE Transactions on Antennas and Propagation*, Vol. 46, No. 6, pp. 853–863, June 1998.

[67]   G. Durgin, N. Patwari, and T. S. Rappaport, "Improved 3D Ray Launching Method for Wireless Propagation Prediction," *Electronics Letters*, Vol. 33, No. 16, pp. 1412–1413, July 1997.

[68]   G. E. Athanasiadou, A. R. Nix, and J. P. McGeehan, "A Microcellular Ray-Tracing Propagation Model and Evaluation of Its Narrow-Band and Wide-Band Predictions," *IEEE Journal on Selected Areas in Communications*, Vol. 18, No. 3, pp. 322–335, Mar. 2000.

[69]   J. P. Rossi, J. C. Bic, A. J. Levy, Y. Gabillet, and M. Rosen, "A Ray Launching Method for Radio-Mobile Propagation in Urban Area," *Proceedings of the IEEE Antennas and Propogation Symposium*, London, Ontario, Canada, Vol. 3, pp. 1540–1543, 1991.

[70]   W. M. O'Brien, E. M. Kenny, and P. J. Cullen, "An Efficient Implementation of a Three-Dimensional Microcell Propagation Tool for

Indoor and Outdoor Urban Environments," *IEEE Transactions on Vehicular Technology*, Vol. 49, No. 2, pp. 622–630, Mar. 2000.

[71] M. F. Catedra, J. Perez-Saez, F. de Adana, and O. Gutierrez, "Efficient Ray-Tracing Techniques for Three-Dimensional Analyses of Propagation in Mobile Communications: Application to Picocell and Microcell Scenarios," *IEEE Antennas and Propagation Magazine*, Vol. 40, No. 2, pp. 15–28, Apr. 1998.

[72] V. Erceg, S. J. Fortune, J. Ling, A. J. Rustako, Jr., and R. A. Valenzuela, "Comparisons of a Computer-Based Propagation Prediction Tool with Experimental Data Collected in Urban Microcellular Environments," *IEEE Journal on Selected Areas in Communications*, Vol. 15, No. 4, pp. 677–684, May 1997.

[73] J. W. Schuster and R. J. Luebbers, "Comparison of GTD and FDTD Predictions for UHF Radio Wave Propagation in a Simple Outdoor Urban Environment," *IEEE Antennas and Propagation Society International Symposium, 1997 Digest*, Vol. 3, pp. 2022–2025, 1997.

[74] G. D. Kondylis, F. DeFlaviis, G. J. Pottie, and Y. Rahmat-Samii, "Indoor Channel Characterization for Wireless Communications Using Reduced Finite Difference Time Domain," *Proceedings of the IEEE Vehicular Technology Conference, 1999*, Vol. 3, pp. 1402–1406, 1999.

[75] A. Lauer, I. Wolff, A. Bahr, J. Pamp, and J. Kunisch, "Multi-Mode FDTD Simulations of Indoor Propagation Including Antenna Properties," in *Proceedings of the IEEE 45th Vehicular Technology Conference*, Chicago, pp. 454–458, 1995.

[76] W. Ying, S. Safavi-Naini, and S. K. Chaudhuri, "A Hybrid Technique Based on Combining Ray Tracing and FDTD Methods for Site-Specific Modeling of Indoor Radio Wave Propagation," *IEEE Transactions on Antennas and Propagation*, Vol. 48, No. 5, pp. 743–754, May 2000.

[77] C. Yang, B. Wu, and C. Ko, "A Ray-Tracing Method for Modeling Indoor Wave Propagation and Penetration," *IEEE Transactions on Antennas and Propagation*, Vol. 46, No. 6, pp. 907–919, June 1998.

[78] B. De Backer, H. Borjeson, F. Olyslager, and D. De Zutter, "The Study of Wave-Propagation through a Windowed Wall at 1.8 GHz," *Proceedings of the IEEE 46th Vehicular Technology Conference: Mobile Technology for the Human Race*, Vol. 1, pp. 165–169, 1996.

[79] C. Yang and B. Wu, "Simulations and Measurements for Indoor Wave Propagation Through Periodic Structures," *Proceedings of the IEEE International Symposium Antennas and Propagation Society*, Vol. 1, pp. 384–387, 1999.

[80] Z. Sandor, L. Nagy, Z. Szabo, and T. Csaba, "3D Ray Launching and Moment Method for Indoor Propagation Purposes," *Proceedings of the 8th IEEE International Symposium on Personal, Indoor and Mobile Radio Communications: Waves of the Year 2000*, PIMRC '97, Vol. 1, pp. 130–134, 1997.

[81]   A. Neskovic and D. Paumovic, "Indoor Electric Field Level Prediction Model Based on the Artificial Neural Networks," *IEEE Communications Letters*, Vol. 4, No. 6, pp. 190–192, 2000.

[82]   K. E. Stocker, B. E. Gschwendtner, and F. M. Landstorfer, "Neural Network Approach to Prediction of Terrestrial Wave Propagation for Mobile Radio," *IEE Proceedings H: Microwaves, Antennas and Propagation*, Vol. 140, No. 4, pp. 315–320, Aug. 1993.

[83]   P. Chang and W. Yang, "Environment-Adaptation Mobile Radio Propagation Prediction Using Radial Basis Function Neural Networks," *IEEE Transactions on Vehicular Technology*, Vol. 46, No. 1, pp. 155–160, Feb. 1997.

[84]   A. A. Zaporozhets, "Application of Vector Parabolic Equation Method to Urban Radiowave Propagation Problems," *IEE Proceedings H: Microwaves, Antennas and Propagation*, Vol. 146, No. 4, pp. 253–256, Aug. 1999.

[85]   C. Brennan and P. J. Cullen, "Application of the Fast Far-field Approximation to the Computation of UHF Pathloss over Irregular Terrain," *IEEE Transactions on Antennas and Propagation*, Vol. 46, No. 6, pp. 881–890, June 1998.

[86]   R. Mazar and A. Bronshtein, "Propagation Model of a City Street for Personal and Microcellular Communications," *Electronics Letters*, Vol. 33, No. 1, pp. 91–92, Jan. 1997.

[87]   N. Blaunstein, "Average Field Attenuation in the Street Waveguide," *IEEE Transactions on Antennas and Propagation*, Vol. 46, No. 12, pp. 1782–1789, Dec.1998.

[88]   B. Chopard, P. O. Luthi, and J. F. Wagen, "Lattice Boltzmann Method for Wave Propagation in Urban Microcells," *IEE Proceedings: Antennas and Propagation*, Vol. 144, No. 4, pp. 251–255, Aug. 1997.

[89]   L. J. Greenstein, D. G. Michelson, and V. Erceg, "Moment-Method Estimation of the Ricean K-Factor," *IEEE Communications Letters*, Vol. 3, No. 6, pp. 175–176, June 1999.

[90]   V. Fung, T. S. Rappaport, and B. Thomas, "Bit Error Simulation for pi/4 DQPSK Mobile Radio Communications Using Two-Ray and Measurement-Based Impulse Response Models," *IEEE Journal on Selected Areas in Communications*, Vol. 11, No. 3, pp. 393–405, Apr. 1993.

[91]   C. C. Hess, *Handbook of Land-Mobile Radio System Coverage*, Artech House, Norwood, MA.

[92]   A. Safak, "Statistical Analysis of the Power Sum of Multiple Correlated Lognormal_Components," *IEEE Transactions on Vehicular Technology*, Vol. 42, No. 1, pp. 58–61, Feb. 1993.

[93]   H. Suzuki, "A Statistical Model for Urban Radio Propagation," *IEEE Transactions on Communications*, Vol. 25, pp. 673–680, 1977.

[94]   A. Krantzik and D. Wolf, "Analysis of a Modified Suzuki Fading Channel Model," *Proceedings of the 1989 International Conference on Acoustics,*

Speech, and Signal Processing, 1989 ICASSP-89, Vol. 4, pp. 2250–2253, 1989.

[95]   M. Patzold, U. Killat, and F. Laue, "A Deterministic Digital Simulation Model for Suzuki Processes with Application to a Shadowed Rayleigh Land Mobile Radio Channel," *IEEE Transactions on Vehicular Technology,* Vol. 45, No. 2, pp. 318–331, May 1996.

[96]   IEEE Vehicular Technology Society Committee on Radio Propagation, "Coverage Prediction for Mobile Radio Systems Operating in the 800/900 MHz Frequency Range," *IEEE Transactions on Vehicular Technology,* Vol. 37, No. 1, pp. 3–72, Feb. 1988.

[97]   K. Yip and T. Ng, "A Simulation Model for Nakagami-m Fading Channels, $m > 1$," *IEEE Transactions on Communications,* Vol. 48, No. 2, pp. 214–221, Feb. 2000.

[98]   F. Vatalaro, "Generalized Rice-Lognormal Channel Model for Wireless Communications," *Electronics Letters,* Vol. 31, No. 22 , pp. 1899–1900, Oct. 1995.

[99]   Y. Karasawa and H. Iwai, "Modeling of Signal Envelope Correlation of Line-of-Sight Fading with Applications to Frequency Correlation Analysis," *IEEE Transactions on Communications,* Vol. 42 No. 6, pp. 2201–2203, June 1994.

[100] T. T. Tjhung and C. C. Chai, "Fade Statistics in Nakagami-Lognormal Channels," *IEEE Transactions on Communications,* Vol. 47, No. 12, pp. 1769–1772, Dec. 1999.

[101] A. Abdi and M. Kaveh, "K Distribution: An Appropriate Substitute for Rayleigh-Lognormal Distribution in Fading-Shadowing Wireless Channels," *Electronics Letters,* Vol. 34, No. 9, pp. 851–852, Apr. 1998.

[102] G. E. Corazza and F. Vatalaro, "A Statistical Model for Land Mobile, Satellite Channels and Its Application to Nongeostationary Orbit Systems," *IEEE Transactions on Vehicular Technology,* Vol. 43, pp. 738–741, Aug. 1994.

[103] P. G. Babalis and C. N. Capsalis, "Impact of the Combined Slow and Fast Fading Channel Characteristics on the Symbol Error Probability for Multipath Dispersionless Channel Characterized by a Small Number of Dominant Paths," *IEEE Transactions on Communications,* Vol. 47, No. 5, pp. 653–657, May 1999.

[104] D. S. Polydorou and C. N. Capsalis, "A New Theoretical Model for the Prediction of Rapid Fading Variations in Indoor Environment," *IEEE Transactions on Vehicular Technology,* Vol. 46, pp. 748–755, Aug. 1997.

[105] R. J. C. Bultitude, P. Melancon, H. Zaghloul, G. Morrison, and M. Prokki, "The Dependence of Indoor Radio Channel Multipath Characteristics of Transmit/Receiver Ranges," *IEEE Journal on Selected Areas in Communications,* Vol. 11, No. 7, pp. 979–990, Sept. 1993.

[106] D. M. J. Devasirvatham, R. R. Murray, and C. Banerjee, "Time Delay Spread Measurements at 850 MHz and 1.7 GHz inside a Metropolitan Office Building," *Electronics Letters,* Vol. 25, No. 3, pp. 194–196, Feb. 1989.

[107] H. Zaghbul, M. Fattouche, G. Morrison, and D. Tholl, "Comparison of Indoor Propagation Channel Characteristics at Different Frequencies," *Electronics Letters*, Vol. 27, No. 22, pp. 2077–2079, Oct. 1991.

[108] G. J. M. Janssen, P. A. Stigter, and R. Prasad, "Wideband Indoor Channel Measurements and BER Analysis of Frequency Selective Multipath Channels at 2.4, 4.75, and 11.5 GHz," *IEEE Transactions on Communications*, Vol. 44, No. 10, pp. 1272–1288, Oct. 1996.

[109] K. Pahlavan, R. Ganesh, and T. Hotaling, "Multipath Propagation Measurements on Manufacturing Floors at 910 MHz," *Electronics Letters*, Vol. 25, No. 3, pp. 225–227, Feb. 1989.

[110] D. A. Hawbaker and T. S. Rappaport, "Indoor Wideband Radiowave Propagation Measurements at 1.3 GHz and 4.0 GHz," *Electronics Letters*, Vol. 26, No. 21, pp. 1800–1802, Oct. 1990.

[111] L. Talbi and G. Y. Delisle, "Experimental Characterization of EHF Multipath Indoor Radio Channels," *IEEE Journal on Selected Areas in Communications*, Vol. 14, No. 3, pp. 431–440, Apr. 1996.

[112] R. J. C. Bultitude and G. K. Bedal, "Propagation Characteristics on Microcellular Urban Mobile Radio Channels at 910 MHz," *IEEE Journal on Selected Areas in Communications*, Vol. 7, No. 1, pp. 31–39, Jan. 1989.

[113] R. J. C. Bultitude, R. F. Hahn, and R. J. Davies, "Propagation Considerations for the Design of an Indoor Broad-Band Communications System at EHF," *IEEE Transactions on Vehicular Technology*, Vol. 47, No. 1, pp. 235–245, Feb. 1998.

[114] S. Ichitsubo, T. Furuno, T. Taga, and R. Kawasaki, "Multipath Propagation Model for Line-of-Sight Street Microcells in Urban Area," *IEEE Transactions on Vehicular Technology*, Vol. 49, pp. 422–427, Mar. 2000.

[115] S. C. Kim, H. L. Bertoni and M. Stern, "Pulse Propagation Characteristics at 2.4 GHz inside Buildings," *IEEE Transactions on Vehicular Technology*, Vol. 45, No. 3, pp. 579–592, Aug. 1996.

[116] T. R. Liu and J. H. Tarng, "Modeling and Measurement of 2.44 GHz Radio Out-Of-Sight Propagation on Single Floors," *Microwave and Optical Technology Letters*, Vol. 14, No. 1 pp. 56–59, Jan. 1997.

[117] R. Ganesh and K. Pahlavan, "Statistics of Short Time and Spatial Variations Measured in Wideband Indoor Radio Channels," *IEE Proceedings H: Antennas and Propagation*, Vol. 140, No. 4, pp. 297–302, Aug. 1993.

[118] M. C. Lawton and J. P. McGeehan, "The Application of a Deterministic Ray Launching Algorithm for the Prediction of Radio Channel Characteristics in Small-Cell Environments," *IEEE Transactions on Vehicular Technology*, Vol. 43, No. 4, pp. 955–969, Nov. 1994.

[119] H. Hashemi, "Impulse Response Modeling of Indoor Radio Propagation Channels," *IEEE Journal on Selected Areas in Communications*, Vol. 11, No. 7, pp.967–978, Sept. 1993.

[120] S. Y. Seidel, T. S. Rappaport, S. Jain, M. L. Lord, and R. Singh, "Path Loss, Scattering and Multipath Delay Statistics in Four European Cities for Digital Cellular and Microcellular Radiotelephone," *IEEE Transactions on Vehicular Technology*, Vol. 40, No. 4, pp. 721–730, Nov. 1991.

[121] T. S. Rappaport and S. Y. Seidel, "900 MHz Multipath Propagation Measurements in Four States' Cities," *Electronics Letters*, Vol. 25, No. 15, pp. 956–958, July 1989.

[122] T. Taga, T. Furuno, and K. Suwa, "Channel Modeling for 2-GHz-band Urban Line-of-Sight Street Microcells," *IEEE Transactions on Vehicular Technology*, Vol. 48, No. 1, pp. 262–272, Jan. 1999.

[123] H. Son and N. Myung, "A Deterministic Ray Tube Method for Microcellular Wave Propagation Prediction Model," *IEEE Transactions on Antennas and Propagation*, Vol. 47, No. 8, pp. 1344–1350, Aug. 1999.

[124] L. J. Greenstein, V. Erceg, Y. S. Yeh, and M. V. Clark, "A New Path-Gain/Delay-Spread Propagation Model for Digital Cellular Channels," *IEEE Transactions on Vehicular Technology*, Vol. 46, No. 2, pp. 477–485, May 1997.

[125] R. Ganesh and K. Pahlavan, "Statistical Modelling and Computer Simulation of Indoor Radio Channel," *IEE Proceedings I: Communications, Speech and Vision*, Vol. 138, No. 3, pp. 153–161, June 1991.

[126] R. A. Valenzula, O. Landron, and D. L. Jacobs, "Estimating Local Mean Signal Strength of Indoor Multipath Propagation," *IEEE Transactions on Vehicular Technology,* Vol. 46, No. 1, pp. 203–212, Feb. 1997.

[127] G. E. Athanasiadou, A. R. Nix, and J. P. McGeehan, "A Microcellular Ray-Tracing Propagation Model and Evaluation of Its Narrow-Band and Wide-Band Predictions," *IEEE Journal on Selected Areas in Communications*, Vol. 18, No. 3, pp. 322–335, Mar. 2000.

[128] Z. Ji, B. H. Li, H. X. Wang, H. Y. Chen, and Y. G. Zhou, "An Improved Ray-Tracing Propagation Model for Predicting Path Loss on Single Floors," *Microwave and Optical Technology Letters*, Vol. 22, No. 1, pp. 39–41, 1999.

[129] W. K. Tam and V. N. Tran, "Multi-ray Propagation Model for Indoor Wireless Communications," *Electronics Letters*, Vol. 32, No. 2, pp. 135–137, 1996.

[130] K. R. Chang and H. T. Kim, "Improvement of the Computation Efficiency for a Ray-Launching Model," *IEE Proeedings: Microwave Antennas and Propagation*, Vol. 145, No. 4, pp. 303–308, 1997.

[131] T. B. Gibson and D. C. Jenn, "Prediction and Measurement of Wall Insertion Loss," *IEEE Transactions on Antennas and Propagation*, Vol. 47, No. 1, pp. 55–57, 1999.

[132] J. A. Kong, *Theory of Electromagnetic Waves*, John Wiley, New York, 1975.

[133] G. Ghobadi, P. R. Shepherd, and S. R. Pennock, "2D Ray-Tracing Model for Indoor Radio Propagation at Millimeter Frequencies, and the Study of

Diversity Techniques," *IEE Proceedings: Microwave Antennas and Propagation*, Vol. 145, No. 4, pp. 349–353, 1998.

[134] W. Honcharenko and H. L. Bertoni, "Transmission and Reflection Characteristics at Concrete Block Walls in the UHF Bands Proposal for Future PCs," *IEEE Transactions on Antennas and Propagation*, Vol. 42, No. 2, pp. 232–239, Feb. 1994.

[135] C. L. Holloway and P. L. Perini, "Analysis of Composite Walls and Their Effects on Short-Path Propagation Modeling," *IEEE Transactions on Vehicular Technology*, Vol. 46, No. 3, pp. 730–738, Aug. 1997.

[136] Z. Ji, B. H. Li, H. X. Wang, H. Y. Chen, and T. K. Sarkar, "Efficient Ray-Tracing Methods for Propagation Prediction for Indoor Wireless Communications," *IEEE Antennas and Propagation Magazine*, Vol. 43, No. 2, pp. 41–49, Apr. 2001.

[137] Z. Ji, T. K. Sarkar, and B. H. Li, "Analysis of the Effects of Walls on Indoor Wave Propagation Using the FDTD Method," *Microwave and Optical Technology Letters*, Vol. 29, No. 1, pp. 19–21, 2001.

[138] C. A. Zelley and C. C. Constantinou, "A Three-Dimensional Parabolic Equation Applied to VHF/UHF Propagation over Irregular Terrain," *IEEE Transactions on Antennas and Propagation*, Vol. 47, No. 10, pp. 1586–1596, 1999.

[139] D. J. Donohue and J. R. Kuttler, "Propagation Modeling over Terrain Using the Parabolic Wave Equation," *IEEE Transactions on Antennas and Propagation,* Vol. 48, No. 2 , pp. 260–277, 2000.

[140] Y. P. Zhang, Y. Hwang and J. D. Parsons, "UHF Radio Propagation Characteristics in Straight Open-Groove Structures," *IEEE Transactions on Vehicular Technology*, Vol. 48, No. 1, pp. 249–254, 1999.

[141] D. Har, H. H. Xia, and H. L. Bertoni, "Path-Loss Prediction Model for Microcells," *IEEE Transactions on Vehicular Technology*, Vol. 48, No. 5, pp. 1453–1462, 1999.

[142] H. L. Bertoni, W. Honcharenko, L. R. Macel, and H. H. Xia, "UHF Propagation Prediction for Wireless Personal Communications," *Proceedings of the IEEE*, Vol. 82, No. 9, pp. 1333–1359, 1994.

[143] D. Ullmo and H. U. Baranger, "Wireless Propagation in Buildings: A Statistical Scattering Approach," *IEEE Transactions on Vehicular Technology*, Vol. 48, No. 3, pp. 947–955, 1999.

[144] S. Kozono and A. Taguchi, "Mobile Propagation Loss and Delay Spread Characteristics with a Low Base Station Antenna on an Urban Road," *IEEE Transactions on Vehicular Technology*, Vol. 42, No. 1, pp. 103–109, 1993.

[145] S. Obayashi and J. Zander, "A Body-shadowing Model for Indoor Radio Communication Environments," *IEEE Transactions on Antennas and Propagation*, Vol. 46, No. 6 , pp. 920–927, 1998.

[146] F. Babich and G. Lombardi, "Statistical Analysis and Characterization of the Indoor Propagation Channel," *IEEE Transactions on Communications*, Vol. 48, No. 3, pp. 455–464, 2000.

[147] H. Hashemi and D. Tholl, "Statistical Modeling and Simulation of the RMS Delay Spread of Indoor Radio Propagation Channels," *IEEE Transactions on Vehicular Technology*, Vol. 43, No. 1, pp. 110–120, 1993.

[148] J. A. Wepman, J. R. Hoffman, and L. H. Loew, "Analysis of Impulse Response Measurements for PCS Channel Modelling Applications," *IEEE Transactions on Vehicular Technology*, Vol. 44, No. 3, pp. 613–620, 1995.

[149] C. L. Holloway, M. G. Cotton, and P. McKenna, "A Model for Predicting the Power Delay Profile Characteristics inside a Room," *IEEE Transactions on Vehicular Technology*, Vol. 48, No. 4, pp. 1110–1120, 1999.

[150] P. E. Driessen, "Prediction of Multipath Delay Profiles in Mountainous Terrain," *IEEE Journal on Selected Areas in Communications*, Vol. 18, No. 3, pp. 336–346, 2000.

[151] Y. Li, "A Theoretical Formulation for the Distribution Density of Multipath Delay Spread in a Land Mobile Radio Environment," *IEEE Transactions on Vehicular Technology*, Vol. 43, No. 2, pp. 379–388, 1994.

9

METHODS FOR OPTIMIZING THE LOCATION OF BASE STATIONS FOR INDOOR WIRELESS COMMUNICATION

SUMMARY

When designing wireless communication systems, it is very important to know the optimum locations for the base station antennas. In this chapter, a mathematical framework has been developed to set up an optimization problem based on the propagation models described in Chapter 8, the solution of which provides information for the optimum location of base station antennas. This methodology is particularly suited for characterizing propagation in an indoor environment. Several methods for optimization of the cost function are presented and the final results are compared with each other. This technique can be applied for the design and planning of optimum locations for base station antennas suitable for an indoor wireless communication environment. Two numerical examples are presented to illustrate the application of this methodology.

9.1 INTRODUCTION

Indoor wireless communication has been a subject of intense investigation. There has been considerable interest in prediction of the propagation mechanisms for an indoor environment [1–4]. Most of the common propagation models currently in vogue have been introduced in Chapter 8. All these models are based on *a priori* knowledge on the location of the base stations or the transmitters.

The infrastructure and planning complexity of an indoor wireless communication system is closely related to the number of base stations required to achieve the desired level of coverage and capacity. Finding the precise location of a base station is normally a heuristic task, and numerous factors must be taken into account. A few examples of such a methodology may be seen in [5–8].

The optimum indoor location of a base station or a transmitter can be posed as a nonlinear programming problem. This nonlinear programming problem can be formulated in the following fashion:

$$\text{Minimize}: \quad f[x] \quad x \in E^n \qquad (9.1)$$

subject to J linear and/or nonlinear equality constraints

$$h_j(x) = 0 \quad j = 1, 2, ..., k \qquad (9.2)$$

and $(p - k)$ linear and/or nonlinear inequality constraints:

$$g_j(x) \geq 0 \quad j = k + 1, k + 2, ..., p \qquad (9.3)$$

where $f([x])$ is the objective or the cost function to be optimized. This describes the salient features of the system and the variables $[x] = [x_1, x_2, ..., x_n]^T$ are the design parameters that need to be optimized. The constraints among the variables are expressed through equations (9.2) and (9.3).

In this chapter we consider the spatial coordinates of the location of the base station as the variables to be optimized. By making use of a propagation model which involves the parameters for the path loss, a cost function showing the coverage of the system of interest can thus be defined. Simulation results are presented utilizing a number of schemes to illustrate the principles of this concept.

9.2   DEFINITION OF THE COST FUNCTION

The cost function to be minimized for the solution of the optimum location of a base station antenna describes the performance of the system of interest and utilizes the spatial locations for the base station as variables. Many factors, such as coverage or path loss [5, 6], quality of service [7], and multiple access capability [8], can be used either by itself or in combination with one another as representative cost functions. Here, we choose the path loss as the main measure of the cost function in our model and we reduce the problem to two dimensions in order to reduce the computational complexity. However, the method can easily be extended to three dimensions.

In our simulations we have chosen the propagation model that has been suggested in [9] as a mathematical function numerically defining the degree of coverage. The path loss at any point d can be calculated from

$$PL(d)[dB] = 10 \log\left(\frac{d}{d_0}\right)^{n_1} U(d_{bp} - d) + 10\left[\log\left(\frac{d_{bp}}{d_0}\right)^{n_1}\right.$$

$$\left. + \log\left(\frac{d}{d_{bp}}\right)^{n_2}\right] U(d - d_{bp}) \qquad (9.4)$$

$$+ \sum_{p=1}^{P} WAF(p) + \sum_{q=1}^{Q} FAF(q)$$

where d_{bp} is the distance of the breakpoint from the transmitter and has been assumed to be 10m. n_1 and n_2 are the path loss exponents on either side of the breakpoint. In general, n_1 should be close to the free-space value of 2.0. n_2 should be greater than 2.0 since it represents propagation in which obstruction occurs in the first Fresnel zone. Here it has been assumed to be 2.5 for all locations. $U(\bullet)$ is the unit step function. $WAF(p)$ and $FAF(q)$ are the attenuation factors for the wall and floor, respectively. Because we only consider the path loss in two dimensions, the attenuation term related to the floor is omitted and the last term in (9.4) is not used in our calculation. WAF is assumed to be 5 dB for hollow plasterboard walls. The path loss is calculated with respect to the variable $d_0 = 1m$. Therefore, the path loss will be the sum of the actual path loss plus the path loss due to the term $d_0 = 1m$, which is assumed to be 37.10 dB.

A threshold for the path loss is introduced as a measure of the quality of the signal. If the path loss at one spatial location is less than an *a priori* specified threshold value, the signal level is considered to be of good quality.

In order to generate a comprehensive model for this problem, a mathematical representation for the radio coverage must be developed over the design space. By considering the various possible and probable scenarios, a convex cost function similar to that defined in [6] can be formed, consisting of the following three terms, which are given by

$$f = \frac{1}{N} \sum_{i=1}^{N} w_i \min_{j=1,\dots,M} (PL_{ij}) + \alpha f_1 + (1-\alpha) f_2 \qquad (9.5)$$

where $f_1 = \frac{1}{N} \sum_{i=1}^{N} w_i p_i D_i$, $f_2 = \max_{i=1,\dots,N} \{w_i p_i D_i\}$, α is a trade-off factor

for f_1 and f_2, $D_i = \max[0, (\min_{j=1,\dots,B} PL_{ij} - t)]$, PL_{ij} is the path loss associated with the ith receiver location related to the jth base station, t is the threshold value of the signal that is acceptable, N is the total number of receivers to be located in the designed areas, B is the number of base stations, and w_i and p_i are the relative priority weight and penalty factor, respectively, associated with the signal at the ith receiver. The weights chosen are a function of the problem.

The first term on the right-hand side of equation (9.5) represents the average weighted path loss of the transmitted signal. The minimization of this term will improve the overall quality of coverage. The minimization of the term f_1 will improve the average path loss that does not meet the corresponding threshold. Optimization of f_2 will ensure that even the worst locations for the antennas enjoy an acceptable level of signal coverage.

9.3   SURVEY OF OPTIMIZATION METHODS

There are many methods that can be used to solve the nonlinear programming problem. Some of these minimization methods require the information for the derivatives of the cost function, while others do not. The cost function to be minimized, in general, is quite complicated and it is difficult to obtain its derivatives analytically. Therefore, a finite-difference approximation is often used to evaluate the derivatives numerically. The various methods that have been applied in our optimization methodology are explained next. To simplify the problem, no constraints have been imposed on the optimization procedure. Details of these optimization methods can be obtained from [10]. One can also get the Fortran computer codes for some of these methods from the accompanying CD associated with this reference. The various optimization methods considered in this section are now summarized.

• **Steepest descent method.** In this method, the gradient of the cost function $f([x])$ at any point $[x]$ is a vector along a local direction of the greatest increase in the value of the function. Clearly, then, one might proceed in the direction opposite the gradient of $f([x])$, that is, in the direction of steepest descent. This method is simple to apply, but sometimes it may converge to the position of a local minimum.

• **BFGS method (quasi-Newton method).** A quasi-Newton method, which was developed by Broyden, Fletcher, Goldfarb, and Shanno, generates a sequence of points which tend to find the local minimum for $f([x])$ at step q in some open set by using the following recipe:

$$x_{q+1} = x_q - [H]_q \nabla f([x]_q) t_q \quad q = 0, 1, \ldots \quad (9.6)$$

where $[H]_q$ is an approximation of the inverse of the Hessian matrix at a local unconstrained minimum and the scalar t_q represents the step size. Different algorithms compute $[H]_q$ in different ways. The BFGS method is one of the effective quasi-Newton methods. The computer program in MATLAB describes the numerical implementation of this technique through the function *fminu*. One can get the details from [11].

• **Simplex method.** By far the most popular direct search method is the simplex technique. A simplex in a ς -dimensional space constitutes a polyhedron with $\varsigma + 1$ vertices. The objective at each iteration is to create a new simplex in which the previous vertex containing an unsuitable numerical value for the cost function is replaced. The simplex is altered by reflection, expansion, or contraction, depending on whether the value for the functional f is improving along the right direction. The description of the method and the computer code for this method is available in MATLAB [11] through the function *fmins*.

• **Hooke and Jeeves's method** [10]. This is another kind of direct search optimization method. This algorithm is a derivative-free multidimensional search technique that employs two types of directions, namely, exploratory directions and pattern-search directions, in order to generate updated values for the parameters to be optimized. Pattern search can be customized to cast a wide net when searching for better points, so that the search may be able to escape a shallow basin that contains a local minimum and find a deeper basin that hopefully converges to the global minimum.

• **Rosenbrock's method** [10]. This method rotates the parameter space (the space for the unknown variables) and aligns one axis with a ridge; all the other axes will remain orthogonal to this axis. This is a form of a conjugate direction method where the path length along the search directions is chosen through trial and error. If the cost function is unimodal and has detectable ridges pointing toward the minimum of the function, this method will proceed with surefooted accuracy towards the minimum of the function. It has turned out that this simple approach is more stable than many sophisticated algorithms, and it requires much less calculations of the objective function over other techniques.

• **Simulated annealing method** [10]. This method is so-named because the optimization process strongly resembles the annealing process for creating a durable metal by systematic cooling and reheating. Simulated annealing can avoid getting stuck in a local minimum. This optimization strategy is based on a stochastic search, and is quite suitable when one is searching a functional which has only one deep minimum. However, when the functional has a number of deep minima, the next algorithm is more suitable.

• **Genetic algorithm** [10]. This is an automatic search technique based on the principles of natural genetics. It begins with a population of encoded individuals. The individual, commonly referred to as a *chromosome*, or a *string*, is a computer-compatible representation of a functional related to an optimization problem. A successive generation of the population is created by applying the genetic operators. The primary operators involved in the methodology of the genetic algorithms are selection and recombination. Mutation is the process of randomly changing encoded bits of information for a newly created population of individuals. Mutation is known as an insurance policy to maintain search diversity within the population. Typically, mutation is performed with a small frequency of occurrence. The process of applying genetic operators continues to create increasing numbers of high-fitness individuals within the population. This process continues for each successive generation. When using a genetic algorithm, the initial values for the parameters to be optimized are not too critical. One may begin by providing a set of random values as an initial guess for the variables. A computer code for this method in Matlab can be downloaded from the Web site: *http://www.ie.ncsu.edu/mirage/GAToolBox/gaot.*

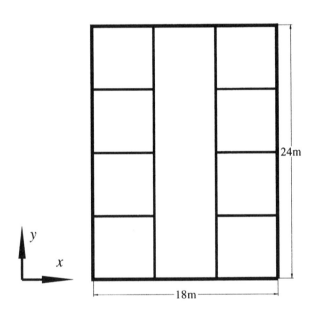

Figure 9.1. Physical layout of the floor for the first example.

9.4 NUMERICAL SIMULATIONS

Two topologies related to the location of base stations have been tested by using the various optimization methods. The first example is illustrated through Figure 9.1. Only one optimum location for the base station is considered in this case. The second example deals with the location of three base stations on the third floor of Link Hall at Syracuse University, as shown in Figure 9.2. For both of these simulations, the initial positions for the base stations have been chosen arbitrarily and are used as the same initial guess for all the optimization methods.

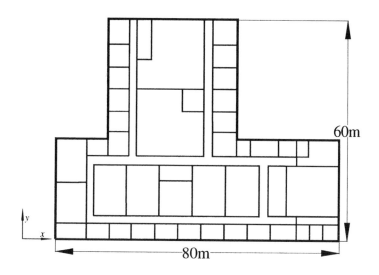

Figure 9.2. Layout of the third floor of Link Hall.

Table 9.1 tabulates the simulation results for both examples when the initial position for the base station has been chosen arbitrarily. In Table 9.1, "Cost F Eval" represents the number of evaluations of the cost function. Hence, it is a good estimate for the computation time for each of the optimization procedures. For the first example, because of the physical symmetry in the layout of the floor, the best location for the base station should be at the center of the floor, whose coordinate locations are given by (9.0, 12.0). It can be concluded from the numerical results seen in Table 9.1 that most of the methods do converge to the correct position, except the method of steepest decent. The solution may have converged to perhaps a local minimum of the functional.

For the second example, the number of search variables increases as the topology becomes a little complex. The simplex, BFGS, and Rosenbrock methods failed to get reasonable results after many iterations. The Hook and Jeeves method, simulated annealing method, and genetic algorithm provided almost the same numerical results.

From these results, we can see that different optimization procedures required different amounts of computation time to yield the same best spatial location for the transmitter, starting with the same initial guess. A suitable initial choice of the parameters to be optimized for some methods, such as simulated annealing, can reduce the number of evaluations of the cost function. It is seen that optimization methods that make use of the information for the derivatives are not necessarily suitable because the cost function is not analytical. Moreover, it may mislead the numerical procedure and converge to a local minimum.

TABLE 9.1
Results of a numerical simulation

Method	First Example		Second Example	
	Optimal Position (x, y)	Cost F Eval.	Optimal Position (3 base stations) $(x_1, y_1); (x_2, y_2); (x_3, y_3)$	Cost F Eval.
Steepest descent	(6.05, 9.12)	146	(2.86, 1.29); (21.11, 39.15); (19.26, 6.28)	311
BFGS	(9.00, 12.00)	1073	(20.87, 22.49); (7.04, 6.09); (17.93 , 8.14)	555
Simplex	(9.00, 12.00)	167	(7.06, 5.34); (48.11, 5.32); (33.24, 46.25)	429
Hook and Jeeves	(8.99, 12.00)	118	(5.45, 8.25); (63.80, 8.62); (22.79, 41.15)	658
Rosenbrock	(9.00, 12.00)	468	(2.06, 4.90); (28.01, 29.55); (9.76, 5.78)	458
Simulated annealing	(9.00, 12.00)	601	(6.07, 6.56); (59.45, 6.67); (22.73, 41.15)	465
Genetic algorithm	(9.00, 12.00)	871	(6.01, 6.73); (60.65, 7.28); (22.34, 40.23)	901

9.5 CONCLUSION

In this chapter we addressed the problem of optimizing the location of the base station in an indoor environment. The cost function is used as a measure of quality radio coverage of the system to be designed. Using an indoor wireless propagation prediction model, several optimization methods have been used to solve the problem. Some of the methods require fewer function evaluations to converge to an optimal solution. However, for the two specific examples presented, the performance of simulated annealing appeared to be the optimum.

REFERENCES

[1]   H. Hashemi, "The Indoor Radio Propagation Channel," *Proceedings of the IEEE*, Vol. 81, No. 7, pp. 943–968, 1993.

[2]   D. Molkdar, "Review on Radio Propagation into and within Buildings," *IEE Proceedings: Microwaves, Antennas and Propagation*, Vol. 138, No. 1, pp. 61–73, 1991.

[3]   S. Y. Seidel and T. S. Rappaport, "914 MHz Path Loss Prediction Models for Indoor Wireless Communications in Multifloored Buildings," *IEEE Transactions on Antennas and Propagation*, Vol. 40, No. 2, pp. 207–17, 1992.

[4]   W. Honcharenko, H. L. Bertoni, and J. Dailing, "Mechanisms Governing Propagation between Different Floors in Buildings," *IEEE Transactions on Antennas and Propagation*, Vol. 41, No. 6, pp. 787–790, 1993.

[5]   S. F. Fortune, D. M. Gay, B. W. Kernighan, and O. Landron, "WISE Design of Indoor Wireless Systems: Practical Computation and Optimization," *IEEE Computation Science and Engineering*, pp. 58–68, Spring 1995.

[6]   H. D. Sherali, C. M. Pendyla, and T. S. Rappaport, "Optimal Location of Transmitters for Micro-cellular Radio Communication System Design," *IEEE Journal on Selected Areas in Communications*, Vol. 14, No. 4, pp. 662–673, 1996.

[7]   K. W. Cheung and R. D. Murch, "Optimizing Indoor Base-Station Locations in Coverage- and Interference-Limited Indoor Environments," *IEE Proceedings: Communications*, Vol. 145, No. 6, pp. 445–450, 1998.

[8]   D. Stamatelos and A. Ephremides, "Spectral Efficiency and Optimal Base Placement for Indoor Wireless Networks," *IEEE Journal on Selected Areas in Communications*, Vol. 14, No. 4, pp. 651–661, 1996.

[9]   K. W. Cheung, J. H. M. Sau, and R. D. Murch, " A New Empirical Model for Indoor Propagation Prediction," *IEEE Transactions on Vehicular Technology*, Vol. 47, No. 3, pp. 996–1001, 1998.

[10]   A. D. Belegundu and T. R. Chandrupatla, *Optimization Concepts and Applications in Engineering*, Chap. 7, Prentice Hall, Upper Saddle River, NJ, 1999.

[11]   Mathworks, *MATLAB*, Version 5.1, Natick, MA, 1997.

10

IDENTIFICATION AND ELIMINATION OF MULTIPATH EFFECTS WITHOUT SPATIAL DIVERSITY

SUMMARY

In wireless communication, fading occurs due to the destructive interference between the signal and its various multipath components that arrive at the receiver with or without significant attenuation and time delay. Channel equalizers through adaptive filters are useful when broadband signals are used for communication. For narrowband signals the effects of multipaths are overcome through diversity, for example through the use of polarization diversity, space diversity, and so on. In this chapter we propose the use of frequency diversity, within the framework of narrowband signals, for identification and elimination of multipaths. This procedure is useful when there is no opportunity to have spatial diversity in the form of a phased array, and in addition, the locations of the transmitter and the receivers are fixed. This may also be useful in antenna pattern measurements in an anechoic chamber or in an open field measurement where the reflected rays from the environment may distort the actual scenario. The procedure consists of sending a set of narrowband tones prior to the transmission of the desired signal, and then the Matrix Pencil method is used to determine the multipath components. Simulation results illustrating the performance of the proposed technique are presented.

10.1   INTRODUCTION

In the area of communication, radar, sonar, processing of seismic data, and so on, one gathers information by transmitting a signal and then uses a receiver to collect the reflected transmitted signal. From the analysis of the reflected signal, it is possible to estimate the properties of the environment. Unfortunately, in many practical situations, the received signal, in general, is accompanied by interferences. The signal that is typically received in a receiving antenna consists of the direct received signal along with some of its replicas, which are attenuated and delayed in the time domain or shifted by a phase component for narrowband signals in the frequency domain. This interference can produce fading, which makes it impossible to extract the correct desired information from this mixture of the signal and its various multipath components. Therefore, it is very important to identify and then eliminate these undesired multipath effects.

An alternative way to solve for wideband channel impairments is through the use of equalizers or in some situations by Rake receivers. Rake receivers attempt to increase the channel gain by combining the various multipaths into one. In this book we are primarily concerned with narrowband systems, as for wideband signals it is necessary to take into account the impulse responses of the transmitting and receiving antennas in order to achieve perfect channel characterization. Moreover, since the impulse responses of the transmitting and receiving antennas are a function of the azimuth angle and physical locations, it becomes too cumbersome to electrically characterize them in any realistic environment. A transmitting antenna of finite size double-differentiates the temporal waveform, whereas a receiving antenna does a single differentiation of the waveshape incident on it. Furthermore, their responses are a function of the azimuth and elevation angles of the waves transmitted and received, respectively. These points have been brought out in Chapter 2. Hence, we will not consider broadband signals, as it is impossible to know *a priori* in a practical situation the values of the azimuth and elevation angles!

If the receiving system has an array of antennas, spatial diversity is provided to combat multipath effects. However, even though there may be some points in space where the interference between the direct and reflected rays are almost destructive, at some other points it may not be so. Hence, through spatial diversity it may be possible to eliminate the effects of multipath components of the signal. But if the receiver consists only of a single antenna, there is no spatial diversity. Then, it is necessary to use alternative methods that can identify and eliminate the multipath effects.

10.2 RECEIVED SIGNAL MODEL WITHOUT SPATIAL DIVERSITY

Consider the case of a wireless communication system as illustrated by Figure 10.1. The transmitter located at position A is transmitting a signal (energy) in all directions. Again we focus our attention on the narrowband case. The receiver is located at position B and receives the direct ray from it and several reflected rays causing interference or fading at the receiver B. The measured signal at B is therefore given by

$$y(f_0) = C \exp\left\{\frac{j 2\pi d f_0}{c}\right\} + \sum_{i=1}^{M} \alpha_i C \exp\left\{\frac{j 2\pi d_i f_0}{c}\right\} + \text{noise} \qquad (10.1)$$

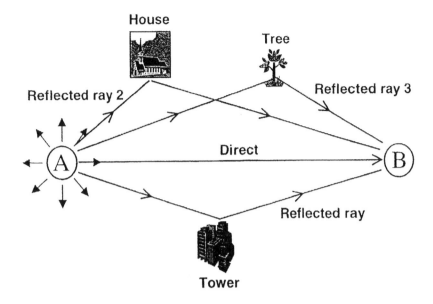

Figure 10.1. Direct and multipath effects in a communication system without spatial diversity.

where $y(f_0)$ is the complex narrowband signal that is received at the receiver, C and f_0 are the amplitude and the frequency of the transmitted signal, respectively. c is the velocity of the light, d is the path length of the direct ray (that corresponds to the distance between the transmitter and the receiver), M is the number of multipaths, and α_i and d_i are the reflection coefficient and the path length for the ith multipath, respectively. The additive noise is assumed to be white, Gaussian, zero mean, with variance σ^2. Next, we describe the technique to identify the multipath effects and then eliminate them.

Before we transmit the desired signal, we first send N tones, which are slightly shifted by Δf. At the receiver we measure the values of the signals at all the frequencies $y\left\{f_0 + \left(q - \dfrac{N-1}{2}\right)\Delta f\right\}$; for $q = 0, 1, \ldots, N-1$ (N is assumed to be an odd number). Here it is assumed that N is greater than twice the number of the multipath components and $N\Delta f$ is still small, so that $f_0 + N\,\Delta f$ is essentially a narrowband signal $\left(1\% \leq \dfrac{N\,\Delta f}{f_0} \leq 10\%\right)$.

The second step consists in a determination of the exponents in equation (10.1). This is carried out by using the Matrix Pencil method [1–4]. It has been shown in [1, 2] that this method has a very low variance for the estimation of the exponents and its residues in the presence of noise. Furthermore, the Matrix Pencil method will work even when the incident signals are correlated [3]. The estimate of the direct path d will be given by the imaginary part of the exponent, which is the least in magnitude. The corresponding weight will be C, the estimate of the complex amplitude of the signal. So once we know C, d and f_0, we can then take out the multipath effects from the signals of interest.

10.3   USE OF THE MATRIX PENCIL METHOD FOR IDENTIFICATION OF MULTIPATH COMPONENTS

The received signal in our case can be written as

$$y_q = \sum_{i=1}^{M+1} \alpha_i C \exp\left\{ \frac{j2\pi d_i}{c} \left(f_0 + \left(q - \frac{N-1}{2} \right) \Delta f \right) \right\} + n_q \quad q = 0, \ldots, N-1$$

(10.2)

with $\alpha_1 = 1$ and $d_1 = d$, for the direct path and n_q is the additive noise component. In this case we have a frequency difference between two consecutive transmitted signals for application of the Matrix Pencil method. An alternative derivation is given in Appendix C.

We now consider two matrices $[Y]_1$ and $[Y]_2$ formed using the received voltages. Consider the matrices $[Y]_1$ and $[Y]_2$ given by

$$[Y]_1 = \begin{bmatrix} y_0 & y_1 & \cdots & y_L \\ y_1 & y_2 & \cdots & y_{L+1} \\ \vdots & \vdots & \vdots & \vdots \\ y_{N-L-1} & y_{N-L} & \cdots & y_{N-2} \end{bmatrix}_{(N-1)\times(L+1)}$$

$$[Y]_2 = \begin{bmatrix} y_1 & y_2 & \cdots & y_{L+1} \\ y_2 & y_3 & \cdots & y_{L+2} \\ \vdots & \vdots & \vdots & \vdots \\ y_{N-L} & y_{N-L+1} & \cdots & y_{N-1} \end{bmatrix}_{(N-1)\times(L+1)}$$

(10.3)

where L is the pencil parameter $\left(\dfrac{N}{3} \le L \le \dfrac{N}{2} \right)$. This parameter should also satisfy $M \le L \le N - M$. Equation (10.2) can be rewritten as follows:

$$y_q = \sum_{i=1}^{M+1} w_i z_i^q \ + \ n_q \qquad q = 0, \ ..., \ N-1 \tag{10.4}$$

where

$$w_i = \alpha_i C \exp\left\{\frac{j2\pi d_i}{c}\left(f_0 - \frac{N-1}{2}\Delta f\right)\right\} \tag{10.5a}$$

and

$$z_i = \exp\left\{\frac{j2\pi d_i}{c}\Delta f\right\} \tag{10.5b}$$

It can be shown that in the noiseless case, the matrices $[Y]_1$ and $[Y]_2$ can be written as

$$[Y]_1 \ = \ [Z]_1[S][Z]_2 \tag{10.6a}$$

$$[Y]_2 \ = \ [Z]_1[S][D][Z]_2 \tag{10.6b}$$

where

$$[S] \ = \ \text{diag } [s_1, s_2, \ ..., \ s_{M+1}]_{(M+1)\times(M+1)} \tag{10.6c}$$

and s_i are the diagonal elements corresponding to the various singular value

$$[D] \ = \ \text{diag } [z_1, z_2, \ ..., \ z_{M+1}]_{(M+1)\times(M+1)} \tag{10.6d}$$

$$[Z]_1 = \begin{bmatrix} 1 & 1 & \cdots & 1 \\ z_1 & z_2 & \cdots & z_{M+1} \\ \vdots & \vdots & \vdots & \vdots \\ z_1^{(N-L-1)} & z_2^{(N-L-1)} & \cdots & z_{M+1}^{(N-L-1)} \end{bmatrix}_{(N-L)\times(M+1)} \tag{10.6e}$$

and

$$[Z]_2 = \begin{bmatrix} 1 & z_1 & \cdots & z_1^{L-1} \\ 1 & z_2 & \cdots & z_2^{L-1} \\ \vdots & \vdots & \vdots & \vdots \\ 1 & z_{M+1} & \cdots & z_{M+1}^{L-1} \end{bmatrix}_{(M+1)\times(L+1)} \tag{10.6f}$$

Then

$$\begin{aligned} [Y]_2 - \lambda[Y]_1 \ &= \ ([Z]_1[S][D][Z]_2) \ - \ \lambda([Z]_1[S][Z]_2) \\ &= \ [Z]_1[S]([D] \ - \ \lambda[I])[Z]_2 \end{aligned} \tag{10.7}$$

Equation (10.7) implies that the nonzero generalized eigenvalues of the Matrix Pencil ($[Y]_2$, $[Y]_1$) are the eigenvalues of the matrix $[D]$. Then since the eigenvalues of $[D]$ are z_i, we have $\lambda_i = z_i$, and then z_i can be estimated from the nonzero eigenvalues λ_i of the Matrix Pencil ($[Y]_2$, $[Y]_1$). The rank of the matrices $[Y]_1$ and $[Y]_2$ is $M + 1$ and we can estimate M from the principal singular eigenvalues of the matrix $[Y]_1$ or $[Y]_2$.

The solution of the generalized eigenvalue problem can be carried out by using the QZ algorithm [5]. However, to avoid instability problems, it is better to convert the problem above to an ordinary eigenvalue problem. This can be done by recognizing the fact that the generalized eigenvalues of the Matrix Pencil ($[Y]_2$, $[Y]_1$) are equal to the eigenvalues of the matrix $[Y]_1^+[Y]_2$, where $[Y]_1^+$ is the pseudo-inverse of the matrix $[Y]_1$. For the noisy case, the pseudo-inverse matrix $[Y]_1^+$ is replaced by the rank $(M + 1)$ truncated pseudo-inverse $[Y]_{1T}^+$ of the matrix $[Y]_1$.

Once the exponents z_i are estimated, the multipath lengths d_i are given by the following equation:

$$d_i = \frac{\arg(z_i)c}{2\pi\,\Delta f} \quad i = 1, \ldots, M + 1 \tag{10.8}$$

As mentioned above, the length of the direct ray path d or the path length of the more dominant multipath component (in the case when no direct ray is present) corresponds to the value of d_i calculated from z_i, which has the minimum magnitude over all the computed values of the exponents.

Finally, the complex amplitudes w_i associated with the direct ray and all the various multipath components can be estimated by solving the following equation:

$$\begin{bmatrix} y_0 \\ y_1 \\ \vdots \\ y_{(N-1)} \end{bmatrix} = \begin{bmatrix} 1 & 1 & \cdots & 1 \\ z_1 & z_2 & \cdots & z_{M+1} \\ \vdots & \vdots & \vdots & \vdots \\ z_1^{(N-1)} & z_2^{(N-1)} & \cdots & z_{M+1}^{(N-1)} \end{bmatrix} \times \begin{bmatrix} w_1 \\ w_2 \\ \vdots \\ w_{M+1} \end{bmatrix} \tag{10.9}$$

Since for the direct ray $\alpha_1 = 1$, the signal amplitude of the direct ray can then be obtained as

$$C = w_1 \exp\left\{ \frac{-j2\pi d}{c}\left(f_0 - \frac{N-1}{2}\Delta f \right) \right\} \tag{10.10}$$

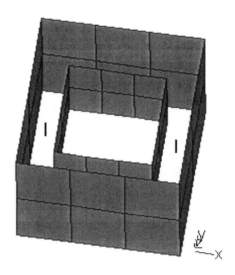

Figure 10.2. Electromagnetic model of the situation similar to that of Figure 10.1 but without the possibility of the existence of any direct ray.

10.4 SIMULATION RESULTS

Computer simulations are carried out to ascertain the performance of the proposed method. A situation similar to the one described in Figure 10.1 is considered and all the electromagnetic effects, including the various multipath components, have been simulated via the WIPL-D electromagnetic analysis code [6]. For this example we have assumed that no direct ray is involved in the communication process. Figure 10.2 illustrates how this situation is modeled. We consider two dipole antennas located in the inner space of two perfectly conducting cubes. Inside the structures there are two dipole antennas and one of them is center fed with an amplitude of 1 V. The other one is the receiving antenna. The length of the edges of the cubes is 1 m and 0.6 m, respectively. The two antennas are considered identical with radius equal to 5 mm and length equal to 0.15 m, so that the resonant frequency of each dipole antenna corresponds to $f_0 = 1$ GHz, and the distance between the two antennas is 0.8 m. Thus, the far-field condition is applicable. We transmit from the first antenna and receive the signal at the second antenna. We also study the effect of noise on the extraction of the signal in this environment through the Cramer–Rao bound (CRB), which provides the least variance for an unbiased estimate of the solution when the data is corrupted by additive zero mean Gaussian noise.

Let us write equation (10.4) in the following form:

$$[Y(t)] = [A(\varphi)][\tilde{S}(t)] + [R(t)] \quad t = 1, 2, ..., P \qquad (10.11)$$

where P is the number of independent simulations and

$$[Y(t)] = [y_0(t) \quad \cdots \quad y_{N-1}(t)]^T, \quad [\tilde{S}(t)] = [w_1(t) \quad \cdots \quad w_M(t)]^T$$

$$[R(t)] = [n_0(t) \quad \cdots \quad n_{N-1}(t)]^T, \quad [\varphi] = [d_1 \quad \cdots \quad d_m]$$

$$[A(t)] = [a(d_1) \quad \cdots \quad a(d_M)] \text{ with } [a(d_i)] = \begin{bmatrix} 1 & z_i & \cdots & z_i^{(N-1)} \end{bmatrix}^T$$

$$(10.12)$$

Here the superscript T denotes the transpose of a matrix.

The CRB for the estimated ray length is given by [7]:

$$\text{CRB} = \sigma^2 \left\{ \sum_{t=1}^{P} \text{Re}\left[[\tilde{S}^H(t)][D]^H \left([I] - [A]([A]^H[A])^{-1}[A]^H\right)[D][\tilde{S}(t)] \right] \right\}^{-1}$$

$$(10.13)$$

where

$$[D] = \begin{bmatrix} \dfrac{\partial a(d_1)}{\partial d_1} & \dfrac{\partial a(d_2)}{\partial d_2} & \cdots & \dfrac{\partial a(d_M)}{\partial d_M} \end{bmatrix} \qquad (10.14)$$

One of the important steps in our simulation is to choose an adequate number of frequencies N and the frequency step Δf in such a way that we will be dealing with essentially narrowband signals. In other words, the total bandwidth (BW) must satisfy $0.01 \leq \text{BW}/f_0 \leq 0.1$ (BW $= N \Delta f$). We must choose N to be large so that it is greater than twice the number of multipaths, and Δf should be small. In addition, care must be taken to ensure that Δf is not too small. Otherwise, the singular values of the data matrix and the values of the exponents would be close to each other, and then it will be difficult to discriminate between the various multipaths. In the simulations presented here, N is chosen to be 9, and we study the influence of Δf, P, and the SNR on the performance of the method.

In the first set of simulations we investigate the robustness of the proposed method by computing the Cramer-Rao bound (CRB). The CRB gives the minimum variance for an unbiased estimate and therefore provide bounds for the best possible solution. Δf is chosen to be 0.01 GHz (BW/$f_0 = 9\%$) and $P = 10$. Figure 10.3 shows the variance of the path length of the dominant ray (d) as a function of the SNR (solid line). The dashed line corresponds to its CRB. For a small number of samples, of course, the CRB is not a good criterion to use, but it is one of the easily computible metric that is available to assess the quality of the solution.

The second result illustrates the influence of Δf and P over the performance of the Matrix Pencil method. Figures 10.4 to 10.6 show the variance of d as a function of SNR for different values of BW/f_0 = 1.8, 2.7, 3.6, 4.5, 5.4, 6.3, 7.2, 8.1, and 9%, and P = 10, 5, and 3, respectively. Better performance is obtained when $BW/f_0 \geq 5\%$. Note that when SNR = 10 dB, a choice of BW/f_0 = 9% gives a better estimation. This is more clear in Figure 10.7, which represents the variance of d as a function of P for SNR = 10 dB. Figure 10.8 shows the variance of d as a function of BW/f_0 for SNR = 10 dB. Here it is also clear that if BW/f_0 is large, the method estimates d better even with very few samples.

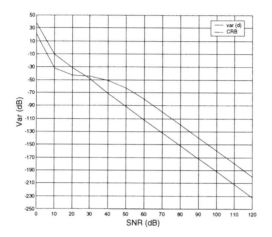

Figure 10.3.   var(d) versus SNR for P = 10 and BW/f = 9%.

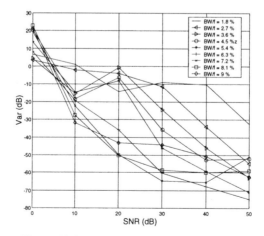

Figure 10.4.   var(d) versus SNR for P = 10.

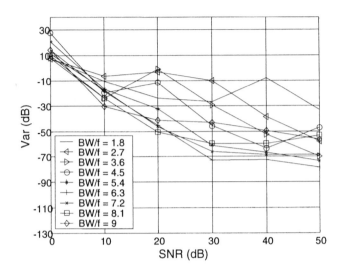

Figure 10.5.   var(d) versus SNR for $P = 5$.

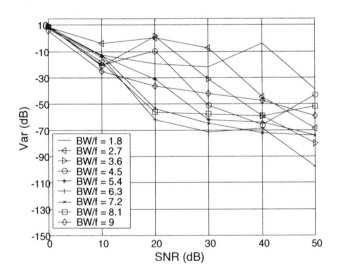

Figure 10.6.   var(d) versus SNR for $P = 3$.

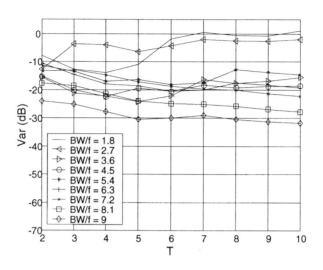

Figure 10.7. var(d) versus P for SNR = 10 dB.

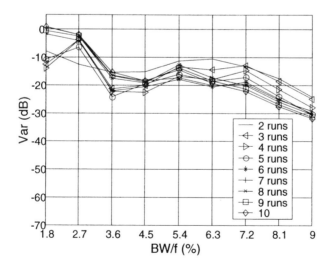

Figure 10.8. var(d) versus BW/f for SNR = 10 dB.

Once d_i is known, the corresponding residue provides the complex amplitude for that multipath component i.

10.5 CONCLUSION

An application of frequency diversity for the identification and elimination of the multipath effects of narrowband signals without spatial diversity has been presented. This is a good method for eliminating the various multipath effects in antenna pattern measurements. With this approach it may be possible to use lower-quality absorbers in anechoic chambers to reduce the cost of building one for practical compact ranges. A similar concept has been used in practice to eliminate the resonances inside an anechoic chamber using this methodology [8]. This frequency diversity procedure consists of sending first a set of narrowband tones prior to the actual signal. The Matrix Pencil method is then used to determine the multipath effects from the received tones. Simulation results have been presented showing a comparison with the CRB and studying the influence of Δf, P, and SNR over the estimation of d_i. It has been shown that for $BW/f_0 = 9\%$, we obtain good estimation even though the tones can be very noisy (SNR = 10 dB).

REFERENCES

[1] Y. Hua and T. K. Sarkar, "Matrix Pencil Method for Estimating Parameters of Exponentially Damped/Undamped Sinusoids in Noise," *IEEE Transactions on Acoustics, Speech, and Signal Processing*, Vol. 38, No. 5, pp. 814–824, May 1990.

[2] T. K. Sarkar and O. Pererira, "Using Matrix Pencil Method to Estimate the Parameters of a Sum of Complex Exponentials," *IEEE Antennas and Propagation Magazine*, Vol. 37, No. 1, pp. 48–55, Feb. 1995.

[3] A. Medouri, A. Gallego, D. P. Ruiz, and M. C. Carrion, "Estimating One- and Two-Dimensional Direction of Arrival in an Incoherent/Coherent Source Environment," *IEICE Transactions on Communications*, Vol. E80-B, No. 11, pp. 1728–1740, Nov. 1997.

[4] Y. Hua and T. K. Sarkar, "Generalized Pencil-of-Function Method for Extracting Poles of an EM System from Its Transit Response," *IEEE Transactions on Antennas and Propagation*, Vol. 37, No. 2, Feb. 1989.

[5] G. H. Golub and C. F. Van Loan, *Matrix Computations*, Baltimore: Johns Hopkins University Press, 1983.

[6] B. M. Kolundzija, J. S. Ognjanovic, and T. K. Sarkar, *WIPL-D: Electromagnetic Modeling of Composite Metallic and Dielectric Structures*, Artech House, Norwood, MA, 2000.

[7] P. Stoica and A. Nehorai, "MUSIC, Maximum Likelihood, and Cramer-Rao Bound," *IEEE Transactions on Acoustics, Speech, and Signal Processing*, Vol. 37, No. 5, pp. 720–741, May 1989.

[8] B. Fourestie, Z. Altman, and M. Kanda, "Efficient Detection of Resonances in Anechoic Chambers Using the Matrix Pencil Method," *IEEE Transactions on Electromagnetic Compatibility*, Vol. 42, pp. 1–5, Feb. 2000.

11

SIGNAL ENHANCEMENT IN MULTIUSER COMMUNICATION THROUGH ADAPTIVITY ON TRANSMIT

SUMMARY

In this chapter a technique is presented on how to enhance the received signals in a multiuser communication environment through the use of adaptivity on transmit. This technique is based on the principle of reciprocity and is independent of the material medium and the near-field environments in which the radio is transmitting or receiving. This assumes that one is using spatial diversity on transmit. The objective here is to select a set of weights to be applied to the input signals fed to each of the transmitting antenna elements, which is a function of the user location. This methodology is not a function of the multipath environment. Furthermore, the transmitted signal may be directed to a particular receiver location, and simultaneously, its strength minimized at other receiver locations while operating on the same frequency. The uniqueness of this approach is that one does not need to know the details of the electromagnetic environment in which they are operating or the spatial locations of the transmitting and receiving antennas. The only information that is necessary in this procedure is the actual received voltages at each of the transmitting antenna elements due to a particular mobile. Numerical simulations have been made to illustrate the novelty of the proposed approach.

11.1 INTRODUCTION

Many methods have been developed and have continued to evolve in recent years, to enhance reception of signals in a multiuser environment [1, 2]. This is necessary for mobile communication in order to enhance the quality of the reception in the presence of multipath fading, near-field scatterers, and so on. In order to have such enhancements it is generally necessary to have spatial diversity. One way to counteract the problem of interference at the receiver due to multipath and the presence of other near-field scatterers, such as buildings, trees, platforms and so on, are to provide spatial diversity as shown in Figure 11.1. However, at a mobile receiver there is generally no spatial diversity, as the footprint of a receiver generally is quite small.

Figure 11.1.   Multiple-user transmit/receive scenario.

On the other hand, it is possible to have multiple transmitting antennas located at a base station. Here, the philosophy is to apply the concept of spatial diversity for narrowband signals on transmit and not on receive. However, in this approach the transmitted energy is directed through spatial diversity and not broadcast indiscriminately through all space. This is in contrast to the frequency diversity scheme on transmit, as explained in Chapter 10. By weighting the signals being fed to each of the antennas, it is possible to provide the necessary spatial diversity on transmit. So the problem we address in this chapter is what can be done in terms of adaptivity on transmit to direct the antenna beam destined for a prespecified mobile receiver, for example, and simultaneously introduce a pattern such that the effective voltage is not induced at the terminals of the other mobiles distributed spatially for which the transmission is not intended. In this way, transmitted energy will be effectively received at the designated receiver while it would cancel out at the other receiver locations for which this transmission is not intended. By providing spatial diversity on transmit, it is possible to mitigate the effects of multipath fading, as the directed energy from the transmitted antennas would combine vectorially at the selected receiving antenna element to produce an effective induced voltage which will be either finite or close to zero.

Many papers and books are available that discuss the problem of spatial diversity for wireless communication [1, 2]. In addition, how to counteract multipath fadings in an adaptive antenna, which can significantly improve the performance of a system, has also been discussed [3–7]. In this chapter we present an alternative way of directing the energy from the transmitter to a specific receiving antenna while simultaneously minimizing the signal directed toward the other receiving antennas. This methodology is based on the reciprocity theorem [8, 9] and is applied to a collection of receiving and transmitting antennas. This transmit-on-diversity scenario has been simulated using the electromagnetic analysis code WIPL-D [10], which can provide a complete description of the voltages and currents induced on various transmit and receive structures due to an incident electromagnetic wave (not necessarily having a plane wavefront) in any environment. This principle of adaptivity on transmit can also be applied to systems operating in any near-field environment. Numerical examples are presented to illustrate this novel methodology.

11.2 DESCRIPTION OF THE PROPOSED METHODOLOGY

In this procedure we simultaneously employ the concept of reciprocity and use the principle of adaptivity on transmit. In this way it is possible to direct the energy transmitted from a base station to a preselected mobile station without worrying about either the presence of other near-field scatterers or the existence of a multipath environment. For example, consider the system represented in Figure 11.1. Here, let us assume, for example, that the transmitter is sending signals using spatial diversity (i.e., using a number of transmitting antennas separated spatially from each other).

Let us assume that the transmitting antenna 1 is transmitting a voltage V_{T_1} at a frequency f_0. The internal impedance of this voltage source is R_{T_1}. At the second transmitting antenna we have a voltage source V_{T_2} transmitting at the same frequency with an internal impedance of R_{T_2}. In addition, there are two receivers, represented by R_1 and R_2. The goal of this transmitting system is that we want to maximize the received power at receiver 1 by exciting transmitting systems 1 and 2 with voltages $V_{T_1}^1$ and $V_{T_2}^1$. Our goal is to minimize the received power at receiver 2 while directing the energy to receiver 1. Similarly, we want to find another set of voltages $V_{T_1}^2$ and $V_{T_2}^2$ which will excite the two transmitting antennas so that they will produce zero voltage at the terminals of receiver 1 (designated by R_1) and induce maximum voltage at receiver 2 (designated by R_2). The characteristics of these two voltages with a superscript of 2, is such that when they are used to excite the transmitting antennas, the electromagnetic signal will be directed to receiver 2 and will induce practically no energy at receiver 1. In this way, we can essentially do adaptivity on transmit. This is ideally suited, for example, in a CDMA environment, where a unique code is assigned to each receiver, so that when we receive any signal we know from which receiver it originated. When we are going to transmit that particularly weighted signal broadcast simultaneously over all the transmitting antennas, we can spatially direct the energy to a prespecified receiver only and simultaneously minimize the received signals at the other receivers for which that particular transmission was not intended. In this way, it is possible to direct the energy to a preselected receiver.

The problem now is how to carry it out! We propose to achieve this objective using the principle of reciprocity. Reciprocity tells us that if we excite transmitting antenna 1 with a voltage V_{T_1} with the other voltages set to zero, such that $V_{T_2} = V_{R_1} = V_{R_2} = 0$, the transmitted signal will induce a current I_{R_1} at the feed point of receiving antenna 1 (i.e., R_1). It will also induce currents at receiver 2 and transmitter 2. However, let us ignore that for the moment. Under the same environment, if we excite receiver 1 with a voltage V_{R_1} with $V_{T_1} = V_{T_2} = V_{R_2} = 0$, it will induce a current, let us say I_{T_1}, at transmitter 1. It will also induce currents at receiver 2 and transmitter 2. However, let us also ignore that for the moment. Then from the principle of reciprocity we know that

$$V_{T_1} I_{T_1} = V_{R_1} I_{R_1} \qquad (11.1)$$

Equation (11.1) represents the reciprocity theorem in the frequency domain. The relationship is much more involved in the time domain, where the reciprocity theorem performs an integration over time as illustrated in Chapter 2. That is why it is possible without violating the principle of reciprocity that a finite length transmitting antenna radiates essentially a double differentiated version of the temporal input waveform applied to it, whereas a receiving antenna differentiates the incident signal in the temporal domain only once.

Here we consider the currents that are induced at the transmitting and receiving antennas which generate induced voltages at the loads which are associated with the feed point of each antenna. The relationship provided by (11.1) is valid irrespective of the shape of the transmitting or receiving antennas. Nor does it depend on the presence of various near-field scatterers such as trees or buildings that are in the immediate neighborhood of the transmitting or receiving antennas, as illustrated in Figure 11.1. One does not even require any knowledge about the spatial locations of the various antennas. The principle of reciprocity can help one to treat such a complex situation without any knowledge of either the electromagnetic environment in which the antennas are communicating or of the spatial locations of objects that influence the electromagnetic coupling mechanisms. Based on the principle of reciprocity, we now implement the system of adaptivity on transmit.

Let us assume that each receiver is given a particular code, as in a CDMA environment. Let us assume that we transmit the code with amplitude 1 V. So when $V_{R_1} = 1$ V with $V_{R_2} = V_{T_1} = V_{T_2} = 0$, it is going to induce currents $I^1_{T_1}$ in transmitting antenna 1 and $I^1_{T_2}$ in transmitting antenna 2. From these induced currents, we know the voltage induced across the loads of the transmitting antennas, as we know the impedance values of the loads. Now what the reciprocity principle tells us is that if we apply 1 V at the feed of transmit antenna 1 so that $V_{T_1} = 1$ and $V_{T_2} = V_{R_1} = V_{R_2} = 0$, it will induce a current $I^1_{R_1}$ into receiver 1 so that

$$I^1_{T_1} = I^1_{R_1} \tag{11.2}$$

Similarly, if we excite receiver 2 only with a voltage of 1 V so that $V_{R_2} = 1$ V and $V_{T_1} = V_{T_2} = V_{R_1} = 0$, we will induce currents $I^2_{T_1}$ and $I^2_{T_2}$ at transmitters 1 and 2, respectively. From reciprocity we have that if we excite transmitter 2 with 1 V at the same frequency f_0, then $V_{T_2} = 1$ and $V_{T_1} = V_{R_1} = V_{R_2} = 0$. Under this situation there will be an induced current $I^2_{R_2}$ in receiver 2. From the principles of reciprocity in the frequency domain, one gets

$$I^2_{T_2} = I^2_{R_2} \tag{11.3}$$

It is important to observe that reciprocity links only the two transmit–receive ports under consideration, namely those related to the transmitter and receiver. The other portions of the electromagnetic network do not come into the picture as long as it remains the same when we switch the voltages on transmitters and receivers to satisfy the relationships of (11.2) and (11.3). Nor is reciprocity dictated by the material medium influencing the propagation path. This implies that the propagation paths, which resulted in the induced currents described by the variables in equations (11.2) and (11.3), can occur even when one is communicating in nonisotropic and heterogeneous mediums.

From the observations above, we can carry out the following procedure:

- **Step 1.** Excite receiver 1 with 1 V with $V_{T_1} = V_{T_2} = V_{R_2} = 0$. This will induce currents $I^1_{T_1}$ and $I^1_{T_2}$ in transmitting antennas 1 and 2, respectively. Let us assume that we can measure these currents.
- **Step 2.** Excite receiver 2 with 1 V with $V_{T_1} = V_{T_2} = V_{R_1} = 0$. This will induce currents $I^2_{T_1}$ and $I^2_{T_2}$ in transmitting antennas 1 and 2, respectively. Let

us assume that we can measure these two currents. The measurement is easy to carry out in a CDMA environment where each receiver carries a unique code.

- **Step 3.** If we now excite transmitter 1 with a voltage W_1^1 and transmitter 2 with a voltage W_2^1, the current induced in receiver 1 will be the sum of the following two terms: $W_1^1 I_{T_1}^1 + W_2^1 I_{T_2}^1$. This is, of course, under the assumption that all the loads terminating the two transmitting and receiving antennas are equal and is numerically equal to $R_{T_1} = R_{T_2} = R_{R_1} = R_{R_2} = 1\Omega$. The induced current at receiver 2 under these conditions will be $W_1^1 I_{T_1}^2 + W_2^1 I_{T_2}^2$. If the signal is to be directed to receiver 1 and not to receiver 2, the weights W_1^1 and W_2^1 should be such that the first sum is unity and the second sum zero. Equivalently, if we now excite transmitter 1 with a voltage W_1^2 and transmitter 2 with a voltage W_2^2, the current induced in receiver 1 will be the sum of these two terms, $W_1^2 I_{T_1}^1 + W_2^2 I_{T_2}^1$. In receiver 2 the induced current will be $W_1^2 I_{T_1}^2 + W_2^2 I_{T_2}^2$. If we were to direct the signal to receiver 2 and not to receiver 1, the first sum should be zero and the second sum should be unity. Therefore, by appropriate choice of the voltages exciting the two transmitting antennas, it is possible to direct the signal to an *a priori* chosen receiver irrespective of the near-field electromagnetic environment in which they may be operating. In summary, the four equations described above which could direct the transmitted energy to a preselected receiver can be written in compact matrix form as

$$\begin{bmatrix} I_{T_1}^1 & I_{T_2}^1 \\ I_{T_1}^2 & I_{T_2}^2 \end{bmatrix} \begin{bmatrix} W_1^1 & W_1^2 \\ W_2^1 & W_2^2 \end{bmatrix} = \begin{bmatrix} 1 & 0 \\ 0 & 1 \end{bmatrix} \tag{11.4}$$

The choice of a different value of the impedance will limit both transmitting and receiving antennas to any values other than 1 Ω. No loss of generality will occur because of such a choice, as it will only affect the scale factor for all four weights.

- **Step 4.** Solution of equation (11.4) provides a set of weights which when used in conjunction with the received signals from the two receivers producing currents at the transmitters will direct the transmitted energy on receiver 1 or receiver 2, depending on the values of the weights chosen. Hence

$$\begin{bmatrix} W_1^1 & W_1^2 \\ W_2^1 & W_2^2 \end{bmatrix} = \begin{bmatrix} I_{T_1}^1 & I_{T_2}^1 \\ I_{T_1}^2 & I_{T_2}^2 \end{bmatrix}^{-1} \begin{bmatrix} 1 & 0 \\ 0 & 1 \end{bmatrix}$$ (11.5)

The caveat here is that if we apply a voltage equal to W_1^1 on transmitting antenna 1 and W_2^1 on transmitting antenna 2 then the received electromagnetic energy from these two transmitting antennas energy will be vectorially additive at the terminals of receiver 1 and would be destructive at the terminals of receiver 2 producing zero current at its loads, whereas if we apply the voltage W_1^2 on transmitting antenna 1 and W_2^2 on transmitting antenna 2, the received electromagnetic energy will be vectorially destructive at the terminals of receiver 1, resulting in practically zero current flowing through its load and will be vectorially additive at the terminals of receiver 2, producing a large value for the induced current.

In summary, by knowing the voltages that are induced at each of the transmitting antennas due to each of the receivers, it is possible to select a set of weights based on reciprocity. This relationship can be applied only at the terminals of the transmitting and receiving antennas. It is independent of the sizes and shapes of the receiving antennas and does not depend on the near-field environments.

Now if there are more receivers than transmitting antennas, it may not be possible to achieve a perfect match (namely, simultaneously directing the received energy on a preselected receiver and practically zero energy induced on the rest of the receivers); but some intermediate solution is still possible. We now illustrate these statements through examples.

11.3   NUMERICAL SIMULATIONS

As a first example, consider three transmitting antennas A_1, A_2, and A_3, located at a base station. The three antennas are considered to be three dipoles of radius 5 mm and their lengths are 15 cm, so that if the operating frequency is 1GHz, the three antennas are a half wavelength long. Next we consider three receiving antennas A_4, A_5, and A_6, as shown in Figure 11.2. They are half-wavelength spaced between themselves and are separated from the transmitters by a distance of 2 m. In this simulation we want to demonstrate that by choosing a proper set of weighted excitations on transmitters A_1, A_2, and A_3, it is possible to direct the electromagnetic energy in such a way that it will be vectorially additive at only one receiver (say, A_4). The vector sum of all the received voltages from these three transmitting antennas will combine in such a way at the other receiving antennas that it will produce practically zero currents at the terminals of the remaining two receivers, A_5 and A_6. In a similar fashion, it is possible to select another set of weights which will result in receiver A_5 developing a large signal at its terminals while the received signals at the terminals of the other two receivers, A_4 and A_6 would be practically zero, and so on.

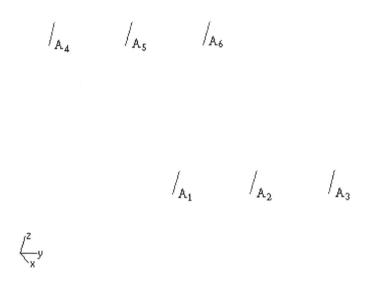

Figure 11.2. Six-antenna system.

Now we illustrate how that can be accomplished based on the principles of reciprocity. First we excite receiver 1 with 1 V and that will induce a current in both receivers A_5 and A_6 and also in transmitters A_1, A_2, and A_3 comprising the base station. These induced currents will generate a voltage across the loads of the remaining five centrally loaded dipole antennas. Let us call the voltages received at the transmitters A_1, A_2, and A_3 due to an excitation voltage of 1 V at receiver A_4 as $V_{A_1}^1$, $V_{A_2}^1$, and $V_{A_3}^1$, respectively. Similarly, if we excite the antenna at the mobile A_5 by 1 V, the voltages received at the three base station antennas would be $V_{A_1}^2$, $V_{A_2}^2$, and $V_{A_3}^2$. Even though voltages are also induced in the other two receivers, they are of no consequence. Finally, if we excite receiver A_6 with 1 V it will generate the following voltages at the three transmitters: $V_{A_1}^3$, $V_{A_2}^3$, and $V_{A_3}^3$. Now the claim is that based on the available information on the three experiments above, we can choose a set of complex voltages W_1^1, W_2^1, and W_3^1, which when fed to the three transmitting antennas will result in an additive vectorial combination of the electromagnetic energy at receiver 1 while simultaneously not inducing practically any currents in the other two receivers. From the principle of reciprocity, we are going to evaluate the induced currents at the loads located at the feed point of the three receivers, when all three transmitting antennas A_1, A_2, and A_3 are excited simultaneously with the voltages W_1^1, W_2^1, and W_3^1. The receiver currents at the three locations would then be

At receiver 1: $W_1^1 I_{A_1}^1 + W_2^1 I_{A_2}^1 + W_3^1 I_{A_3}^1$

At receiver 2: $W_1^1 I_{A_1}^2 + W_2^1 I_{A_2}^2 + W_3^1 I_{A_3}^2$

At receiver 3: $W_1^1 I_{A_1}^3 + W_2^1 I_{A_2}^3 + W_3^1 I_{A_3}^3$

The objective now is to select the weights W_i^j in such a fashion that the received voltage would be finite at receiver 1 and zero at receivers 2 and 3. This is enforced through the equation

$$\begin{bmatrix} I_{A_1}^1 & I_{A_2}^1 & I_{A_3}^1 \\ I_{A_1}^2 & I_{A_2}^2 & I_{A_3}^2 \\ I_{A_1}^3 & I_{A_2}^3 & I_{A_3}^3 \end{bmatrix} \begin{bmatrix} W_1^1 \\ W_2^1 \\ W_3^1 \end{bmatrix} = \begin{bmatrix} 1 \\ 0 \\ 0 \end{bmatrix} \tag{11.6}$$

Now if the three transmitting antennas are excited with the following three complex voltages obtained from the solution of (11.6),

$$W_1^1 = -0.415 + j0.366; \; W_2^1 = -0.679 - j0.446; \; W_3^1 = 0.845 + j0.062$$

then the electromagnetic energy will be vectorially additive at the terminals of receiver 1 and will cancel each other, resulting in practically zero currents at the terminals of the other two receivers. Based on these three voltages, we can carry out an electromagnetic simulation using WIPL-D [10] so as to obtain the following three induced currents in the three receivers (with $R_{T_1} = R_{T_2} = R_{T_3} = 50\,\Omega$):

$$I_{T_1} = 1.0 - j\,0.007; \; I_{T_2} = -0.003 + j0.001; \; I_{T_3} = 0.001 - j0.002$$

(All the currents are in mA.) This clearly shows that for this particular choice of excitation voltages, the electromagnetic environment will be such that it will induce a large current at the load of receiver 1 and practically zero currents at the terminals of the other two receivers. This in no way implies that the current distribution on the receiving antenna is zero. It is not. It is practically zero only at the feed terminals and nowhere else! This is because the principle of reciprocity holds only at prespecified terminals or ports of a network.

Similarly, using the procedure above, one can show that if the excitation voltages are chosen for the three transmitting antennas as follows:

$$W_1^2 = -0.679 - j0.446; \; W_2^2 = 0.137 + j0.011; \; W_3^2 = -0.679 - j0.446$$

then the currents induced at the terminals of the three receivers will be given by

$$I_{T_1} = -0.003 - j0.0003; \; I_{T_2} = 1.0 - j0.01; \; I_{T_3} = -0.003 + j0.0003$$

(All the currents are in mA.) This clearly shows that the induced energy can be vectorially additive only at receiver 2, while producing no appreciable currents at the terminals of the other two receivers, 1 and 3.

Finally, if the excitation voltages at the three transmitting antennas are chosen as

$$W_1^3 = 0.845 + j0.062; \; W_2^3 = -0.679 - j0.446; \; W_3^3 = -0.415 + j0.366$$

then one can show that the currents induced at the terminals of the three receivers will be

$$I_{T_1} = -0.0012 - j0.0025; \; I_{T_2} = -0.003 + j0.001; \; I_{T_3} = 1.0 - j0.007$$

(All the currents are in mA.) This clearly demonstrates that by appropriately choosing simultaneously the three complex amplitudes of the excitations at different transmitting antennas it is possible to create an electromagnetic environment such that the induced electromagnetic energy will be vectorially additive at any one of the terminals of the preselected receivers and will try to cancel the energy at the terminals of the other receiving antennas, resulting in practically zero current. Such a procedure can be carried out without knowing precisely the near-field environment in which the transmitter and receiver are operating, nor is it necessary to know the precise spatial locations of the transmitting or receiving antennas.

In summary, the steps involved in this new method are as follows:

• **Step 1.** Excite each of the mobile receiving antennas and measure the currents that are induced in the base station transmitting antennas.

• **Step 2.** Using these induced currents, one can generate a matrix, the inversion of which will contain the information necessary to excite the transmitting antennas.

• **Step 3.** Then depending on the choice of which of the mobile receiving antennas the signal is meant for, a set of weights can be solved for, which will result in a vectorial addition of all the incident electromagnetic energy at the terminals maximizing the induced currents. This would produce the maximum voltage at that pre-determined receiver and practically zero induced currents at the terminals of the other receivers.

By following the procedure above one can carry out adaptivity on transmiting based on the principle of reciprocity. This methodology works even when there are near-field scatterers present in the vicinity of the transmitter or the receiver. From our numerical simulations using WIPL-D code [10], we have observed that this principle can be applied over a relatively wide band of frequencies. We choose the same six antennas as described in the previous example. In this case, we now sweep the frequency from 0.9 to 1.1 GHz, generating a 20% bandwidth. Even when we use a 10% bandwidth, the currents induced at elements other than the targeted receivers are down by about −15 dB. For a 20% bandwidth centered at 1 GHz, the currents induced at the other receivers are down by about −5 dB over that of the desired receiver. The results are illustrated in Figures 11.3 to 11.5.

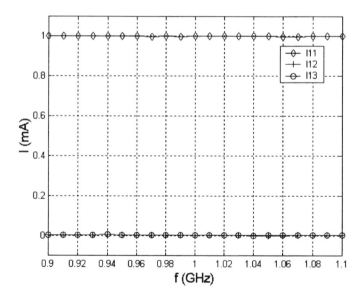

Figure 11.3. Magnitude of the measured currents at the three receivers with W_i^1 as the weight vector.

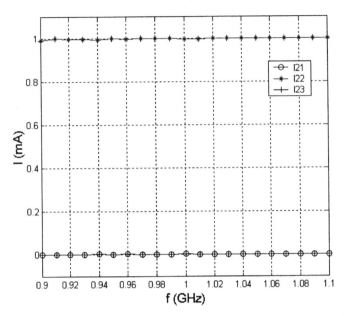

Figure 11.4. Magnitude of the measured currents at the three receivers with W_i^2 as the weight vector.

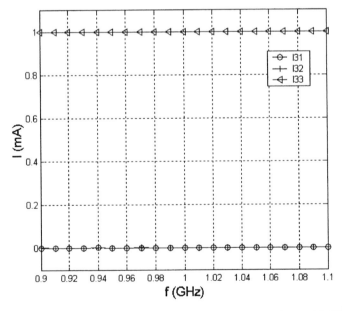

Figure 11.5. Magnitude of the measured currents at the three receivers with W_i^3 as the weight vector.

As a third example, we put six antennas inside two concentric perfectly conducting cylinders, as shown in Figure 11.6. The dimension of the outer conducting cylinder is 1 m × 1 m × 0.5 m. The transmitting and receiving antenna sets are separated by 0.8 m and they are spaced 0.25 m from each other. In this case, one can also choose a set of weights so that the energy is directed to a particular receiver. For the three sets of weights we see that the energy can be directed to a preselected receeiver as shown by the results of Table 11.1.

Figure 11.6. Six antennas located inside a concentric cube.

TABLE 11.1
Magnitude of the currents measured at the three receiving antennas 1, 2, and 3 for three different choices of excitations

$\left|I_{R_1}\right|$	$\left|I_{R_2}\right|$	$\left|I_{R_3}\right|$
0.99	0.014	0.006
0.0012	1.0	0.0012
0.0057	0.014	1.0

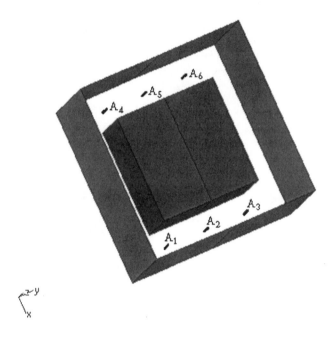

Figure 11.7. Simulation situation when the conducting inner cylinder is replaced by a dielectric cube.

Next we replace the inner conducting cylinder by a dielectric cube of relative permittivity $\varepsilon_r = 4$, as shown in Figure 11.7. All the other dimensions remain the same. In this case it is also possible to choose a set of appropriate weights W such that the energy can be directed to a particular receiver. The magnitude of the currents in this case for the three different types of excitations are given in Table 11.2.

TABLE 11.2
Magnitude of the currents measured at the three receiving antennas 1, 2, and 3 for three different choices of excitations

$\left|I_{R_1}\right|$	$\left|I_{R_2}\right|$	$\left|I_{R_3}\right|$
0.99	0.002	0.001
0.003	1.0	0.0008
0.001	0.004	1.0

The three different rows correspond to three different excitations when applied to each of the transmitters will direct the energy to the terminals of a particular receiver while simultaneously inducing g no currents at the remaining receivers.

As the final example, we consider a situation where the number of receiving antennas is greater than the number of transmitting antennas. First let us consider three transmitting antennas and six receiving stations. The antennas are all dipoles of length 15 cm and radius 5 mm centrally loaded with 50 Ω and operating at 1 GHz. In this case it will not be possible to maximize the received energy at the terminals of a particular receiver and simultaneously minimize it at the terminals of all the other receivers. Since the number of degrees of freedom is much less than the number of variables that we need to control, we wanted to observe the nature of the performance of this methodology for the underdetermined case. Table 11.3 presents the magnitude of the current at the six receiving antennas when three transmitting antennas are simultaneously radiating energy.

TABLE 11.3

Magnitude of the received currents at the terminals of each one of the six receiving antennas

$\left|I_{R_1}\right|$	$\left|I_{R_2}\right|$	$\left|I_{R_3}\right|$	$\left|I_{R_4}\right|$	$\left|I_{R_5}\right|$	$\left|I_{R_6}\right|$
0.45	0.30	0.01	0.14	0.06	0.37
0.30	0.50	0.33	0.04	0.21	0.06
0.01	0.33	0.54	0.34	0.04	0.14
0.14	0.04	0.34	0.54	0.33	0.01
0.06	0.21	0.04	0.33	0.50	0.30
0.37	0.06	0.14	0.01	0.30	0.45

From the table it is seen that there is approximately a −3.5 dB difference in the signal levels between the desired receiver and the neighboring ones.

Next, we increase the number of transmitters from three to four, keeping the numbers of receivers fixed. Here we choose a set of excitation voltages for each of the transmitting antennas so that the energy could be directed to the terminals of a particular receiver. The results of the simulation are given in Table 11.4. From the table it is seen that at best there is a −4 dB isolation between the receivers when the transmitter directs the energy to a particular receiver.

Finally, when we increase the number of transmitting antennas to five, the principle of directing electromagnetic energy to a prespecified receiver is further highlighted. The results of the simulation are presented in Table 11.5. Of course, when we increase the number of antennas from three to four and finally to five, we have better control over directing the energy to a preselected receiver, as illustrated by the data in Tables 11.3, 11.4, and 11.5.

TABLE 11.4
Magnitude of the received currents at the terminals of each one of the six receiving antennas

$|I_{R_1}|$	$|I_{R_2}|$	$|I_{R_3}|$	$|I_{R_4}|$	$|I_{R_5}|$	$|I_{R_6}|$
0.86	0.20	0.08	0.10	0.21	0.14
0.20	0.57	0.34	0.08	0.20	0.22
0.08	0.33	0.56	0.33	0.08	0.09
0.09	0.08	0.33	0.56	0.33	0.08
0.22	0.20	0.08	0.34	0.57	0.20
0.14	0.21	0.10	0.08	0.20	0.86

TABLE 11. 5
Magnitude of the induced currents at the terminals of each one of the six receiving antennas

$|I_{R_1}|$	$|I_{R_2}|$	$|I_{R_3}|$	$|I_{R_4}|$	$|I_{R_5}|$	$|I_{R_6}|$
0.95	0.08	0.13	0.13	0.08	0.05
0.08	0.86	0.24	0.21	0.19	0.08
0.12	0.21	0.69	0.31	0.21	0.12
0.12	0.21	0.31	0.69	0.21	0.12
0.08	0.14	0.21	0.21	0.86	0.08
0.05	0.08	0.13	0.13	0.08	0.95

11.4   CONCLUSION

A novel method based on the principle of reciprocity is presented for maximizing the induced current at the terminals of a particular receiving antenna by choosing an appropriate set of weights for the transmitting antennas, thereby resulting in adaptivity on transmit. In this procedure it is not necessary to have any knowledge of the near/far-field electromagnetic environments in which the transmitting and receiving antennas are operating. Nor is it necessary to know the spatial coordinates of any of the antennas. By following the prescribed procedure based on reciprocity, it is possible to have the transmitted energy from all the antennas combine vectorially at the terminals of a prespecified receiver in the presence of coherent multipaths and near-field scatterers. The same principle can be used to simultaneously cancel the induced electromagnetic energy at the terminals of the other receiving antennas. Sample numerical results using an electromagnetic simulation tool have been presented to illustrate the applicability of this novel approach based on reciprocity.

REFERENCES

[1]   G. B. Giannakis, Y. Hua, P. Stoica, and L. Tong, *Signal Processing Advances in Wireless and Mobile Communications*, Vol. 1, Prentice Hall, Upper Saddle River, NJ, 2000.

[2]   L. Setian, *Antennas with Wireless Applications*, Prentice Hall, Upper Saddle River, NJ, 1998.

[3]   T. K. Sarkar, S. Park, J. Koh, and R. A. Schneible, "A Deterministic Least Squares Approach to Adaptive Antennas," *Digital Signal Processing,* Vol. 6, pp. 185-194, 1996.

[4]   S. Choi and D. Yun, "Design of Adaptive Antenna Array for Tracking the Source of Maximum Power and Its Applications to CDMA Mobile Communications," *IEEE Transactions on Antennas and Propagation*, Vol. 45, No. 9, Sept. 1997.

[5]   S. Choi, D. Shim, and T. K. Sarkar, "A Comparison of Tracking-Beam Arrays and Switching-Beam Arrays Operating in a CDMA Mobile Communication Channel," *IEEE Antennas and Propagation Magazine*, Vol. 41, No. 6, pp. 10–22, Dec. 1999.

[6]   R. C. Qui and I. T. Lu, "Multipath Resolving with Frequency Dependence for Wide-Band Wireless Channel Modeling," *IEEE Transactions on Vehicular Technology*, Vol. 48, No. 1, Jan. 1999.

[7]   G. D. Durgin and T. S. Rapport, "Theory of Multipath Shape Factors for Small-Scale Fading Wireless Channels," *IEEE Transactions on Antennas and Propagation*, Vol. 48, No. 5, May 2000.

[8]   R. F. Harrington, *Time-Harmonic Electromagnetic Fields*, McGraw-Hill, New York, 1961.

[9]   K. F. Lee, *Principles of Antenna Theory*, Wiley, New York, 1984.

[10]   B. M. Kolundzija, J. S. Ognjanovic, and T. K. Sarkar, *WIPL-D: Electromagnetic Modeling of Composite Metallic and Dielectric Structures, Software and User's Manual*, Artech House, Norwood, MA, 2000.

12

DIRECT DATA DOMAIN LEAST SQUARES SPACE-TIME ADAPTIVE PROCESSING

SUMMARY

We start this chapter by presenting a direct data domain least squares (D^3LS) approach to space-time adaptive processing (STAP) for enhancing signals in a nonhomogeneous environment. The nonhomogeneous environment may consist of nonstationary clutter and could include blinking jammers. The D^3LS method is applied to data collected by an antenna array utilizing space and in time (Doppler) diversity. Conventional STAP generally utilizes statistical methodologies based on estimating a covariance matrix of the interference using data from secondary range cells. As the results are derived from ensemble averages, the two-dimensional filter in space and time (optimum in a probabilistic sense) used to extract the signal of interest is obtained for the operational environment. The environment is assumed to be a wide-sense stationary process. However, for highly transient and nonhomogeneous environments the conventional statistical methodology is difficult to apply. Hence, the D^3LS method is presented, as it analyzes the data in space and time over each range cell separately. The D^3LS is a deterministic method. From an operational standpoint, an optimum method could be a combination of these two diverse—deterministic and statistical—methodologies, as illustrated by some examples. Initially, we describe several new D^3LS techniques. One is based on the computation of a generalized eigenvalue for the strength of the signal of interest (SOI) and the others are based on the solution of a set of block Hankel matrix equations. Since the matrix of the system of equations to be solved has a block Hankel structure, the conjugate gradient method and the fast Fourier transform (FFT) can be utilized for efficient solution of the adaptive problem. Illustrative examples presented in this chapter use measured data from the multichannel airborne radar measurements (MCARM) database to detect a Saberliner in the presence of urban, land, and sea clutter. An added advantage of the D^3LS in solving real-life problems is that many realizations can be obtained simultaneously for the same solution for the SOI. The degree of variability among the results obtained through separate independent solution methodologies can provide a confidence level for the processed results. Finally, a knowledge-based STAP methodology is described which may effectively combine a number of the algorithms in an optimum manner to produce an engineering solution.

12.1   INTRODUCTION

Building on the one-dimensional adaptive processing presented in Chapter 4, this chapter presents a two-dimensional deterministic adaptive signal processing technique for nulling interferers and estimating the signal of interest (SOI). The expansion into an additional domain (time) is necessitated by the fact that the interfering signals are two-dimensional [1] in nature. Unlike stochastic methods, that rely on training data to estimate the statistics of the interference in order to null interferers, this approach operates on a snapshot-by-snapshot basis to determine the adaptive weights. For this two-dimensional case, a snapshot is defined as the space-time data corresponding to a single range cell. Due to the reduced data set, and because of the techniques employed, this approach can be implemented in real time using real signal processing systems. An additional advantage of this method over the stochastic methods is that it can operate effectively in highly nonstationary environments, and can null noncoherent interferers as well as coherent interferers. In Appendices A and B, some of the rationals for using a D^3LS method over the conventional statistical methodology have been outlined. This procedure has applications in many signal-processing fields, including communications, sonar, radar, and medical imagining. In developing this two- dimensional technique, the application that will be addressed here is that of an airborne radar scenario, a description of which can be found in [1, 2].

Prior to discussing the direct data domain least squares space-time adaptive processing (D^3LS STAP) algorithms, we will touch on some of the issues associated with an airborne radar application and provide a description of the signals that is related to the information about the system. This is followed by a discussion of other approaches which have been pursued in this field, and finally, the D^3LS STAP techniques are presented, first for uniformly spaced linear arrays and then for circular arrays.

12.2   AIRBORNE RADAR

For airborne radars it is necessary to detect targets in the presence of clutter, jammers, and thermal noise. The airborne radar scenario has been described in [1–6] and is summarized here for completeness. This scenario is depicted in Figure 12.1. It is necessary to suppress the levels of the undesired interferers well below the weak desired signal. The problem is complicated due to the motion of the platform as the ground clutter received by an airborne radar is spread out in range, spatial angle, and over Doppler.

AIRBORNE RADAR SCENARIO

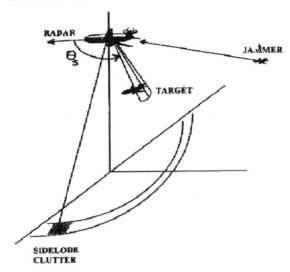

Figure 12.1. Airborne radar scenario.

So for airborne radar surveillance systems the detection of airborne and ground targets is complicated by many factors in the radar signal environment, as illustrated by Figure 12.2 [2]. Here the target signal competes with the sidelobe clutter at the same Doppler frequency, main lobe clutter at the same angle, and jammers (which arrive from a set angle but exist over a wide frequency extent) also at the same angle. Changes in the altitude and velocity of the aircraft also affect system performance. Due to the geometry, as the elevation of the platform increases, the amount of clutter entering the system increases (clutter extends in range). The Doppler spread of this clutter is a function of the platform velocity and the cosine of the direction of arrival (DOA) of the clutter energy (with respect to the velocity vector of the platform). Therefore, an increase in the velocity of the platform reduces the clutter-free spectrum of the system and increases the minimum discernible velocity (MDV), which complicates the detection of slow-moving ground targets. Compounding this are the near-field airframe effects and the size and weight limitations imposed by the platform, which limit radar power and aperture size. In addition to the platform challenges, the radar system must contend with a severely nonstationary environment (consisting of jammers, multipath, and clutter discretes) while attempting to detect such targets as low observables (LOs) or targets employing concealment, camouflage, and deception (CCD) [7].

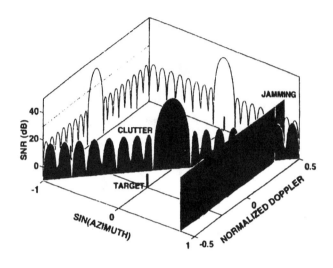

Figure 12.2. Airborne radar signal environment.

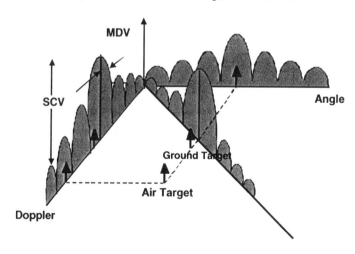

Figure 12.3. Angle-Doppler structure of clutter.

In airborne or space radar, the clutter in a given range cell has a structure determined by the motion of the aircraft platform, as shown in Figure 12.3. The slope of the clutter ridge in angle-Doppler space is determined by the speed of the aircraft. In an airborne moving target scenario (AMTI) case, the threat target is widely spaced from main lobe clutter in the Doppler domain, and it is possible to use Doppler processing to separate targets from clutter. The limitation on target detection is determined by the subclutter visibility (SCV). For the ground moving target indicator (GMTI) case, the problem is more difficult since the target is close to the main beam clutter in Doppler. Placing a null on main beam clutter reduces the gain on target and hence detection performance. The goal of GMTI is to reduce the minimum detectable velocity (MDV), the lowest velocity where a target can be separated from clutter.

One way to detect small signals of interest in such a noisy environment is to have a large array providing sufficient power and a large enough aperture to achieve narrow beams. In addition, the array must have extremely low sidelobes simultaneously on transmit and receive. This is very difficult and expensive to achieve in practice. An electrically large aperture will provide a narrow beam on transmit and receive with which to search, and the low sidelobes would help keep interferers from entering the system through the sidelobes. The problem with this solution is that manufacture of the array would be very difficult, as extremely tight mechanical tolerances would be necessary and thus will be expensive to build. In addition, the real estate on airborne platforms is limited. This is the situation with the early warning system airborne platforms called AWACS (Airborne Early Warning and Control System). Another solution, the one we discuss here, is to use space-time adaptive processing to suppress the interferers and enable the system to detect potentially weak target signals.

Therefore, instead of using a high-gain antenna with very low sidelobes, we plan to achieve the same goal through space-time adaptive processing (STAP). STAP is carried out by performing two-dimensional filtering on signals which are collected by simultaneously combining signals from the elements of an antenna array (the spatial domain) as well as from the multiple pulses from a coherent radar (the temporal domain). The data collection mechanism is shown in Figure 12.4. The temporal domain thus consists of multiple pulse repetition periods of a coherent processing interval (CPI). By performing simultaneous multidimensional filtering in space and time, the goal is not only to eliminate clutter that arrives at the same spatial angle as the target but also to remove clutter that comes from other spatial angles which has the same Doppler frequency as the target. Hence, STAP provides the necessary mechanism to detect low observables from an airborne radar.

The goal of adaptive processing is to weight the received space-time data vectors as seen in Figure 12.4 to maximize the output signal-to-interference plus noise ratio (SINR). Traditionally, the weights are determined based on an estimated covariance matrix of the interference. The weights maximize the gain in the look direction, while placing pattern nulls along the directions of the interferers. This interference plus noise is a combination of clutter, electronic countermeasures (ECMs), and thermal noise.

12.3 SIGNALS AND INFORMATION

We consider a pulsed Doppler radar situated on an airborne platform which is moving at a constant velocity. The radar consists of an antenna array where each element has its own independent receiver channel. The linear antenna array has $N+1$ elements uniformly spaced by a distance Δ, as shown in Figure 12.5. In this configuration the received voltage at each of the elements represents the combination of voltages from each of the column elements of the two-dimensional array. They are added vectorially before entering the receive channel and can easily be extended to the case where there is a receive channel behind every element of a planar array.

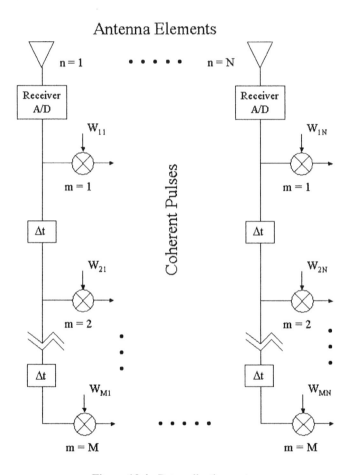

Figure 12.4. Data collection system.

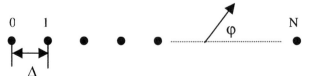

Figure 12.5. Uniform linear array

We also assume that the system processes M coherent pulses within a coherent processing interval (CPI) (i.e., the radar transmits a coherent burst of M pulses at a constant pulse repetition frequency) where each pulse repetition interval consists of the transmission of a pulsed waveform of finite bandwidth and the reception of reflected energy captured by the aperture and passed through a receiver with a bandwidth equal to that of the pulse. In the receive chain the signal is downconverted, matched filtered, sampled, and digitized, and the baseband samples are stored. In this manner complex samples are generated at R range bins for M pulses at N elements. To facilitate working with the array output, the baseband samples can be arranged into a three-dimensional matrix commonly referred to as a data cube, as shown in Figure 12.6. The three axes of the data cube correspond to the pulse (M), element (N), and range (R) dimensions. At a particular range r_i, the sheet or slice of the data cube is referred to as a space-time snapshot, indicated by the shaded plane in Figure 12.6.

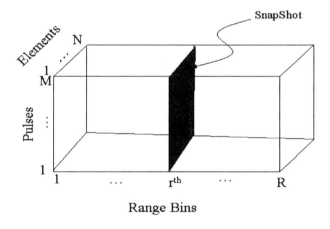

Figure 12.6. Representative data cube.

Therefore, with M pulses and N antenna elements, each having its own independent receiver channel, the received data for a coherent processing interval consists of RMN complex baseband samples. These samples, often referred to as the *data cube*, consist of $R \times M \times N$ complex baseband data samples of the received pulses. The data cube then represents the voltages defined by $V(m; n; r)$ for $m = 1$, ..., M; $n = 1$, ..., N; and $r = 1$, ..., R. These complex baseband measured voltages contain the signal of interest (SOI), jammers, and clutter, including thermal noise. A space-time snapshot then is referred to as MN samples for a fixed range gate value of r.

We assume that the signals entering the array are narrowband and consist of the signal of interest (SOI) and interference plus noise. The noise (thermal noise) originates in the receiver and is assumed to be independent across elements and pulses. The interference is external to the receiver and consists of clutter (reflection of the transmitted electromagnetic energy from the earth), jammers, mutual coupling, and multipath (due to the SOI, clutter, and/or jamming). We assume that for each jammer, the energy impinging on the array is confined to a particular direction of arrival (DOA) and is spread in frequency. The jammers may be blinking or stationary. From the data cube shown in Figure 12.6, we focus our attention to the range cell r and consider the space-time snapshot for this range cell.

We assume that the SOI for this range cell r is incident on the uniform linear array from an angle φ_s and is at Doppler frequency f_d. φ_s is measured from the tail of the aircraft, as shown in Figure 12.5. Our goal is to estimate its amplitude, given φ_s and f_d only. In a surveillance radar, φ_s and f_d set the look directions and a SOI (target) may or may not be present along this look direction and Doppler. Let us define $S(p; q)$ to be the complex voltage received at the qth antenna element corresponding to the pth time for the same range cell r. We further stipulate that the voltage $S(p; q)$ is due to a signal of unity magnitude incident on the array from the azimuth angle φ_s corresponding to Doppler frequency f_d. Hence, the signal-induced voltage under the assumed array geometry and a narrowband signal is a complex sinusoidal given by

$$S(p, q) = \exp\left[j \left(\frac{2\pi \Delta q}{\lambda} \cos\varphi_s + \frac{2\pi f_d\, p}{f_r} \right) \right]$$

$$\text{for } p = 1, ..., M; \; q = 1, ..., N \tag{12.1}$$

where λ is the wavelength of the radio-frequency radar signal and f_r is the pulse repetition frequency.

Let $X(p; q)$ be the actual measured complex voltages that are in the data cube of Figure 12.6 for the range cell r. The actual voltages X will contain the signal of interest of amplitude α (α is a complex quantity), jammers which may be due to coherent/incoherent multipaths of the radiated signal, and clutter which is the reflected electromagnetic energy from the ground which has been transmitted through both the main lobe and sidelobes. The interference competes with the SOI at the Doppler frequency of interest. There is also a contribution to the measured voltage from the receiver thermal noise. Hence the actual measured voltages $X(p; q)$ are

$$X(p, q) = \alpha \exp\left[j \left(\frac{2\pi \Delta q}{\lambda} \cos\varphi_s + \frac{2\pi f_d\, p}{f_r} \right) \right]$$

$$+ \text{ clutter } + \text{ jammer } + \text{ thermal noise} \tag{12.2}$$

$$= V(p; q; r)$$

The goal is to extract the SOI, α, given these voltages X, the direction of arrival for the SOI, φ_s, and the Doppler frequency, f_d.

12.4 PROCESSING METHODS

Having stored the complex space-time-range samples in the data cube, adaptive signal processing routines can now operate on the data. Traditionally, the statistical approach to adaptive array processing has been taken [1–8]. Earlier approaches adjusted the spatial weights iteratively in an attempt to either maximize SNR [7] or minimize the mean-squared-error [7]. Then Brennan et al. [3] showed that in the case of Gaussian interference the space-time filter weights that maximize the probability of detection, for a fixed false alarm rate, are given by

$$[W] = \gamma \, [\Re]^{-1} \, [s]^* \qquad (12.3)$$

where γ is a complex constant, $[\Re]$ is the covariance matrix of the interference, and $[s]$ is a steering vector. The superscript * represents the complex conjugate of a quantity. $[s]$ sets the look direction, the direction in angle and Doppler being tested for the presence of a target. Note that $[s]$ sets the look direction only, while the actual target may be at a different angle-Doppler point close to the look direction. The covariance matrix, $[\Re]$, is estimated by averaging over *secondary data* chosen from range cells close to the range cell of interest (the primary range cell) and is written as

$$[\hat{\Re}] = \frac{1}{K} \sum_{r=1}^{K} [X_r][X_r^*] \qquad (12.4)$$

where K is the number of range samples and X_r is the received voltage vector for the rth range cell. This is shown pictorially in Figure 12.7. Here it is assumed that the training data does not contain any signal energy. This is the key feature possessed by the statistical algorithms. If this assumption that the secondary cells do not contain any signal energy is violated, the methods fail to provide quality results, as we shall observe later.

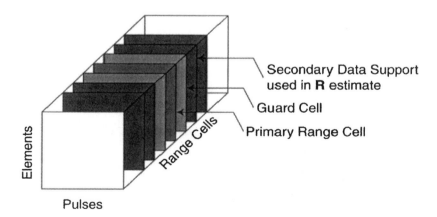

Figure 12.7. Estimating the space-time interference covariance matrix.

Equations (12.3) and (12.4) illustrate the two main difficulties in applying the fully adaptive procedure: the number of degrees of freedom and the assumption of homogeneous data. Underlying several STAP approaches is a third problem, ignoring array effects, such as mutual coupling between the antenna elements and coupling with the main air frame.

In (12.3) the number of unknowns and size of the covariance matrix directly determine the degrees of freedom. The total computational load rises as the third power of the number of unknowns. Choosing this parameter is therefore crucial to a practical implementation of STAP. In the fully adaptive approach, the number of unknowns is the number of antenna subarrays (N) times the number of pulses (M) in the data cube. The algorithm estimates the NM-dimensional covariance matrix of the interference. In practice, an accurate estimate requires about $2NM$ to $3NM$ independent and identically distributed (i.i.d.) secondary data samples [9]. This number is very large, making it impossible to evaluate the covariance matrix and the adaptive weights in a reasonable computation time. The goal of STAP research has therefore been to reduce the number of unknowns, namely the adaptive weights, while retaining the quality of performance.

Equation (12.4) estimates the covariance matrix using K secondary data vectors from range bins close to the range cell of interest. The inherent assumption is that the statistics of the interference in the secondary data is the same as that within the primary range cell (i.e., the data is assumed homogeneous). K must be greater than twice the number of unknowns, between $2NM$ and $3NM$ in the fully adaptive case [7]. It is impossible to obtain a large number of i.i.d. homogeneous secondary data vectors. No clutter scene is perfectly homogeneous and most, if not all, land clutter is sufficiently non-homogeneous to affect performance. In addition, some regions are worse than others, as for example, due to the nonhomogeneity of the environments due to urban clutter, or land/sea interfaces, as shown in Figure 12.8. This leads to severely degraded performance.

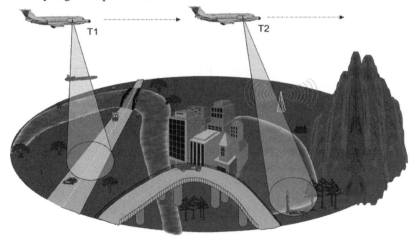

Figure 12.8. Clutter nonhomogeneity.

Traditionally STAP algorithms were developed for proof of concept, assuming the receiving antenna array is a linear array of isotropic point sensors. In practice, such an array is not feasible and the elements must be of some physical size. This implies that the array not only receives, but also scatters the incident fields, leading to mutual coupling between elements. Additionally, near-field scattering off the aircraft body has a significant impact on how the array receives incident signals. Ignoring array effects leads to significantly degraded STAP performance.

This joint domain (element–pulse) optimum processor requires knowledge of the statistics of the interference. Since this is not known *a priori* in a real system, especially one in which the statistics are rapidly changing, a suboptimum approach is needed. To overcome the drawbacks of the fully adaptive algorithm, researchers have limited the number of adaptive weights, which are often referred to as the degrees of freedom (DOFs) to reduce the problems associated with sample support and computation expense. Wang and Cai [9] introduced the joint domain localized (JDL) algorithm, a post-Doppler, beamspace approach that adaptively processes the radar data after transformation to the angle-Doppler domain. Adaptive processing is restricted to a localized processing region (LPR) in the transform domain, significantly reducing the number of unknowns while retaining maximal gain against thermal noise. The reduced DOF leads to corresponding reductions in required sample support and computation load. The JDL algorithm for the case of an ideal linear array of equispaced, isotropic, point sensors transforms the space-time data to the angle-Doppler domain using a two-dimensional fast Fourier transform (FFT). Under certain restrictions, this approach is valid because the spatial and temporal steering vectors form the Fourier coefficients [2, pp. 12–17]. In order to highlight the restrictions placed on the algorithm by the original formulation, this section clarifies the original development of Wang and Cai [9].

By observing that for an ideal array the spatial and temporal steering vectors are identical to the Fourier coefficients, the transformation to the angle-Doppler domain can be simplified under two conditions.

1.) If a set of angles are chosen such that $d/\lambda \sin \varphi$ is spaced by $1/N$ and a set of Doppler frequencies are chosen such that f/f_R is spaced by $1/M$, the transformation to the angle-Doppler domain is equivalent to the computation of a two-dimensional discrete Fourier transform (2D-DFT).

2.) If the angle φ corresponds to one of these angles and the Doppler f corresponds to one of these Doppler frequencies, the steering vector is a column of the 2D-DFT matrix and the angle-Doppler steering vector is localized to a single angle-Doppler bin, as shown in Figure 12.9.

The JDL algorithm as originally developed in [9] assumes that both these conditions are met. This simplification is possible only in the case of the ideal, equispaced, linear array of Figure 12.5. Owing to beam mismatch, the localization to a single point in angle-Doppler space is exact only for the look steering vector.

As shown in Figure 12.9, a LPR centered about the look angle-Doppler point is formed and interference is suppressed in this angle-Doppler region only. The LPR covers η_a angle bins and η_d Doppler bins. The choice of η_a and η_d is independent of N and M (i.e. the localization of the target to a single angle-Doppler bin decouples the number of adaptive degrees of freedom from the size of the data cube while retaining maximal gain against thermal noise). The covariance matrix corresponding to this LPR is estimated using secondary data from neighboring range cells. The adaptive weights are then calculated using (12.3). The estimated covariance matrix $[\hat{\Re}]$ is replaced with $[\tilde{\Re}]$, the estimated angle-Doppler covariance matrix corresponding to the LPR of interest. The steering vector $[s]$ is replaced with the angle-Doppler steering vector $[\tilde{s}]$:

$$[\tilde{w}] = [\tilde{\Re}]^{-1} [\tilde{s}] \tag{12.5}$$

The number of adaptive unknowns is equal to $\eta_a \eta_d$. The steering vector for the adaptive process is the space-time steering vector $[s]$ of (12.3) transformed to the angle-Doppler domain. Under the two conditions listed above, $[\tilde{s}]$ is given by the length $\eta_a \eta_d$ vector:

$$[\tilde{s}] = [0, 0, ..., 0, 1, 0, ..., 0, 0]^T$$

It must be emphasized that this simple form of the steering vector is valid only because the DFT is an orthogonal transformation. The space-time steering vector is transformed to angle-Doppler domain using the same transformation as that used for the data.

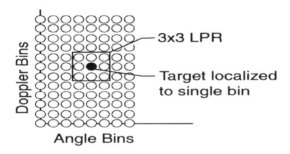

Figure 12.9. Localized processing regions (LPR) in JDL for $\eta_a - \eta_d = 3$.

When applying the JDL algorithm to measured data, a crucial assumption in the development of [9] is invalid. The elements of a real array cannot be point sensors. Owing to their physical size, the elements of the array are subject to mutual coupling. Furthermore, the assumption of a linear array is restrictive. A planar array allows for DOF in azimuth and elevation. Therefore, the Fourier coefficients do not form the spatial steering vector, and a DFT does not transform the spatial data to the angle domain. In this case, a DFT is mathematically feasible but has no physical meaning.

In a physical array, the spatial steering vectors must be measured or obtained using a numerical electromagnetic analysis. These steering vectors must be used to transform the space domain to the angle domain. This transformation is necessarily nonorthogonal with a corresponding spread of target information in the angle-Doppler domain. JDL applied to a real array ignored the nonorthogonal nature of the measured spatial transform [10].

Factored or cascaded signal processing approaches were developed as an alternative to the joint domain optimum (JDO) approach. In these approaches either the element pulse data is beamformed along the elements and then the beam-pulse data is Doppler processed, or the data is Doppler processed along the pulses, then the element-Doppler data is beamformed. These approaches perform one-dimensional filtering, which results in nulls extending across the entire space of the other dimension. The taxonomy of the various space-time processing methods has been identified in [2]. There has been some debate over which factored approach is superior. In the three different cases evaluated in [10–16], the factored approaches either performed equally to the joint domain suboptimum approach or significantly poorer. This performance by the factored approaches, which still require a significant amount of secondary data, is why Wang and Cai sought and developed an algorithm which operated in the joint domain across a smaller dimensional space. This algorithm, the joint domain locally (JDL) optimized processor, was presented in [10–16] and showed significant improvement over the factored approaches. This processor still requires training data, although significantly less than other suboptimum approaches, and nonstationary environments also affect its performance significantly. The shortcomings of these approaches include the inability to perform in real time, as well as their poor performance in nonstationary environments. Therefore other signal processing methods are needed.

While many researchers were investigating suboptimum space-time adaptive approaches which relied on estimating the underlying statistics, other researchers investigated deterministic ways to suppress the interference. Luthra [17] proposed an algorithm which could null $M - 1$ coherent jammers, using data from $2M$ elements (spatial information) while maintaining a fixed gain in the look direction. This approach was improved upon in [18], where Sarkar and Sangruji used a new, simpler data matrix and solved for the weights using the conjugate gradient method. Further improvement was made in [19, 20], where Schneible implemented multiple constraints to preserve an SOI which might arrive slightly off the look direction. Then Sarkar et al. [21, 22] presented two more variations of this approach, the backward procedure and the forward-backward procedure. These algorithms where then extended to the space-time domain, where Park [23] presented a generalized eigenvalue processor, and Sarkar et al. [21] implemented the direct data domain least squares (D^3LS) processor, described in Chapter 4.

12.5   DIRECT DATA DOMAIN SPACE-TIME APPROACH

12.5.1 Space-Time Processing

The adaptive algorithms presented in Chapter 4 used data from a space snapshot, which consists of samples from across the array at an instant in time (a given pulse at a given range bin). In this section we present four algorithms that operate across pulses and elements, increasing the degrees of freedom over that of the element domain alone. The first processor described implements a generalized eigenvalue equation [21, 22], while the last three processors implement a least squares solution to a linear matrix equation [21, 22].

Using the linear array defined earlier (Figure 12.5), we now take advantage of the temporal diversity provided by the coherent pulses within a CPI as well as the spatial diversity provided by the elements of the array. Using this data from a given range bin r, shown in Figure 12.6, the adaptive weights can be obtained deterministically, on a snapshot-by-snapshot basis, and implemented in the architecture shown in Figure 12.4. Assuming a narrowband signal, the complex envelope of the received SOI with unity amplitude for the mth pulse and nth element can be described through (12.1). The signals entering the array consist of the SOI plus interference (clutter, jammers, multipath, etc.) and noise. For the nth element and mth pulse, at the rth range bin, the complex envelope of the received signal is

$$X(m, n) \;=\; \alpha S(m, n) \;+\; \text{interference} \;+\; \text{noise} \tag{12.6}$$

where α is the amplitude of the SOI entering the array.

In the D^3 procedures to be described, the adaptive weights are applied to the single space-time snapshot for the range cell r. Here a two-dimensional array of weights numbering $N_a N_p$ is used to extract the SOI for the range cell r. Hence the weights are defined by $w(p; q; r)$ for $p = 1, ..., N_p < M$ and $q = 1, ..., N_a < N$ and are used to extract the SOI at the range cell r. Therefore, for the D^3LS method we essentially perform a high-resolution filtering in two dimensions (space and time) for each range cell.

12.5.2 Two-Dimensional Generalized Eigenvalue Processor

For each pulse-element cell (given range bin r) the difference equation

$$X(m, n) \;-\; \alpha S(m, n) \tag{12.7}$$

removes the SOI from the nth element and mth pulse sample, leaving noise plus interference. It is important to note that α, the amplitude of the SOI, is still an unknown quantity. Based on (12.7) a two-dimensional matrix pencil can be created whose solution will result in a weight vector which will null out the interferers and extract the SOI. The elements of this matrix pencil can be constructed by sliding a window (box) over the space-time snapshot data, as shown by the shaded plane in Figure 12.6. By creating a vector using the elements in the window, each window position generates a row in the S and X matrices as shown next:

$$
\begin{array}{c}
\text{Box 1} \longrightarrow \\
\text{Box 2} \longrightarrow \\
S = \\
\text{Box 3} \longrightarrow
\end{array}
\begin{pmatrix}
S_{11} & \cdots & S_{1Na} & \cdots & S_{Np1} & \cdots & S_{NpNa} \\
S_{12} & \cdots & S_{1(Na+1)} & \cdots & S_{Np2} & \cdots & S_{Np(Na+1)} \\
\vdots & & \vdots & & \vdots & & \vdots \\
S_{21} & \cdots & S_{2Na} & \cdots & S_{(Np+1)} & \cdots & S_{(Np+1)Na} \\
\vdots & & \vdots & & \vdots & & \vdots \\
S_{(M\text{-}Np)(N\text{-}Na)} & \cdots & S_{(M-Np)Na} & \cdots & S_{M(N-Na)} & \cdots & S_{MNa}
\end{pmatrix}
$$

$$(12.8a)$$

$$
\begin{array}{c}
\text{Box 1} \longrightarrow \\
\text{Box 2} \longrightarrow \\
X = \\
\text{Box 3} \longrightarrow
\end{array}
\begin{pmatrix}
X_{11} & \cdots & X_{1Na} & \cdots & X_{Np1} & \cdots & X_{NpNa} \\
X_{12} & \cdots & X_{1(Na+1)} & \cdots & X_{Np2} & \cdots & X_{Np(Na+1)} \\
\vdots & & \vdots & & \vdots & & \vdots \\
X_{21} & \cdots & X_{2Na} & \cdots & X_{(Np+1)} & \cdots & X_{(Np+1)Na} \\
\vdots & & \vdots & & \vdots & & \vdots \\
X_{(M\text{-}Np)(N\text{-}Na)} & \cdots & X_{(M-Np)Na} & \cdots & X_{M(N-Na)} & \cdots & X_{MNa}
\end{pmatrix}
$$

$$(12.8b)$$

The window size along the element dimension is N_a and is N_p along the pulse dimension. Selection of N_a determines the number of spatial degrees of freedom, while N_p determines the temporal degrees of freedom. Typically, for a single domain processing, N_a and N_p must satisfy the equations

$$N_a \leq \frac{N+1}{2} \qquad (12.9)$$

$$N_p \leq \frac{M+1}{2} \qquad (12.10)$$

The advantage of a joint domain processing is that either of these bounds can be relaxed (i.e., one can exchange spatial degrees of freedom with the temporal degrees of freedom). So, indeed, it is possible to cancel a number of interferers which is greater than the number of antenna elements in a joint domain processing. The total number of degrees of freedom, Q, for any method is

$$Q = N_a \times N_p \qquad (12.11)$$

Given the system constraints, most airborne radar systems contain more temporal degrees of freedom than spatial (i.e., $N \ll M$). Therefore, since the terms $X(p, q) - \alpha S(p, q)$ eliminate the SOI, these elements represent the contribution due to the unwanted signal multipaths, jammers, unwanted signals at the same Doppler, and receiver thermal noise. In D^{3}LS adaptive processing, the goal is to take a weighted sum of these matrix elements defined in (12.7) and extract the SOI, which is going to be α for the range cell r.

The total number of degrees of freedom (DOFs = Q) then represents the total number of weights, and this is the product $N_a N_p$, where N_a is the number of spatial DOFs and N_p is the number of temporal DOFs. Next it is illustrated how a D^{3}LS approach is taken for the extraction of SOI.

The least squares procedure for the 1D case (i.e., with only the subscript q and no p) is available in [20, 21]. Here the same least squares procedure is extended to two dimensions. Consider the two matrices C_1 and C_2. The elements of C_1 and C_2 are formed by

$$C_1(x\,;y) \; = \; S(g + h - 1;\quad d + e - 1) \tag{12.12}$$

where

$$C_2(x\,;y) \; = \; X(g + h - 1;\quad d + e - 1) \tag{12.13}$$

$$1 \le x = g + (d-1)N_p \tag{12.14}$$

$$1 \le y = h + (e-1)N_p \le N_a N_p \tag{12.15}$$

$$1 \le d \le N - N_a \tag{12.16}$$

$$1 \le e \le N_a \tag{12.17}$$

$$1 \le g \le M - N_p \tag{12.18}$$

$$1 \le h \le N_p \tag{12.19}$$

so that $1 \le x, y \le N_a N_p$. Now if we consider a matrix pencil of size $N_a N_p$,

$$[C_2]_{N_a N_p \times N_a N_p} \; - \; \alpha [C_1]_{N_a N_p \times N_a N_p} \tag{12.20}$$

this represents the contribution of the unwanted signals, as the desired components have been canceled out. The elements of the matrices $[C_2]$ and $[C_1]$ are created out of the data matrices $[X(p, q)]$ and $[S(p, q)]$, respectively, as defined by (12.2) and (12.1).

Now in the STAP processing, the elements of the weight vector $[W]$ are chosen in such a way that the contribution from the jammers, clutter, and thermal noise is zero. Hence, if we define the generalized eigenvalue problem,

$$[R][W] \; = \; \{[C_2] - \alpha [C_1]\}[W] \; = \; 0 \tag{12.21}$$

then α, the strength of the signal, is a generalized eigenvalue and the weights are given by the corresponding generalized eigenvector. Here W is a column vector of length $N_a N_p$ for the range cell r. Since we have assumed that only the SOI is arriving from φ_s corresponding to the Doppler f_d, the matrix $[C_1]$ is of rank one, and hence the generalized eigenvalue equation has only one nonzero eigenvalue, which provides an estimate of the complex amplitude of the signal.

Alternatively, one can view the left-hand side of (12.21) as the total noise signal at the output of the adaptive processor due to jammer, clutter, and thermal noise. One is therefore trying to reduce the noise voltage at the output of the adaptive processor, which is given by

$$N_{\text{out}} = [R][W] = \{[C_2] - \alpha[C_1]\}[W] \tag{12.22}$$

The total output noise power then can be obtained as

$$N_{\text{power}} = [W]^H \{[C] - \alpha[C_1]\}^H \{[C_2] - \alpha[C_1]\}[W] \tag{12.23}$$

where H represents the conjugate transpose of a matrix. Our objective is to set the noise power as small as possible by selecting $[W]$ for a fixed signal strength α. This is done by differentiating the real quantity N_{power} with respect to the elements of $[W]$ and setting each component equation to zero. This yields equation (12.21).

The total number of DOFs, $N_a N_p$, is determined by both M and N. Clearly, we need $N_p < M$ and $N_a < N$ so that enough equations can be generated to form equation (12.21). Generally, $M >> N$, and therefore there are a larger number of temporal DOFs than spatial DOFs. The goal therefore is to extract the SOI at a given Doppler and angle of arrival in a given range cell r by using a two-dimensional filter of size $N_a N_p$. The filter is going to operate on the data snapshot depicted in Figure 12.6 of size NM to extract the SOI.

In real-time applications, it is difficult to solve numerically for the generalized eigenvalue problem in sufficient time, particularly if the value $N_a N_p$ representing the total number of weights is large and the matrix $[C_2]$ is highly rank deficient. For this reason, we convert the solution of a generalized eigenvalue problem given by (12.21) to the solution of a linear matrix equation.

12.5.3 Least Squares Forward Processor

The formulation of the direct data domain least squares space-time algorithm [21] can be obtained through extension of the one-dimensional case. As has been done in Chapter 4, we start by developing the forward case and then present the backward and forward-backward algorithms. As before, the matrix equation to be solved can be defined as

$$[T][W] = 0 \tag{12.24}$$

where $[T]$ is the system matrix and $[W]$ is the vector of space-time weights, which when determined will null the interferers.   The system matrix, $[T]$, contains the angle-Doppler look direction of the SOI as well as the cancellation rows, which contain the angle-Doppler information on the interferers. This interferer information is obtained through difference equations similar to equation (4.46), where the contribution of the SOI is removed, leaving information of interferers only. In the two-dimensional case these difference equations are performed with elements offset in space only, time only, and space and time.   Define the element-to-element offset of the SOI in space and time, respectively, as

$$Z_1 \ = \ e^{j2\pi \frac{d}{\lambda} \cos(\varphi_s)} \tag{12.25}$$

$$Z_2 \ = \ e^{j2\pi \frac{f_d}{f_r}} \tag{12.26}$$

Again, SOI has an angle of arrival of φ_s and a Doppler frequency of f_d. The three types of difference equations are then given by

$$X(n, m) \ - \ X(n + 1, m) \, Z_1^{-1} \tag{12.27}$$

$$X(n, m) \ - \ X(n, m + 1) \, Z_2^{-1} \tag{12.28}$$

$$X(n, m) \ - \ X(n + 1, m + 1) \, Z_1^{-1} \, Z_2^{-1} \tag{12.29}$$

Note that in (12.27), the signal component (SOI) is canceled from samples taken from different antenna elements at the same time. Similarly, (12.28) represents signal cancellation from samples taken at the same antenna elements at different time.   Finally, (12.29) represents signal cancellation from neighboring samples in both space and time.   Therefore, we are performing a filtering operation simultaneously using $N_a N_p$ samples of the space-time data. The cancellation rows of the matrix $[T]$ can now be formed using (12.27)–(12.29) and the windowing in Figure 12.10. In this case the dots in Figure 12.10 represent the induced voltages, $X(m, n)$ as defined in (12.2), for a given element-pulse location. Just as was done for the generalized eigenvalue algorithm for the 1D case in Chapter 4, a space-time window is passed over the data.

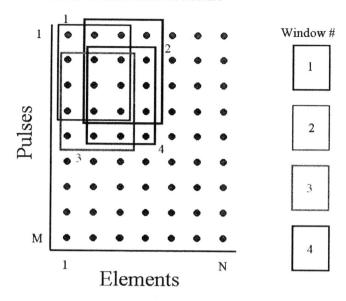

Figure 12.10. Space-time data.

For each given location of the window function, three rows in matrix [T] are formed by implementing (12.27)–(12.29), which removes the SOI. The rows are formed by performing an element-by-element subtraction between the elements of the windows and then arranging the resulting data into a row vector, as shown in Figure 12.11. The window is then slid one space to the right and three more rows are generated, and so on. After this window has reached the second column to the far right and three rows are generated, the window is lowered a row and shifted back to the left side of the data array, and the generation of rows continues. This is repeated until $Q - 1$ cancellation rows have been formed. The elements of this row can be obtained by placing an $N_a \times N_p$ window, such as window 1 in Figure 12.10, over data. In order to restore the signal component in the adaptive processing, we fix the gain of the subarray (in both space and time) formed by fixing the first row of the matrix [T]. The elements of the first row are given by

$$T(1; y) = Z_1^{(e-1)} Z_2^{(h-1)}$$ (12.30)

where y, e, and h are given by equations (12.15), (12.17), and (12.19), respectively. By fixing the gain of the system for the given Doppler and the direction of arrival, the following row vector can be generated:

$$[1 \quad Z_1 \quad Z_1^2 \quad \cdots \quad Z_1^{Na-1} \quad Z_2 \quad Z_1Z_2 \quad Z_1^2Z_2 \quad \cdots \quad Z_1^{Na-1}Z_2 \quad Z_2^2 \quad Z_1Z_2^2 \quad \cdots \quad Z_1^{Na-1}Z_2^{Np-1}]$$

As was done for the 1D case, the first row of the system matrix is used to set the gain of the system in the look direction. For the 2D case, the look direction is specified by the angle of arrival, φ_s, and the Doppler frequency, f_d, of the SOI.

Example: Computing a Cancellation Row Using , $N_a = 3$ and $N_p = 4$

$$X_{\text{window1}} - X_{\text{window2}} \bullet Z_1^{-1},$$

$$\Rightarrow \begin{pmatrix} x_{11} & x_{12} & x_{13} \\ x_{21} & x_{22} & x_{23} \\ x_{31} & x_{32} & x_{33} \\ x_{41} & x_{42} & x_{43} \end{pmatrix} - \begin{pmatrix} x_{12} & x_{13} & x_{14} \\ x_{22} & x_{23} & x_{24} \\ x_{32} & x_{33} & x_{34} \\ x_{42} & x_{43} & x_{44} \end{pmatrix} \bullet Z_1^{-1}$$

$$= \begin{pmatrix} x_{11} - x_{12}z_1^{-1} & x_{12} - x_{12}z_1^{-1} & x_{13} - x_{12}z_1^{-1} \\ x_{21} - x_{12}z_1^{-1} & x_{22} - x_{12}z_1^{-1} & x_{23} - x_{12}z_1^{-1} \\ x_{31} - x_{32}z_1^{-1} & x_{32} - x_{12}z_1^{-1} & x_{33} - x_{12}z_1^{-1} \\ x_{41} - x_{42}z_1^{-1} & x_{42} - x_{12}z_1^{-1} & x_{43} - x_{12}z_1^{-1} \end{pmatrix}$$

$$= \begin{pmatrix} \underline{x}_{11} & \underline{x}_{12} & \underline{x}_{13} \\ \underline{x}_{21} & \underline{x}_{22} & \underline{x}_{23} \\ \underline{x}_{31} & \underline{x}_{32} & \underline{x}_{33} \\ \underline{x}_{41} & \underline{x}_{42} & \underline{x}_{43} \end{pmatrix} \quad \textit{Converting to a row vector}$$

$$\left(\underline{x}_{11} \ \ \underline{x}_{12} \ \ \underline{x}_{13} \ \ \underline{x}_{21} \ \ \underline{x}_{22} \ \ \underline{x}_{23} \ \ \underline{x}_{31} \ \ \underline{x}_{32} \ \ \underline{x}_{33} \ \ \underline{x}_{41} \ \ \underline{x}_{42} \ \ \underline{x}_{43} \right)$$

Figure 12.11. Creating a cancellation row.

By setting the product of $[T]$ and $[W]$ equal to a column vector $[Y]$ the matrix equation is completed and it becomes a square system. The first element of $[Y]$ consists of the constraint gain C, and the remaining $Q - 1$ elements are set to zero in order to complete the cancellation equations. The resulting matrix equation is then given by

$$[T][W] = [Y] = \begin{bmatrix} C \\ 0 \\ \vdots \\ 0 \end{bmatrix} \tag{12.31}$$

where C is a complex constant. In solving this equation one obtains the weight vector $[W]$, which places space-time nulls in the direction of the interferers while maintaining gain in the direction of the SOI. The amplitude of the SOI can be estimated using

$$\hat{\alpha} = \frac{1}{C} \sum_{m=1}^{N_p} \sum_{n=1}^{N_a} W(n+m-1) X(n,m) \tag{12.32}$$

The analysis above was conducted for a single constraint. As in the 1D case, the SOI in the 2D case could arrive at the array slightly off the look direction, either in angle or Doppler or both. In order to keep the processor from nulling the SOI, multiple constraints can be implemented. The added constraints would reduce the number of degrees of freedom, but given the antenna beam width and Doppler filter width of a real system, the constraints could help maintain the system gain over this finite look direction extent. In a manner similar to the single constraint, L constraints can be implemented using L row constraints where the look direction of the ℓth row is determined by φ_ℓ and f_ℓ. For the ℓth constraint,

$$Z_{\ell 1} = e^{j\,2\pi \frac{d}{\lambda} \cos(\varphi_\ell)} \tag{12.33}$$

$$Z_{\ell 2} = e^{j\,2\pi \frac{f_\ell}{f_r}} \tag{12.34}$$

and the ℓth row of $[T]$, denoted as $T(\ell, :)$, becomes

$$T(\ell, :) = [1 \quad Z_{\ell 1} \quad (Z_{\ell 1})^2 \quad \cdots \quad (Z_{\ell 1})^{N_a-1} \quad Z_{\ell 2} \quad Z_{\ell 1} Z_{\ell 2} \quad (Z_{\ell 1})^2 Z_{\ell 2} \quad \cdots$$
$$(Z_{\ell 1})^{N_a-1} Z_{\ell 2} \quad (Z_{\ell 2})^2 \quad Z_{\ell 1}(Z_{\ell 2})^2 \quad \cdots \quad (Z_{\ell 1})^{N_a-1} \quad (Z_{\ell 2})^{N_p-1}] \tag{12.35}$$

The L constraints provide a more accurate solution when there is some uncertainty associated with either the Doppler or the direction of arrival.

12.5.4 Least Squares Backward Processor

A second direct least squares space-time processor can be implemented by conjugating the element-pulse data and processing this data in reverse. It is well known in the parametric spectral estimation literature that a sampled sequence consisting of a sum of complex exponentials can be estimated by observing it in either the forward or reverse direction. If we now conjugate the data and form the reverse sequence, we obtain an equation similar to (12.31) for the weights. In this case the first rows of $[T]$ and $[Y]$ are the same as before, as in (12.31). The remaining equations of (12.31) now have to be modified. Under the present circumstances, one would obtain the three consecutive rows of the $[T]$ matrix by taking a weighted difference between the neighboring elements to form

$$T(x, y) = X^*(M - h - g + 2 ; \ N - d - e + 2)$$
$$- Z_1^{-1} X^*(M - h - g + 2 ; \ N - d - e + 1) \tag{12.36}$$

$$T(x + 1 ; y) = X^*(M - h - g + 2 ; \ N - d - e + 2)$$
$$- Z_2^{-1} X^*(M - h - g + 1 ; \ N - d - e + 2) \tag{12.37}$$

$$T(x + 2 ; y) = X^*(M - h - g + 2 ; \ N - d - e + 2)$$
$$- Z_1^{-1} Z_2^{-1} X^*(M - h - g + 1 ; \ N - d - e + 1) \tag{12.38}$$

for any row number x, and h, g, d, and e have been defined in (12.16)–(12.19).The row number increases by multiples of three, and

$$1 \leq y = N_p(e - 1) + h \ \leq N_a N_p \tag{12.39}$$

The form of this linear matrix equation is similar to that of the forward algorithm, resulting in

$$[B][W] = [Y] \tag{12.40}$$

where the matrices $[B]$, $[W]$, and $[Y]$ are $Q \times Q$, $Q \times 1$, and $Q \times 1$ matrices, respectively. The constraint rows in $[B]$ are implemented in the same manner as the constraint rows in $[T]$. The difference between $[B]$ and $[T]$ is in the cancellation equations. For the backward method, these equations are formed by first conjugating the space-time snapshot given in Figure 12.11. Then using a windowing procedure similar to the forward case, three cancellation rows are generated for each position of the window, except now the window starts in the lower right corner of the space-time snapshot, as shown in Figures 12.10 and 12.11. This window is then moved to the right and up the snapshot. The three difference equations that are used to cancel the SOI are given by

$$X^*(n, m) \ - \ X^*(n - 1, m) \, Z_1^{-1} \tag{12.41}$$

$$X^*(n, m) \ - \ X^*(n, m - 1) \, Z_2^{-1} \tag{12.42}$$

$$X^*(n, m) \ - \ X^*(n - 1, m - 1) \, Z_1^{-1} \, Z_2^{-1} \tag{12.43}$$

Using equations (12.41)–(12.43), the SOI is removed from the windowed data. Once the weights are solved for by solving a system of equations similar to (12.40), the strength of the desired signal at range cell r is estimated from [19]

$$\alpha \approx \frac{1}{C} \sum_{e=1}^{N_a} \sum_{h=1}^{N_p} W\{N_p(e-1) + h\} X^*(M - h + 1 ; N - e + 1)   \quad (12.44)$$

Thus the backward procedure provides a second independent realization of the same solution. In a practical environment, where the real solution is unknown, generation of two independent sets of solutions may provide some degree of confidence in the final results.

For systems where the DOA and Doppler frequency of the SOI are not known exactly, but are known approximately (e.g., within the mainbeam of the antenna), multiple constraints can be implemented to preserve the SOI. The procedure for doing this is identically to that of the forward method. For each additional constraint a constraint equation replaces a cancellation equation in [B] and the corresponding amplitude is placed in [Y]. The constraint equations are determined using (12.35).

12.5.5 Least Squares Forward-Backward Processor

A system of equations may be formed by combining the forward and backward solution procedure as described for the 1D case [18, 21]. Since, in this process, one is doubling the amount of data available by considering it in both the forward and reverse directions, one can essentially do one of the following two things: either increase the number of equations and solve an equation similar to (12.31) or (12.40) in a least squares fashion, or equivalently, increase the number of weights and hence the number of degrees of freedom (DOFs) by as much as 50%. The second alternative is extremely attractive if one is processing the data on a snapshot-by-snapshot basis due to a highly nonstationary environment. In this way one can effectively deal with a situation where the number of data samples may be too few to perform any other processing.

Since in this case the total number of data points is $2MN$, the number of degrees of freedom can be increased from the two cases presented earlier. Hence the number of degrees of freedom in this case will be $N_a' N_p'$ where $N_a' > N_a$ and $N_p' > N_p$. The increase in the number of degrees of freedom depends on the number of antennas N and the time samples M. But clearly, $N_a' N_p'$ is significantly greater than $N_a N_p$. This increase may be by a factor of approximately 2 when dealing with a data cube where $N = 22$ (the number of antennas in the array) and $M = 128$ (the number of time samples). By using the samples from N antenna elements and M pulses, we formulate the following matrix equation:

$$[FB][W] = [Y]   \quad (12.45)$$

Here again the system matrix [FB] consists of constraint rows and cancellation rows. The constraint rows preserve the SOI during the adaptive process. There is at least one constraint row, while multiple constraints may be used just as in (12.35) to maintain the gain of the array toward the SOI, which may possess a slightly different look direction φ_s and Doppler frequency f_d along the look direction. The remaining rows in [FB] consist of cancellation equations that are formed in both the forward and backward directions.

We now apply the various STAP algorithms described so far to real experimental data to study the performance of each.

Figure 12.12. MCARM testbed.

12.6   DESCRIPTION OF THE DATA COLLECTION SYSTEM

The Multichannel Airborne Radar Measurement (MCARM) program had as its objective the collection of multiple spatial channel airborne radar data for the development and evaluation of STAP algorithms for future Airborne Early Warning (AEW) systems. The airborne MCARM testbed, a BAC1-11 aircraft, used for these measurements is shown in Figure 12.12. The phased array is hosted in an aerodynamic cheek-mounted, mounted just forward of the left wing of the aircraft. The L-band (1.24 GHz) active array consists of 16 columns, with each column having two four-element subarrays, shown in Figure 12.13. The elements are vertically polarized, dual-notch reduced-depth radiators. These elements are located on a rectangular grid with azimuth spacing of 4.3 inches and elevation spacing of 5.54 inches. There is a 20-dB Taylor weighting across the eight elevation elements, resulting in a 0.25-dB elevation taper loss for both transmit and receive. The total average radiated power for the array was approximately 1.5 kW. A 6-dB modified trapezoid weighting for the transmit azimuthal illumination function is used to produce a 7.5° beamwidth pattern along the boresight with −25 dB RMS sidelobes. This pattern can be steered up to ±60°. Of the 32 possible channels, only 24 receivers were available for the data collection program. Two of the receivers were used for analog sum and azimuthal difference beams. There are therefore 22 ($N = 22$) digitized channels, which in this work are arranged as a rectangular 2×11 array. Each CPI comprises 128 ($M = 128$) pulses at a pulse repletion interval of 1984 Hz.

Figure 12.13. MCARM antenna array.

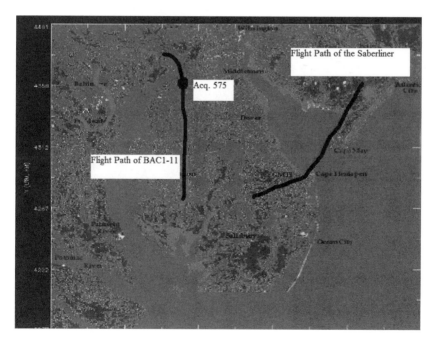

Figure 12.14. Flight paths of the BAC1-11 and the Saberliner over the Delmarva peninsula.

In the following examples the beam was pointed downward in elevation by about 5°. The flight path of the phased array was over the Delmarva peninsula (as defined by the landmass between the Chesapeake Bay and the Atlantic Ocean). In this experiment, the phased array on the BAC1-11 is trying to locate a Sabreliner approaching the BAC1-11 in the presence of sea, urban, and land clutter. The data cubes generated from these measurements, which are available from the AF Research Laboratory Web site (*http://sunrise.deepthought.rl.af.mil*), are used to analyze the validity of the algorithms presented in this chapter. The details are available in [24]. The geographical region for the flight is shown in Figure 12.14 and the regions of ground clutter return in Figure 12.15. This indicates that there is possible to have urban, land, and sea clutters simultaneously.

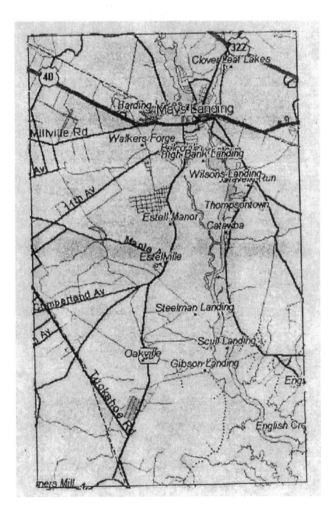

Figure 12.15. Scene of the region of ground clutter.

In the ideal case of a linear array, the magnitude of the steering vector is constant at each element. Figure 12.16 shows the variation in magnitude for the MCARM array. The magnitude varies by as much as 4 dB over the 32 elements. This variation is due to the mutual coupling between the elements of the antenna array. The data was first preprocessed to compensate for the mutual coupling effects between the elements before the space-time signal processing techniques were applied.

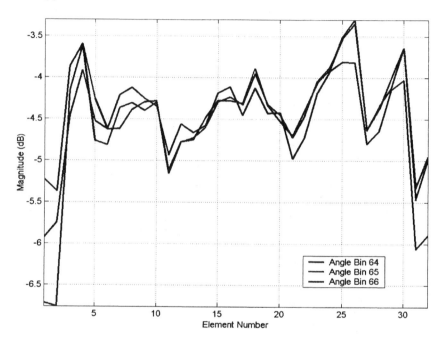

Figure 12.16 Magnitude of MCARM steering vectors.

12.7   NUMERICAL EXAMPLE

We next apply the D^3LS techniques to the analysis of MCARM data set RL050575.dat. This deals with an actual target buried in clutter. The data was collected by an airborne antenna array. The antenna array had 22 channels in addition to the sum and the difference channel. For each channel, the data in the time domain was sampled at 1984 Hz and there are 128 time samples ($M = 128$). The third dimension of the data set corresponds to the range profile and there are 630 range bins. The 3-dB beamwidth of the antenna is approximately $7.8°$. The data was gathered over a flight path over the Delmarva peninsula. The flight path of the down-looking phased array of the BAC1-11 is shown by the left curve in Figure 12.14, on which the particular data set was collected. Its position when it took the data is marked by the circle on the left-hand side. In addition, there is a Sabreliner flying toward the BAC1-11 in a slanted fashion, as shown by the second curve in Figure 12.14. The position of the Sabreliner is marked on the right-hand side. From data collected from geostationary satellites, it appears that the target is at 91° in azimuth and corresponds to the range cell at 318 (this is the second ambiguous range cell, namely 630 + 318) and corresponding to a Doppler frequency of approximately $f_d = 520$ Hz. Also, this data set contains received land, sea, and urban clutter from the regions shown in Figure 12.16. It also had signal return from highways which may have had some cars traveling by at that time. In the current analysis it is assumed that the signal is coming from $\varphi = 90°$ (i.e., broadside).

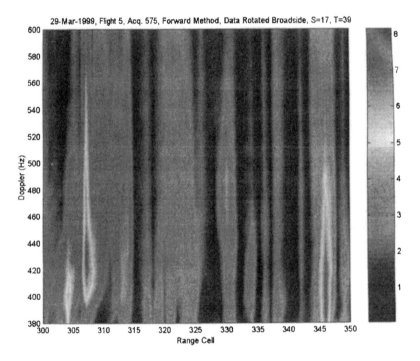

Figure 12.17. Application of the forward method to RL050575.dat.

We next apply the superresolution D^3LS analysis where a two-dimensional filtering technique is applied to each range cell as described by the forward method. The order of the filter required to identify the signals in the presence of clutter consisted of 17 weights or filter taps in space ($N_a = 17$) and a 39-order filter in time ($N_p = 39$), so that the total number of degrees of freedom is N_aN_p and is 663. This filter was applied to each range bin, corresponding to the signal of arrival from the broadside direction (i.e., $\varphi = 90°$) and the Doppler frequency was swept from 380 to 600 Hz in steps of 10 Hz. The range cells are swept from 300 to 350. Figure 12.17 represents the contour plot of the estimated signal return utilizing the forward method with weights of $N_a = 17$ and $N_p = 39$. It is seen that there are some activities around Doppler 500 Hz near the range cells of 308, 330, and 347. Next the backward method is used to analyze the same data set using the same number of degrees of freedom ($17 \times 39 = 663$). The results are shown in Figure 12.18. Again large returns around 500 Hz Doppler frequency are observed in the range cells of 305, 320, 330, and 338. The application of the forward-backward method with the following weights of order ($N_a = 19$, $N_p = 61$) results in signal returns as shown in Figure 12.19. This is nearly a two fold increase in the number of degrees of freedom used in either the forward or backward method. It is seen that the return is dominant at the range cell of 330. By comparing the results of the three graphs, one could say with confidence which strong signal is a true return corresponding to a particular Doppler and range cell. This is because the three methods are analyzing the same data set in three independent ways, and hence it makes sense to compare the three results. Therefore, simultaneous use of all three methods would provide a reliable estimate of the signal and will minimize the probability of false alarm.

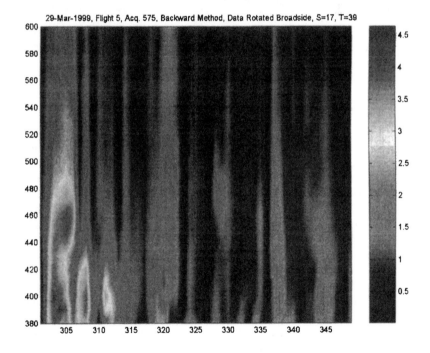

Figure 12.18. Application of the backward method to RL050575.dat.

Typical running time for each data point using the forward or backward method is less than a minute on a Pentium PC with a clock of 450 MHz. The forward-backward method takes slightly more time than either the forward or backward method. It is important to note that each range cell/Doppler/look angle can be

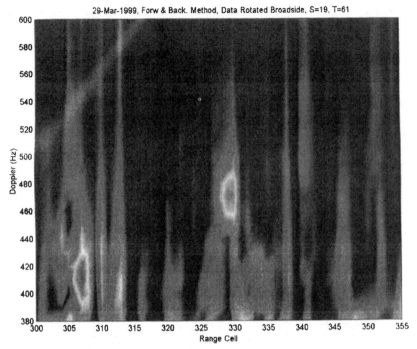

Figure 12.19. Application of the forward-backward method to RL050575.dat.

processed in parallel. Hence the computational requirements are very modest for real-time applications. In all the computations, the value of the gain factor C in equations (12.31), (12.40), and (12.45) has been chosen as unity.

Next the conventional stochastic method is used to estimate the signal strengths. The application of a 9×9 covariance matrix utilizing a JDL stochastic approach [9] is also used to estimate the signals. This is after the data has been Fourier transformed into the Doppler domain utilizing a Kaiser–Bessel window and a 128-point FFT. The result is shown in Figure 12.20, where some weak activity can be seen in range cells 308 and 330 around the Doppler of 500 Hz. However, without any "ground truth," it is difficult to predict which signal return is actually the Saberliner in all of these! This is because there are channel mismatches in the measurements and various uncertainties, such as the crab angle of the two aircrafts. The actual results show some deviations from the theoretical estimates. There may be several factors of certainty in the measured data such as the velocity of each aircraft, its elevation, and its direction of travel. However, all the methods predicted returns around the Doppler of 500 Hz in the range bin of 330. This slight discrepancy in Doppler and range can happen due to various factors, as outlined. Some shift may occur due to the matched filter processing if the target is not exactly at $\varphi_s = 90°$, or due to errors in the array calibration introduced by mutual coupling and near-field coupling with the aircraft frame [25] which were not fully accounted for in the analysis.

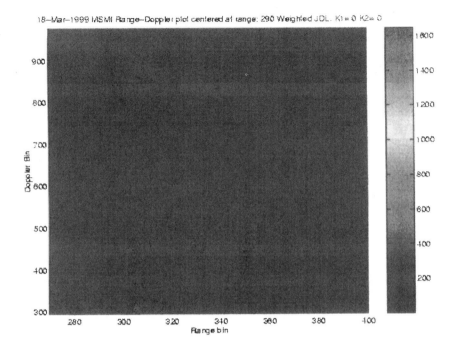

Figure 12.20. Application of the stochastic method to RL050575.dat.

12.8   SPACE-TIME ADAPTIVE PROCESSING USING CIRCULAR ARRAYS

In this section we describe a D^3LS space-time adaptive processing (STAP) approach for adaptively enhancing signals in a nonhomogeneous environment of jammers, clutter, and thermal noise utilizing a circular antenna array. Limited examples are presented to illustrate the application of this approach. Using a single snapshot of data a circular array cannot discriminate signals at the same Doppler but with different angles of arrival. However, if we use the transformation presented in Chapter 6, we can then use a single space-time snapshot which corresponds to a single range cell. The circular array that we are going to use in this section consists of a total of E elements equally distributed in the azimuth angle, as shown in Figure 12.21. The angular separation φ_e between each of the elements is

$$\varphi_e = \frac{2\pi}{E} \qquad (12.46)$$

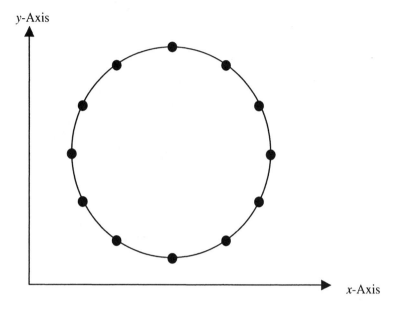

Figure 12.21. Geometry of a circular array.

Let the spatial coordinate of the nth element be $(x_n; y_n)$, which will be oriented along φ_n with respect to the x-axis. The angle is given by

$$\varphi_n = \frac{2\pi}{E}\,(n-1) \qquad (12.47)$$

If ϑ is the radius of the circular array, then

$$x_n = \vartheta \cos \varphi_n \qquad (12.48)$$

$$y_n = \vartheta \sin \varphi_n \qquad (12.49)$$

Here it is assumed that the elements of the array are omnidirectional point radiators. However, it is quite straightforward to take the mutual coupling between the elements into account as well as the electromagnetic coupling between the phased array and the airborne platform.

Let us assume that at any time only Γ of the E elements are active. The radar transmits a coherent burst of M pulses at a constant pulse repetition frequency. The time interval over which the received pulses are collected in the array is the coherent processing interval. The pulse repetition interval is the inverse of the pulse repetition frequency. A pulsed waveform of finite duration and (approximately finite bandwidth) is transmitted. On receive, at any of the given Γ elements, matched filtering is done where the receiver bandwidth is equal to the transmit bandwidth. Matched filtering is carried out separately on each pulse return, after which the signals are digitized and stored. So for each pulse repetition interval, R time samples are collected to cover the desired range interval. Hence, we term R the number of range cells. Therefore, with M pulses and Γ antenna elements, each having its own independent receiver channel, the received data for a coherent processing interval consists of $RM\Gamma$ complex baseband samples. These samples are often referred to as the *data cube*, consisting of $R \times M \times \Gamma$ complex baseband samples of the received pulses. The data cube then represents the voltages defined by $V(m; n; r)$ for $m = 1$, ..., M; $n = 1$, ..., Γ; and $r = 1$, ..., R, as shown in Figure 12.21. These measured voltages contain the signal of interest (SOI), jammers, and clutter, including thermal noise. A space-time snapshot is then referred to as MN *sample* for a fixed range gate value of r.

In the D^3LS procedure to be described, adaptive weights are applied to the space-time snapshot for the range cell r. Here a two-dimensional array of weights numbering $N_a N_p$ is used to extract the SOI for the range cell r. N_a is always taken to be $\Gamma - 1$. Hence the weights are defined by $w(p; q; r)$ for $p = 1$, ..., $N_p < M$ and $q = 1$, ..., $N_a = \Gamma - 1$ and are used to extract the SOI at the range cell r. Note that in this system the number of time samples M must be greater than $N_a N_p$. In this procedure, we essentially perform a high-resolution filtering in two dimensions for each range cell.

Here we use the D^3LS approach to deal with STAP for circular arrays. In this alternative approach the joint space-time (multidimensional) filtering is carried out for each range cell separately, and hence we process the data dealing with each space-time snapshot individually. No secondary data is required in this methodology. The STAP procedure for circular arrays is described next.

12.9   DIRECT DATA DOMAIN LEAST SQUARES STAP FOR CIRCULAR ARRAYS

From the data cube shown in Figure 12.6, we focus our attention on the range cell r and consider the space-time snapshot of MN data for this range cell [26]. We assume that the SOI for this range cell r is incident on the circular array from an azimuth angle φ_s from the x-axis and is at Doppler frequency f_d. Our goal is to estimate its amplitude, given φ_s and f_d. Let us define $S(p; q)$ to be the complex voltage received at the qth antenna element corresponding to the pth time instance for some range cell r. We further stipulate that the known voltages $S(p; q)$ are due to a signal of unity magnitude incident on the array from the azimuth angle φ_s

corresponding to Doppler frequency f_d. Hence, the signal-induced voltages under the assumed array geometry and a narrowband signal comprise a complex sinusoidal given by

$$S(p;q) = \exp\left[j2\pi\left(\frac{x_q\cos\varphi_s + y_q\sin\varphi_s}{\lambda} + p\times f_d\times pri \right)\right] \qquad (12.50)$$

$$\text{for } q = 1, ..., \Gamma; \quad p = 1, ..., M$$

where λ is the wavelength of the radio-frequency radar signal, *pri* is the pulse repetition interval, and x_q and y_q are the spatial coordinates of the antenna elements of the circular array. Let $X(p; q)$ be the actual measured complex voltages that are in the data cube of Figure 12.6 for the range cell r. The actual voltages X will contain the signal of interest of amplitude α (α is a complex quantity); jammers, which may be due to coherent multipaths in both the mainlobe and the sidelobe; and clutter, which is the reflected electromagnetic energy from the ground, which will compete with the SOI at the Doppler frequency of interest. There is also a contribution to the measured voltage for the range cell r from receiver thermal noise. Hence the actual measured voltages $X(p; q)$ are

$$X(p;q) = \alpha \exp\left[j2\pi\left(\frac{x_q\cos\varphi_s + y_q\sin\varphi_s}{\lambda} + p\times f_d\times pri \right)\right]$$

$$+ \text{ clutter } + \text{ jammer } + \text{ thermal noise} \qquad (12.51)$$

$$= V(p;q;r)$$

Now if one forms the following difference in the signals from range cell r, the elements of the matrix pencil,

$$[X]_{N_a-N_p} - \alpha[S]_{N_a-N_p} \qquad (12.52)$$

represent the contribution due to the unwanted signal multipaths, jammers, and unwanted signals at the same Doppler and receiver thermal noise. In the adaptive processing, the goal is to take a weighted sum of the matrix elements defined by (12.52) and extract the SOI for the range cell r. The total number of degrees of freedom then represents the total number of weights, and this is the product $N_a N_p = N_p (N - 1)$, where N_a is the number of spatial degrees of freedom (and is always equal to $N - 1$ in this case) and N_p is the number of temporal degrees of freedom. In this formulation it is necessary that the total number of time samples M be greater than $(N - 1) N_p$: that is,

$$M \geq (\Gamma - 1)N_p \qquad (12.53)$$

Next we show illustrated how a D³LS approach is taken for the extraction of SOI.

Following the procedures outlined in Section 12.5, one can set up an eigenvalue equation to solve for the SOI. For computational reasons we transform it to a matrix equation. Let

$$Z_k = \exp\left[j2\pi \left(\frac{x_k \cos \varphi_s + y_k \sin \varphi_s}{\lambda} \right) \right] \tag{12.54}$$

$$\beta = \exp[j\, 2\pi\, f_d \cdot \text{pri}] \tag{12.55}$$

and we form a reduced rank matrix $[T]$ of dimension $(N_a N_p - 1) \times (N_a N_p)$ from the elements of the matrix $[X]$, where

$$T(x; y) = \frac{X(g+h-1;\ k-1)}{\beta^h Z_{k-1}} - \frac{X(g+h-1;\ k)}{\beta^h Z_k} \tag{12.56}$$

$$T(x+1; y) = \frac{X(g+h-1;\ k-1)}{\beta^h Z_{k-1}} - \frac{X(g+h;\ k-1)}{\beta^{h+1} Z_{k-1}} \tag{12.57}$$

$$T(x+2;\ y) = \frac{X(g+h-1;\ k-1)}{\beta^h Z_{k-1}} - \frac{X(g+h;\ k)}{\beta^{h+1} Z_k} \tag{12.58}$$

and

$$1 \le k \le N_a = N - 1$$
$$1 \le y = N_p\, (k-1) + h \le N_a N_p \tag{12.59}$$

for any row x (its value is between 1 and $N_a N_p$), the following variables take values between

$$1 \le g \le N_p N_a$$
$$1 \le h \le N_p \tag{12.60}$$

If we consider the three consecutive rows of the matrix $[T]$, we observe that they have been formed by taking a weighted difference of two-dimensional block of data of size $N_a N_p$. The weighted difference is taken using the elements of matrix $X(p; q)$. Therefore, the elements $T(x; y)$ are formed by writing the $N_a N_p$ difference matrix of equations (12.56)–(12.58) as three consecutive rows of the matrix. The weighted differences forming $N_a N_p$ elements occupy one row of the matrix. The elements of the matrix [T] are thus obtained by a weighted subtraction of the induced voltages from the neighboring elements (either in space or in time) so that in these elements the desired signals are canceled out and the elements of [T] contain no components of the signal corresponding to the Doppler f_d and the direction of arrival φ_s. We choose the weights [W] such that

$$[T][W] = 0 \tag{12.61}$$

where the column matrix [W] has been generated by arranging the $N_a N_p$ weights from $w(m; n; r)$ in a linear column array corresponding to the processing of data from range cell r. In order to restore the signal component in the adaptive processing, we fix the gain of the subarray (in both space and time) formed by fixing the first row of the matrix T. The elements of the first row are given by

$$T(1; y) = 1 \tag{12.62}$$

where

$$y = N_p(e-1) + h \tag{12.63}$$

$$\left.\begin{array}{l} 1 \le e \le N_a \\ 1 \le h \le N_p \end{array}\right\} \tag{12.64}$$

We set the gain of the system along the direction φ_s of the arrival of the signal and corresponding to the Doppler frequency f_d, at some constant C and so let

$$Y(1) = C \tag{12.65}$$

The final equation is formed by combining (12.61), (12.62), and (12.65), resulting in the matrix equation

$$[T][W] = [Y] = \begin{bmatrix} C \\ \vdots \\ 0 \\ 0 \\ 0 \end{bmatrix} \tag{12.66}$$

Once the weight [W] is known from (12.66), the signal strength for the range cell r is estimated from the measured voltages as

$$\alpha = \frac{1}{C} \sum_{e=1}^{N_a} \sum_{h=1}^{N_p} W\{N_p(e-1) + h\} \frac{V(h; e; r)}{\beta^h Z_e} \tag{12.67}$$

12.10   NUMERICAL EXAMPLE USING A CIRCULAR ARRAY

As an example,consider a 40-element circular array. The elements are distributed evenly along the arc of the circle of radius seven wavelengths. Only 11 contiguous elements of the 40 elements array are active at one time, so that it scans over a 45° sector. Then the next 11 elements are excited, and so on. First let us assume that the 11 elements comprising the sector 0 to 45° are active. The signal of interest is coming from 20° and its level is considered to be 0 dB. The SOI has a Doppler of 560 Hz. In addition, we consider two other strong targets located very close to the SOI. One of the targets is 37 dB stronger than the signal. It is arriving from an azimuth of 28° and it is at a Doppler of 555 Hz. The second interference is arriving from an azimuth of 30° and is 39.5 dB stronger than the SOI. It has a Doppler of 565 Hz. In addition, there are two other strong interferers located at the periphery of the active sector of 45°. One of them is arriving from 47° and is at a Doppler of −570 Hz. It is 39 dB above the signal. The second one is arriving from 5° at a Doppler of −550 Hz and is 38.3 dB stronger than the signal. The signals are all sampled at 1950 Hz at all the elements. In addition, there is thermal noise in all the antenna elements for all times. It is 24.7 dB below the signal level at each antenna element. A data cube over a range cell is generated for this scenario. There are 128 time samples at each antenna element, and there are 11 elements. To this data cube we applied the direct data domain least squares method. In this case there were 10 spatial taps and 25 temporal taps. This is equivalent to filtering the data cube by a two-dimensional filter of order 250. The output signal-to-noise ratio for this scenario was estimated to be 5 dB. This results in a signal-to-interference plus noise enhancement from −39 dB to +5 dB.

12.11   HYBRID STAP METHODOLOGY

Performance degradation of statistical-based STAP algorithms due to nonhomogeneous data occurs in two forms. In one form the secondary data is not i.i.d., leading to an inaccurate estimate of the covariance matrix. For example, the clutter statistics in urban environments fluctuate rapidly with range. To minimize the loss in performance due to nonhomogeneous sample support, a nonhomogeneous detector (NHD) may be used to identify secondary data cells that do not reflect the statistical properties of the primary data. These data samples are then eliminated from the estimate of the covariance matrix. The second form of performance loss is due to a discrete nonhomogeneity within the primary range cell. For example, a large target within the test range cell but at a different angle and/or Doppler appears as a false alarm at the look angle-Doppler domain. Other examples include a strong discrete nonhomogeneity, such as a large building (corner reflector), in the primary range cell. These false alarms appear through the sidelobes of the adapted beam pattern. The secondary data cells do not carry information about the discrete nonhomogeneity, and hence a statistical algorithm cannot suppress discrete (uncorrelated) interference within the range cell under test.

The inability of statistical STAP algorithms to counter nonhomogeneities in the primary data motivates research in the area of nonstatistical D^3LS algorithms. These algorithms use data from the range cell of interest only, eliminating the sample support problems associated with statistical approaches.

The main contribution of this section is the introduction of a two-stage hybrid STAP algorithm combining the benefits of both nonstatistical and statistical methods. The hybrid approach uses the nonstatistical algorithm as a first-stage filter to suppress discrete interferers present in the range cell of interest. This first stage serves as an *adaptive transform* from the space-time domain to the angle-Doppler domain and is followed by JDL processing in the second stage. The adaptive transform replaces the steering vector-based nonadaptive transform used in JDL. The second stage is designed to filter out the residual correlated interference [12]. This hybrid approach is useful when there is a lot of clutter in the mainlobe region. As outlined in Appendix A, the advantage of the statistical algorithms lies in the fact that when the signal and interference spectra overlap, it is not possible to separate them by direct filtering. Other information, such as the correlation function of the interferers or their power spectral density, may be used in addition to extract the SOI. That is why for many practical applications this hybrid approach may be more effective.

Consider the general framework of any STAP algorithm. The algorithm processes received data to obtain a complex weight vector for each range bin and each look angle/Doppler. The weight vector then multiplies the primary data vector to yield a complex number. The process of obtaining a real scalar from this number for threshold comparison is part of the post-processing and not inherent in the algorithm itself. The adaptive process therefore *estimates the signal component in the look direction*, and hence the adaptive weights can be viewed in a role *similar to the nonadaptive steering vectors* used to transform the space-time data to the angle-Doppler domain.

The JDL processing algorithm begins with a transformation of the data from the space-time domain to the angle-Doppler domain. This is followed by statistical adaptive processing within a LPR in the angle-Doppler domain. The hybrid approach uses the D^3LS weights, replacing the nonadaptive steering vectors used earlier. By choosing the set of look angles and Dopplers to form the LPR, the D^3LS weights perform a function analogous to the nonadaptive transform. The D^3LS algorithm serves as a first-stage *adaptive* transformation from the space-time to the angle-Doppler domain. JDL statistical processing in the angle-Doppler domain forms the second stage of adaptive processing to filter the residual correlated interference. The D^3LS algorithm is used repeatedly with the η_a look angles and η_d look Doppler frequencies to form the LPR. The space-time data is transformed to the LPR in the angle-Doppler domain using these adaptive weights. Using the D^3LS weights, the discrete interferers are suppressed. The advantages associated with the JDL algorithm, such as reduction in the required secondary data support, carry over to the hybrid algorithm. Unlike the JDL algorithm, in this hybrid algorithm a transformation matrix is used which changes from range cell to range cell. The hybrid algorithm therefore has a significantly higher computation load than the JDL algorithm. The hybrid algorithm forms the adaptive transformation matrix for each range cell and then transforms this primary and *associated* secondary data to the angle-Doppler domain. This process is repeated for each range cell [27].

12.11.1 Applying the Hybrid Algorithm to Measured Data

We present two examples of the application of the hybrid algorithm to measured data. The examples use data from the MCARM database. The examples use two acquisitions (acquisitions 575 and 152 on flight 5) [24] to illustrate the suppression of discrete interference in measured data.

Before the hybrid algorithm can be applied to the MCARM database, array effects must be accounted for. The D^3LS method is implemented using an equispaced, linear array of point sensors. This allowed for the assumption of no mutual coupling between the elements and for the fact that for each pulse, the target signal advances from one element to the next by a constant spatial multiplicative factor z_s. This, in turn, allowed for the crucial assumption of the elimination of the target signal in the entries of the interference matrix. In addition, the MCARM antenna is an array of 22 elements arranged in a rectangular 2×11 grid. For a rectangular array, these assumptions are invalid. The effects of the mutual coupling have been taken care of through the measured steering vectors.

12.11.1.1   Injected Target in MCARM Data.   In this example, a discrete non-homogeneity is introduced into the data by adding a strong fictitious target at a single range bin, but not at the look angle-Doppler. Two cases are considered within this example: no injected target and an injected weak target. The first case illustrates the suppression of the discrete nonhomogeneity. In the second case, a weak target is injected at the same range bin as the nonhomogeneity, but at the look angle and Doppler. This case illustrates the ability of the hybrid algorithm to detect weak targets in the presence of strong discrete nonhomogeneities. In this case, only 22 of the 128 pulses in the CPI are used (i.e., $N = 22$, $M = 22$). The details of the injected nonhomogeneity and weak target are shown in Table 12.1.

TABLE 12.1
Parameters for injected nonhomogeneity and target in MCARM data

Parameter	Nonhomogeneity	Target
Amplitude	0.0241	0.000241
Angle bin	35	65 (broadside)
Doppler bin	-3	-2
Range bin	290	290

The hybrid algorithm is applied to the data from the range bin with the nonhomogeneity and surrounding range bins. The output from the modified sample matrix inversion (MSMI) output from the second stage of the hybrid algorithm is plotted as a function of range. In this example, five Doppler bins and five angle bins form the LPR for both the JDL algorithm and the second stage of the hybrid algorithm. One hundred secondary data vectors are used to estimate the 25×25 covariance matrix. For the case without an injected target, Figure 12.22a compares the output from the JDL algorithm with the output of the hybrid algorithm. As can be seen, the JDL algorithm indicates the presence of a large target in the look direction (angle bin 65). This is because the large nonhomogeneity at angle bin 35 and Doppler bin −3 is not suppressed by the statistical algorithm, leading to false alarms in the look direction. On the other hand, the hybrid algorithm shows no target at broadside. The nonhomogeneity is suppressed in the first D^3 stage, and residual clutter is suppressed in the second JDL stage.

A synthetic target injected at the look direction and Doppler illustrates the performance sensitivity of the hybrid algorithm to weak targets. The parameters of the weak target are listed in Table 12.1. Figure 12.22b compares the output of the two algorithms in the case of a strong nonhomogeneity and a weak target. The JDL algorithm again shows the presence of a strong target in the look direction. However from Figure 12.22a we know that the strength of the statistic is caused by the nonhomogeneity. On the other hand, the plot for the hybrid algorithm shows that the statistic at the target range bin is 6.9 dB above the next highest peak.

This example shows that the hybrid algorithm may be used to detect a weak target in the presence of a discrete nonhomogeneity within the range cell of interest. This is a unique capability compared to all other STAP approaches.

12.11.1.2 Moving Target Simulator Tones in the MCARM Data. Certain acquisitions within the MCARM database include signals from a moving target simulator (MTS) at known Doppler shifts. In acquisition 152 on flight 5 [24], the MTS tones occur in angle bin 59. In this example, the look direction is set to angle bin 85 for a mismatch (with the MTS direction), and the JDL and hybrid algorithms are applied to the same acquisition. For this look direction, the MTS tones at angle bin 59 act as strong targets at a different angle bin (i.e., discrete nonhomogeneities). As in the previous example, two different cases are considered: no injected target and a weak injected target. The first case illustrates suppression of the MTS tones acting as discrete, strong nonhomogeneities. The second case illustrates the sensitivity of the hybrid algorithm to weak targets. This example uses all 128 pulses in the CPI (i.e., $N = 22$, $M = 128$).

In this acquisition, the MTS tones are in range bin 449–450 with the strongest tone at a Doppler corresponding to bin −53 and angle bin 59. The example focuses on the suppression of this tone. Figure 12.23a plots the MSMI statistic of the two algorithms for the case without any artificial injected targets. The JDL algorithm detects a large target at range bins 449 and 450. This false alarm is due to the strong MTS tone at angle bin 59 even though the look direction is set at angle bin 85. The hybrid algorithm, however, suppresses the strong MTS tone, showing no activity at range bins 449 and 450.

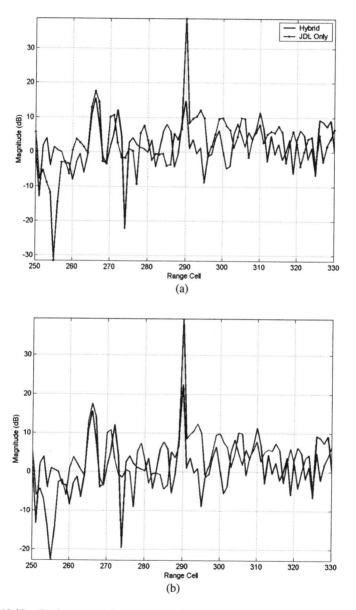

Figure 12.22. Performance of the hybrid algorithm in countering nonhomogeneities: injected target (a) with nonhomogeneity and no target and (b) with nonhomogeneity and target.

Figure 12.23b plots the results of using the two algorithms to detect a weak target injected into range bin 450. The parameters of the weak target are: magnitude: 0.0001, Doppler bin: -53, angle bin: 85. This weak target is easily detected by the hybrid algorithm with the statistic at the target range bin 9.8 dB above the next highest peak.

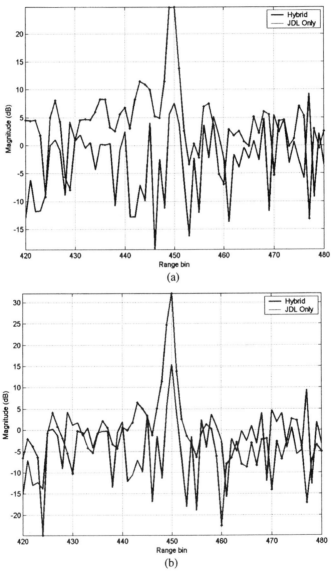

Figure 12.23. Performance of the hybrid algorithm in countering nonhomogeneities: MTS data (a) with nonhomogeneity and no target and (b) with nonhomogeneity and target.

The beam patterns associated with the two algorithms illustrate the improvement in using the D^3LS algorithm as the first stage of a two-stage hybrid method. Figure 12.24 plots the spatially adapted beam pattern at the look Doppler frequency for the JDL and hybrid algorithms. The plot for the hybrid algorithm shows the deep null in the adapted pattern of the hybrid algorithm near angle bin 59, while the JDL pattern does not show such a null. In applying the JDL algorithm to the MCARM data acquisition with MTS tones, the strong tones leak through the sidelobes of the adapted pattern, leading to false alarms.

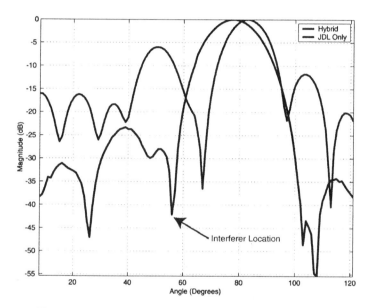

Figure 12.24. Beam pattern associated with the hybrid and JDL methods.

In summary, we have presented a hybrid algorithm, developed specifically for the nonhomogeneous data case. Statistical algorithms cannot suppress discrete nonhomogeneities because the secondary data possesses no information regarding such interference. The D^3LS method, however can suppress such discrete interference. The proposed two-stage hybrid algorithm alleviates this drawback by implementing a second stage of statistical processing after using the D^3LS algorithm as an adaptive transform to the angle-Doppler domain. This algorithm combines the advantages of the statistical and nonstatistical approaches. The D^3LS method is particularly effective at countering nonhomogeneous interference. The statistical STAP algorithm then improves on suppression of the residual correlated interference. Even with *ad hoc* compensation for mutual coupling, the hybrid algorithm shows a significant improvement over statistical methods in the suppression of discrete nonhomogeneities. We anticipate that a true evaluation of the mutual coupling would improve the performance of the hybrid algorithm.

12.12   KNOWLEDGE-BASED STAP PROCESSING

The field of space-time adaptive processing has received much interest in the past 30 years. The sum total of the research is extensive, with several classes of algorithms, some practical and others not so practical. In addition, interesting new algorithms [28] and algorithms that address particular interference situations [29] are being continually developed. What is clear is that there is no single algorithm that is optimal in all interference scenarios. All statistical algorithms require the estimation of a covariance matrix. In a nonhomogeneous scene, the choice of the secondary data has a huge impact on the performance of the algorithm. It is essential that in a real-world situation, the secondary data be chosen properly.

This research therefore heads toward the concept of *knowledge-based STAP* (KB-STAP). KB-STAP [30] chooses the best of several possible STAP algorithms for detection with knowledge-based control of algorithm parameters and selection of secondary data using NHDs. The basic elements of a comprehensive KB-STAP formulation are shown in Figure 12.25. Any comprehensive algorithm for practical implementation of STAP requires at least three elements: a nonhomogeneity detector to separate the received data into homogeneous and nonhomogeneous sectors, a statistical algorithm for use within the homogeneous sectors, and a hybrid algorithm for use within the nonhomogeneous sector. This section presents the performance improvements possible using such a combined scheme [31].

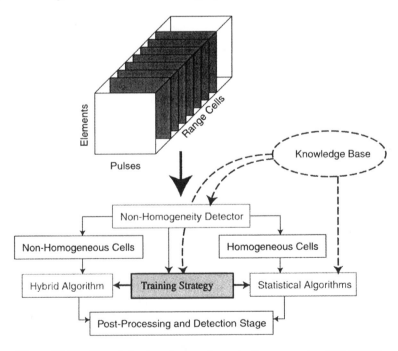

Figure 12.25.   Knowledge-based space-time adaptive processing (KB-STAP).

The combined approach is tested using data from the MCARM database. As described earlier, each CPI comprises the data corresponding to 22 digitized channels and 128 pulses at a PRF of 1984 Hz. The data cube comprises 630 range cells, sampled at 0.8 μs. Each range bin, therefore, corresponds to 0.075 mile. The array operates at a center frequency of 1.24 GHz. Included with each CPI is information regarding the position, aspect, velocity, and main beam transmit direction. This information is used to correlate target detections with ground features.

The example illustrates the following issues addressed here: namely, nonhomogeneities and the use of the appropriate processing algorithm in appropriate portions of the radar data cube. Nonhomogeneity detection is accomplished using JDL assuming homogeneous data. Any range bin with a statistic above a chosen threshold is considered nonhomogeneous. The statistical algorithm is JDL again, although in the second stage, only homogeneous data is used in the sample support. The hybrid algorithm also uses only homogeneous data for sample support in the second stage.

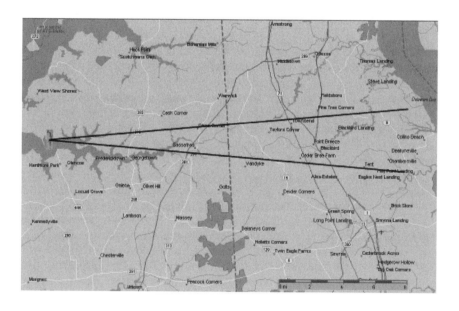

Figure 12.26. Location and transmit direction of the MCARM aircraft during acquisition.

This example uses data from acquisition 575 on flight 5 [24]. While taking this acquisition the radar platform was at latitude–longitude coordinates of $(39.379°, -75.972°)$, placing the aircraft close to Chesapeake Haven, Maryland, near the Delmarva Peninsula. The plane was flying mainly south with velocity 223.78 mph and east with velocity 26.48 mph. The aircraft location and transmit main beam are shown in Figure 12.26. The main beam is close to broadside. Note that the main beam illuminates several major highways. In addition to the targets on the roadways illuminated by the array, we inject two artificial targets at closely spaced range bins to illustrate the effects of nonhomogeneities in secondary training data. Based on the measured steering vectors and chosen Doppler shifts, the response of the two simulated targets may be calculated. The artificial targets are injected in range bins 290 and 295. In this acquisition the transmit pulse is zero-shifted to range bin 74 (i.e., the targets are at ranges of 16.2 miles and 16.575 miles, respectively). The parameters of the injected targets are given in Table 12.2. Note that the two targets are at the same Doppler frequency and the second target is 20 dB stronger than the first.

TABLE 12.2
Parameters of the injected targets

Target 1		Target 2	
Amplitude	1×10^{-4}	Amplitude	1×10^{-3}
Range bin	290	Range bin	295
Doppler bin 9	137.5 Hz	Doppler bin 9	137.5 Hz
Angle	$1°$	Angle	$1°$

This example uses three angle bins and three Doppler bins (a 3×3 LPR) in all stages of adaptivity, including JDL-NHD. Thirty-six secondary data vectors are used to estimate the 9×9 angle-Doppler LPR covariance matrix. In addition, two guard cells are used on either side of the primary data vector. Based on these numbers, without a NHD stage, range bin 295 would be used as a secondary data vector for detection within range bin 290. The example compares the results of using the JDL algorithm without nonhomogeneity detection and the combined approach illustrated in Figure 12.25. Figure 12.27 presents the results of using the JDL algorithm without any attempt to remove nonhomogeneities from the secondary data support. The range-Doppler plot is of the MSMI statistic after applying a threshold. In producing this figure, a threshold of 40 is used [i.e., any Doppler-range bin with a MSMI statistic greater than 40 (amplitude not in dB) is said to contain a target, while any Doppler-range bin with a statistic below 40 is declared target free]. The plot is for adaptive processing between range bins 150 and 350 (i.e., ranges between 5.7 and 20.7 miles and all 128 Doppler bins). Due to platform motion the radar is approaching the declared targets at a speed of 26.48 mph.

As is shown later, certain range bins that are declared to contain a target can be correlated with the map in Figure 12.26 as corresponding to roadways. However, this approach results in several false alarms, including several at extremely high radial velocities. In addition, the first injected target at range bin 290 is not detected. This is because of the presence of the larger target at range bin 295 in the secondary data when range bin 290 is in the primary data set.

Figure 12.27 clearly illustrates the need for a stage to identify non-homogeneities and eliminate them from the secondary data support. Applying STAP to measured data results in several false alarms and the possibility of targets in the secondary data, masking weak targets. The processing structure of Figure 12.25 addresses this need.

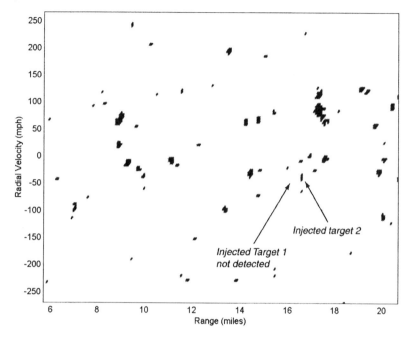

Figure 12.27. JDL processing without accounting for nonhomogeneities.

In the implementation used here, a JDL-NHD (nonhomogeneity detector) is used to identify non-homogeneous range cells. A range cell is considered to be nonhomogeneous if the JDL-MSMI statistic is above 18.52, significantly lower than the threshold of 40 used to generate Figure 12.27. Assuming Gaussian interference, using 36 secondary data vectors to estimate a 9×9 covariance matrix to obtain an MSMI statistic, this threshold corresponds to a false alarm rate of $P_{fa} = 10^{-4}$. Note that the true false alarm rate using measured data is significantly higher.

The combined algorithm uses JDL processing in those cells declared homogeneous and hybrid processing in those cells declared nonhomogeneous. Again, a 3×3 LPR is used, both in the JDL algorithm and in the hybrid algorithm. In the second application of the JDL algorithm in homogeneous range cells, only other homogeneous cells are used for sample support. Within the non-homogeneous cells, a hybrid algorithm is used (i.e., the D^3LS algorithm is used nine times for three-angle and three-Doppler look directions, *using the same primary data*). The angle-Doppler data so obtained is used for further JDL processing. Homogeneous cells are used to obtain sample support for second-stage JDL processing.

Figure 12.28 shows the result of using the combined approach. Notice the significantly fewer false alarms than in Figure 12.27 when using a purely statistical algorithm without nonhomogeneity detection. In essence, the hybrid algorithm is applied to all those range/Doppler bins where the JDL-MSMI statistic is greater than 18.52. Use of the hybrid algorithm suppresses the nonhomogeneities, thereby significantly reducing the false alarms.

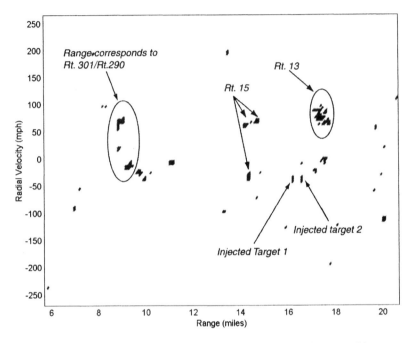

Figure 12.28. Combined processing accounting for nonhomogeneities.

In addition, the weaker injected target is detected since the stronger target at range bin 295 is eliminated from the sample support for range bin 290. Furthermore, the range bins of most target detections can be directly correlated with the state highways in Maryland and Delaware. Routes 299 and 301 in Maryland are closely spaced at a range of 9.0 and 9.8 miles. The aircraft was moving due east at a speed of 26.48 mph. The ground speed of the targets is therefore approximately 50 mph toward and away from the aircraft.

The range of the several target detections at the far range shown in the plot, approximately 20 miles, is not immediately attributable to Route 9 in Delaware. At broadside, Route 9 is at a range of 21 miles. The detected targets are between 19.4 and 20.4 miles. However, note that Route 9 is north-south at broadside curves and has a short east-west section within the 3-dB main beam. The distance to this section is between 19.1 and 20.6 miles. These targets are detected at these range bins and are present in both Figures 12.27 and 12.28. At a range of approximately 11 miles is a strong detection. Accounting for the aircraft motion, this detection has zero ground velocity. This corresponds to the town of Van Dyke.

This section has presented a comprehensive approach to STAP incorporating the essential elements of a practical scheme: nonhomogeneity detection, a statistical algorithm for STAP in homogeneous cells, and a hybrid algorithm for STAP in nonhomogeneous cells. The example illustrates the importance of these concepts to the ground moving target indicator case. This scheme yields huge performance improvements over the traditional STAP algorithm as applied to real measured data.

12.13   CONCLUSION

A direct data domain least squares approach has been presented to carry out space-time adaptive processing using both linear and circular arrays. Even though the circular array can resolve signals which are close in azimuth and Doppler, special attention must be paid to this analysis when the signals are at the same Doppler. Limited examples have been presented to illustrate the applicability of this technique to deal with real airborne platform data. In addition, we present a hybrid methodology that can combine the best of the direct data domain approach and of the stochastic methodology. These combined procedures can be automated through the use of knowledge-based STAP, which is termed KB-STAP.

12.14   REFERENCES

[1]   R. Klemm, "Adaptive Clutter Suppression for Airborne Phased Array Radar," *IEE Proceeding: Radar Signal Processing*, Vol. 130, pp. 125–131, Feb. 1983.

[2]   J. Ward, *Space Time Adaptive Processing of Airborne Radar*, Technical Report 1015, Lincoln Laboratory, Lexington, MA, Dec. 1994.

[3]   L. E. Brennan and I. S. Reed, "Theory of Adaptive Radar," *IEEE Transactions on Aerospace and Electronic Systems*, Vol. 9, pp. 237–252, Mar. 1973.

[4]   L. E. Brennan, J. D. Mallet, and I. S. Reed, "Adaptive Arrays in Airborne MTI Radar," *IEEE Transactions on Antennas and Propagation*, Vol. 24, pp. 605–615, Sept. 1976.

[5]   I. S. Reed, J. D. Mallett, and L. E. Brennan, "Rapid Convergence Rate in Adaptive Arrays," *IEEE Transactions on Aerospace and Electronic Systems*, Vol. 10, pp. 853–863, Nov. 1974.

[6]   E. C. Banle, R. C. Fante, and J. A. Torres, "Some Limitations on the Effectiveness of Airborne Adaptive Radar," *IEEE Transactions on Aerospace and Electronic Systems*, Vol. 28, pp. 1015–1032, Oct. 1992.

[7]   M. C. Wicks, "A Comparative Study of Space–Time Processing for Airborne Radar," Ph.D. dissertation, Syracuse University, Syracuse, NY, May 1995.

[8]   D. H. Johnson and D. E. Dudgeon, *Array Signal Processing*, Prentice Hall, Englewood Cliffs, NJ, 1993.

[9]   H. Wang and L. Cai, "On Adaptive Spatial–Temporal Processing for Airborne Surveillance Radar Systems," *IEEE Transactions on Aerospace and Electronic Systems*, Vol. 30, No. 3, pp. 660–669, July 1994.

[10] W. L. Melvin and B. Himed, "Comparative Analysis of Space–Time Adaptive Algorithms with Measured Airborne Data," *Proceedings of the 7th International Conference on Signal Processing Applications and Technology*, Boston, Oct. 1996.

[11] R. S. Adve, T. B. Hale, and M. C. Wicks, "Joint Domain Localized Adaptive Processing in Homogeneous and Non-homogeneous Environments, Part I, Homogeneous Environments," to appear in *IEE Proceedings on Radar, Sonar and Navigation*. Accepted for publication Nov. 1999.

[12] R. S. Adve, T. B. Hale, and M. C. Wicks, "Joint Domain Localized Adaptive Processing in Homogeneous and Non-homogeneous Environments, Part II, Non-homogeneous Environments," to appear in *IEE Proceedings on Radar, Sonar and Navigation*. Accepted for publication Dec. 1999.

[13] F. R. Dickey, Jr., M. Labitt, and F. M. Standaher, "Development of Airborne Moving Target Radar for Long Range Surveillance," *IEEE Transactions on Aerospace and Electronic Systems*, Vol. 27, pp. 959–971, Nov. 1991.

[14] J. Ender and R. Klemm, "Airborne MTI via Digital Filtering," *IEEE Proceedings*, Vol. 136, Pt. F, No. 1, pp. 22–29, Feb. 1989.

[15] R. S. Adve and M. C. Wicks, "Joint Domain Localized Processing Using Measured Steering Vectors," *Proceedings of the IEEE National Radar Conference*, Dallas, TX, 1998.

[16] R. L. Fante, "Cancellation of Specular and Diffuse Jammer Multipath Using a Hybrid Adaptive Array," *IEEE Transactions on Aerospace and Electronic Systems*, Vol. 27, No. 5, pp. 823–837, Sept. 1991.

[17] A. Luthra, "A Solution to the Adaptive Nulling Problem with a Look-Direction Constraint in the Presence of Coherent Jammers," *IEEE Transactions on Antennas and Propagation*, Vol. 34, pp. 702–710, May 1986.

[18] T. K. Sarkar and N. Sangruji, "An Adaptive Nulling System for a Narrowband Signal with a Look Direction Constraint Utilizing the Conjugate Gradient Method," *IEEE Transactions on Antennas and Propagation*, Vol. 37, pp. 940–944, July 1989.

[19] R. Schneible, "A Least Squares Approach for Radar Array Adaptive Nulling," Ph.D. dissertation, Syracuse University, Syracuse, NY, May 1996.

[20] T. K. Sarkar, S. Park, J. Koh, and R. A. Schneible, "A Deterministic Least Square Approach to Adaptive Antennas," *Digital Signal Processing: A Review Journal*, Vol. 6, pp. 185–194, 1996.

[21] T. K. Sarkar, J. Koh, R. S. Adve, R. A. Schneible, M. C. Wicks, M. Salazar-Palma, and S. Choi, "A Pragmatic Approach to Adaptive

Antennas," *IEEE Antennas and Propagation Magazine*, Vol. 42, No. 2, pp. 39–55, Apr. 2000.

[22] T. K. Sarkar et al., "A Deterministic Least-Squares Approach to Space–Time Adaptive Processing (STAP)," *IEEE Transactions on Antennas and Propagation*, Vol. 49, pp. 91–103, Jan. 2001.

[23] S. Park, "Estimation of Space–Time Parameters in Non-homogeneous Environments," Ph.D. dissertation, Syracuse University, Syracuse, NY, Aug. 2000.

[24] D. Sloper, D. Fenner, J. Arntz, and E. Fogle, *Multichannel Airborne Radar Measurement (MCARM), MCARM Flight Test*, Technical Report RL-TR-96-49, Vol. I, Rome Laboratory, Rome, NY, Apr. 1996.

[25] R. S. Adve and T. K. Sarkar, "Compensation for the Effects of Mutual Coupling on Direct Data Domain Adaptive Algorithms," *IEEE Transactions on Antennas and Propagation*, Vol. 48, No. 1, pp. 86–94, Jan. 2000.

[26] T. K. Sarkar and R. S. Adve, "Space Time Adaptive Processing Using Circular Arrays," *IEEE Antennas and Propagation Magazine*, Vol. 43, No. 1, pp. 138–143, Feb. 2001.

[27] R. Adve et al., *Knowledge-Base Application to Ground Moving Target Detection*, AFRL-SN-RS-TR-2001-185, Air Force Research Laboratory, Sensors Directorate, Rome Research Site, Rome, NY, Sept. 2001.

[28] J. S. Goldstein, J. R. Guerci, and I. S. Reed, "Advanced Concepts in STAP," *Proceedings of the IEEE International Radar Conference*, Washington, DC, pp. 699–704, May 2000.

[29] J. R. Guerci, J. S. Goldstein, P. A. Zulch, and I. S. Reed, "Optimal Reduced-Rank 3D STAP for Joint Hot and Cold Clutter Mitigation," *Proceedings of the IEEE National Radar Conference*, Boston, pp. 119–124, Apr. 1999.

[30] P. Antonik, H. Schuman, P. Li, W. Melvin, and M. Wicks, "Knowledge-Based Space–Time Adaptive Processing," *Proceedings of the IEEE National Radar Conference*, Syracuse, NY, May 1997.

[31] R. S. Adve, T. B. Hale, M. C. Wicks, and P. Antonik, "Ground Moving Target Indication Using Knowledge Based Space Time Adaptive Processing," *Proceedings of the IEEE International Radar Conference*, Washington, DC, pp. 735–740, May 2000.

APPENDIX A

THE CONCEPT OF A RANDOM PROCESS AND ITS PHILOSOPHICAL IMPLICATIONS IN ANALYZING COMMUNICATION SYSTEMS

SUMMARY

When trying to analyze a complex communication system, scientists often apply concepts from stochastic modeling and analysis to obtain a description of the system, frequently assuming that this will supplement our knowledge and improve our understanding. The philosophy is to obtain a result that would occur on the average when this system is working under normal conditions. However, we must consider the fact that the introduction of probability in communication system analysis often involves invoking certain assumptions and additional information about the system which may not be valid. Hence, under these circumstances one may obtain a result that may not be commensurate with the conceived communication system. The objective of this appendix is to highlight the basic assumptions that are invariably associated with the signal analysis in a system using stochastic analysis and the introduction of probabilistic methods. Surprisingly, in many cases, analysis using stochastic methods may provide results equivalent to those obtained using deterministic methods. Examples are presented to illustrate our approach, explain the basic assumptions, and formulate the mathematical framework associated with a stochastic analysis. We also demonstrate the equivalence between a random and a deterministic process and under what conditions they approach the Cramer–Rao bound. Analysis using stochastic models to describe a system may be superior to a deterministic description. However, such a characterization comes with a large cost, namely one must have more definitive knowledge about the system, knowledge that is often unavailable. For convenience, with the application of a random model, the concepts of stationarity and ergodicity are used to simplify the mathematical analysis of measured data. It is shown here that the introduction of ergodicity in probability is similar to a deterministic analysis of a single waveform, and in both cases characterizes the entire underlying mathematical agenda. An example is presented to illustrate the salient features of an ergodic process as opposed to a deterministic process. It is seen that for practical problems it might be easier and more relevant to introduce a deterministic model and then carry out a stochastic analysis, but this may not be practical since the underlying ensemble is not available, nor are its probability density functions. Moreover, a deterministic solution may present the best solution for a given data set, whereas the stochastic approach yields an "average" solution for all the waveforms in the ensemble. Hence the stochastic solution may not be the desired one for the given data set. However, when accurate statistics are available, a better solution may be obtained using probabilistic methods. A probabilistic approach tends to be more useful for analysis, whereas a deterministic approach tends to be more useful for system design.

A.1   INTRODUCTION

Often when dealing with large communication system analysis problems, concepts of probability theory are introduced either to represent uncertainty in the model or to supplement what is known about the system. In addition, when a problem is too large or too complex to solve by any other method, one might take recourse in a probabilistic approach. One point to be made in this appendix is that a probabilistic model cannot be used to compensate for insufficient or inadequate knowledge about the system to be analyzed. The mere introduction of a probabilistic model will not produce more knowledge about a system. Besides, the probabilistic analysis may not be an appropriate one for a given problem. In order to carry out a probabilistic analysis and to correctly interpret its results, the underlying assumptions must be understood. It must also be clear whether one is interested in a performance average or is concerned with specific signals or events. For example, if the adaptive antenna in a mobile system is transmitting your bank balance, you may want your result to be absolutely correct for that time instance and not on the average. In fact, what happens on the average is immaterial!

It is true that 40 years ago when the concepts of adaptive antenna and related signal processing techniques were introduced, life was analog in nature. However, we have a mature digital society and therefore it is important for us to rethink the basic philosophy involved with the introduction of probability theory based techniques into such signal analysis applications, even though noise is still considered to be analog in nature. The goal of this appendix is to initiate such a dialogue.

The principles of probabilistic modeling and the mathematical concepts of probability and probability space are summarized in Section A.2. In Section A.3 the discussion is extended to random variables. Random processes are introduced in Section A.4 and it is shown how the concept of stationarity can simplify a random process. Further simplification results from the property of ergodicity, if applicable. This is explained in Section A.5 and it is shown that a deterministic description can be used for an ergodic process. An example based on the filtering of a signal in the presence of white noise is presented in Section A.6 to bring out the different mathematical assumptions that often exist in applying a stochastic model as opposed to a deterministic model. Through this filter example, it is demonstrated that a probabilistic model can generate a solution that cannot be obtained from a deterministic analysis, provided that the assumptions made are actually applicable to the analysis of a system. However, if that is not the case, the probabilistic model may yield inferior or incorrect results.

A.2   PROBABILITY THEORY

Probability theory is a branch of mathematics that has been developed for more effective mathematical treatment of situations in the real world that involve uncertainty, or randomness, in some sense. It turns out, however, that intensive study of the theory alone can still leave the would-be practitioner peculiarly inept in its application [1–4]. This is because correct application of probability mathematics requires that the problem at hand be formulated in a suitable way, as outlined below. It needs to be mentioned that there exist various approaches to, and interpretations of, probability. However, in engineering, the frequency interpretation of probability is used, and this is what is discussed here [1, 2].

Before mathematics is applied to a real-world problem, an important first step is the establishment of a suitable model, or idealization, of the real world. This involves fitting the real-world situation into a particular way of thinking about things—into a *conceptual model*. When using most kinds of mathematics, this occurs almost automatically. But the conceptual models used with probability theory are considerably more involved than those needed for most other branches of applied mathematics. The establishment of adequate models is therefore an essential ingredient in tackling real-world problems by means of probability theory. If one neglects this point, it may result in a confused and erroneous application of the theory. The conceptual world always forms the bridge between a real-world experience and mathematical analysis.

When probability theory is to be used, an appropriate type of conceptual model is a *probabilistic experiment*, which has four distinct steps [4, 5]:

1.) Statement of purpose. This is an expression of the intent to make specific real-world observations under specific circumstances. It includes a description of all results that might arise. These are the *possible outcomes*.
2.) Description of an experimental procedure.
3.) Execution of the experimental procedure.
4.) Noting of results.

Consider the experiment of throwing a simple die. The purpose of the experiment is to obtain one of the six possibilities, or possible outcomes, which are observable when throwing the die: "one dot faces up," "two dots face up," "three dots face up," "four dots face up," "five dots face up," "six dots face up." The procedure here is that the experimenter is to throw the die in the customary manner on a reasonably smooth horizontal surface and to observe the number of dots facing up when it comes to rest. When the procedure is carried out correctly, the die comes to rest at a certain spot on the surface and a certain number of dots face up. This number is noted by the experimenter. One of the possible outcomes must actually be observed upon completion of a probabilistic experiment, and this is called the *actual* outcome of that probabilistic experiment [5]. It turns out that probability mathematics are not really concerned with completed probabilistic experiments—only with probabilistic experiments prior to their execution.

It is important to recognize that the procedure is strictly deterministic, whereas the final result of the procedure brings into evidence occurrences and properties about which there is initial uncertainty. Furthermore, the probabilistic experiment cannot have as its purpose determination of the probability or the degree of uncertainty. However, using the data from repetitions of such an experiment, one can assign probabilities. A question such as "Will hydrochloric acid produce a precipitate in an unknown liquid sample?" or "Will it rain tomorrow?" has no answer from a probabilistic point of view [5–9], unless a suitable probabilistic experiment can be defined. Probability theory may be of no help in dealing with a problem where all the possible final outcomes of the experiment are not catalogued. Once a problem situation has been formulated as a probabilistic experiment, probability theory can be applied.

According to the development presented in [5–9], the mathematical formulation of probability theory begins with a set of *elements*, denoted S, where the elements represent all possible outcomes for the situation at hand. Consider again the experiment of throwing a normal six-faced die. When one throws the die, one can have any of the six specified possible outcomes. In the die-throwing experiment, therefore, the set S consists of six elements, representing the six possible outcomes. So a *set* is a collection of things called elements. A *subset* of S is a set made up of some of the elements of S.

Next we define an event. An *event* is identified with some subset of S. Therefore, we say that a particular event has occurred if the actual outcome, upon performing the experiment, is a member of the corresponding subset. If the experiment again is the throw of a die, an example of an event is "an even number of dots face up" or "four dots face up" or "six dots face up." The event is represented by the subset of the set S that consists of three elements: namely, those elements that correspond to the three possible outcomes just mentioned. So we see that an event represents some collection of possible outcomes. If the experiment is performed and its actual outcome belongs to a specified event, that event is said to have occurred.

The relationship between the conceptual world and the mathematical world can be visualized as follows:

CONCEPTUAL WORLD	$\leftrightarrow$	MATHEMATICAL WORLD
• Possible outcomes	$\leftrightarrow$	• Elements
• The collection of all possible outcomes of an experiment	$\leftrightarrow$	• The set S
• Event	$\leftrightarrow$	• Subset of S
• Performing an experiment	$\leftrightarrow$	• *Selection* of an element of the sets S

If the connection and correspondence between the conceptual and mathematical worlds is made carelessly, inconsistencies may arise which will result in paradoxical results. Such difficulties have existed throughout much of the history of probability. These problems are not inherent to probability theory but are due to a careless analysis.

The collection of all possible outcomes of a probabilistic experiment is also called an *ensemble*. The ensemble corresponds to the set S of the mathematical world and is assumed to be completely specified. Therefore, the mathematics of probability is deterministic. Uncertainty exists only regarding which of the specified possible outcomes is going to occur when the experiment is carried out. Therefore, the application of probability theory to a given problem requires that we know a priori all the various possible outcomes that can arise. What we do not know is whether or not a specific possible outcome will occur at a particular instance, but probability theory allows us to determine with what degree of likelihood it occurs.

Making use of the correspondence between the conceptual world and the mathematical world, we have so far build up the mathematical models to the point of defining the set S and subsets of S. But this is not enough to carry out probability calculations. What is needed next is to associate probabilities with various events. Mathematically, this amounts to establishing a *probability space* for the experiment that is under consideration [2, 4–6]. This means that all the mathematical "givens" are stated. They are of three different kinds: the set S, which we have already mentioned; the collection of all subsets of S, which is denoted by A; and the probability function P, which assigns a probability to each of the members of A. Thus, the probability space is characterized by a triple: (S, A, P). Let us consider once more the example of throwing a normal balanced die. Then the set S consists of six possible outcomes and therefore has six elements ξ_k, for $k = 1, 2, 3, 4, 5, 6$. This is illustrated in Figure A.1. The collection A contains all subsets of S such as "a 1 or 2 is thrown," "an even number is thrown," and so on. The probability function is very simple in this case. It specifies the probability of any subset as the number of elements making up that subset, multiplied by 1/6.

Another feature, not of the mathematics of probability but of the conceptual model, is that there must be *statistical regularity*. This means that the probabilistic experiment pertaining to any given problem situation must be repeatable many times, without one execution of the experiment influencing the results of later executions. Frequently, statistical regularity can be inferred from the physical nature of a real-world problem, without actually performing experiments. If statistical regularity holds for a given experiment and the experiment is performed a large number of times N, the fraction of the executions in which a particular event occurs is called the *relative frequency* of occurrence of the event.

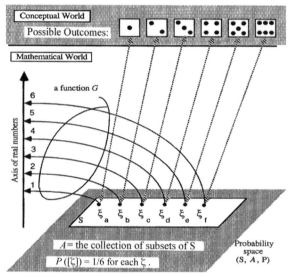

Figure A.1. Experiment of throwing a die.

In the limit as $N \rightarrow \infty$, this becomes the *probability* that this event occurs. In this way, probabilities can be estimated experimentally. But in many cases, probabilities can be inferred from the physical features of the experiment. Thus, when a well-made balanced die is thrown from a cup and allowed to roll, each of the six faces of the die can be assumed to have the same probability of facing up when the die comes to rest. So we write $P\{\xi_k\} = 1/6$ for $k = 1, 2, ..., 6$, where $\{\xi_k\}$ is the subset which says that "k dots are showing" when the die is thrown. However, in many practical situations of uncertainty, it may not be possible to establish all the possible outcomes that can result when an experiment is performed, or with what frequency or probability these possible outcomes occur.

A.3   INTRODUCTION OF RANDOM VARIABLES

In many probability problems, a numerical value or set of values is associated with each of the possible outcomes of an experiment [1, 2, 4–9]. Even data that is not numerical may often be coded numerically for convenience. This requires an extension of the mathematical model as discussed so far. With each element ξ_k of the probability space (S, A, P) gets associated a real number (or a vector of real numbers). The result is a function, or mapping, from the probability space to the space $\Re$ of real numbers. A significant feature of such a function is that it also maps the probabilities from the given probability space to the subsets of $\Re$ [5]. So the probability assignment is carried over into the realm of real numbers as illustrated in Figure A.1 for the die-throwing experiment. In this way, the probability of any given set of real numbers becomes defined as the probability of the subset in A which gets mapped into that set of real numbers.

This kind of function, which is able to map probabilities from a probability space to the space $\Re$, is called a *random variable*. One convenience of a random variable is that it allows probabilities to be displayed graphically, along the real axis. In Figure A.1, for instance, we show a random variable G which assigns to the element ξ_1 the number 1, to ξ_2 the number 2, and so on. The probabilities which have been mapped by G are then denoted by P_G. Figure A.2 shows how a bar graph can be drawn which displays the probabilities assigned to the sets $\{1\}$, $\{2\}$, ..., $\{6\}$.

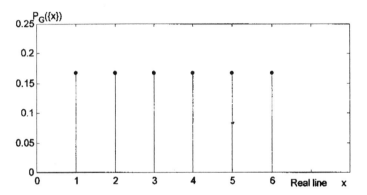

Figure A.2. Probability distribution of the die-throwing experiment.

The term *random variable* is misleading. There is nothing random associated with the random variable. The only randomness is associated with the occurrence or nonoccurrence of a well-defined event in an experiment. Nor is it a variable. It is a real-valued point function defined on a probability space. Its relation to the conceptual model can be summarized as follows:

CONCEPTUAL WORLD		MATHEMATICAL WORLD
• Probabilistic experiment	↔	• Probability space (S, A, P)
• Each possible outcome has associated with it a numerical value.	↔	• A numerical value is associated with each element $\xi \in S$ by defining a random variable X on (S, A, P).
• The experiment is performed, resulting in an actual outcome, and the associated numerical value.	↔	• One of the elements $\xi \in S$ is selected. With it the number $X(\xi)$ is selected.

As another example, consider the experiment in which "a coin is tossed 10 times" and the result of each toss is recorded. The possible outcomes of this experiment can be expressed as 10-tuples, where each component in the 10-tuple can be either "head" or "tail." The sample space S consists of 2^{10} equally likely possible outcomes. Here we can define the random variable B, which expresses the "number of heads obtained in 10 tosses." This random variable will simply associate one of the integers from 0 to 10 with each of the possible outcomes, and the probability distribution P_B assigns positive probability to each of these 11 integers. The magnitude of each probability for $m = 0, 1, ..., 10$ is given by the

binomial distribution $P_B\left(\{m\}\right) = \binom{10}{m} / 2^{10}$,   and a bar graph for P_B is shown in

Figure A.3 [5].

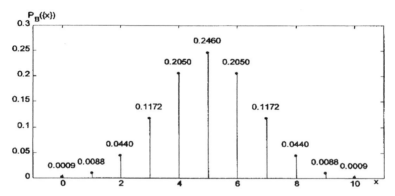

Figure A.3. Probability density function for the coin-tossing experiment.

The random variables G and B mentioned above are called discrete random variables because they assign positive probability only to isolated points in $\Re$. If a random variable X (abbreviated as r.v.) is not discrete, it induces a probability distribution P_X on $\Re$ which no longer can be described by a bar graph such as used in the previous examples. We then have a continuous r.v., because it assigns probability over a continuum. For a continuous r.v. X, the probability distribution P_X is described by the cumulative distribution function $F_X(x)$, which expresses for every point $x \in \Re$ "how much probability lies to the left of it and directly on it." In other words, as x increases, $F_X(x)$ "accumulates" all probability that it encounters. Hence, the cumulative distribution function must be 0 at $-\infty$, since no probability can be assigned there, and it must be 1 when it reaches $+\infty$, since all probability must be assigned below $+\infty$. More convenient is often the probability density function $f_X(x)$, which is the derivative of $F_X(x)$.

Frequently in probability problems it happens that a random variable gets defined but is not known in complete detail. Nevertheless, what is known about the r.v. may be sufficient to arrive at the desired solution. In complicated problems, it also saves time and effort if an r.v. is not specified completely. Next, we see how that works.

A sum of real numbers, where each term is weighted by a probability, is called a *statistical average*. The statistical average of all the values that can be assigned by a discrete r.v. X is called the *expected value*, or *expectation*, of the r.v. and is denoted by $\mathcal{E}[X]$. It is given by

$$\mathcal{E}[X] = \Sigma_x \, x \, P_X(\{x\}) \tag{A.1}$$

where the summation is over all real numbers x belonging to the range of X. If the range for x is infinite, (A.1) is an infinite series, which must converge absolutely, otherwise, $\mathcal{E}[X]$ is undefined. For a continuous r.v. X, the expectation is defined by

$$\mathcal{E}[X] = \begin{cases} \int_{-\infty}^{\infty} x f_x(x)\, dx & \text{if the integral converges absolutely} \\ \\ \text{otherwise undefined} \end{cases} \tag{A.2}$$

The expected value of a r.v. is also called the first moment m_1 of the probability distribution which that r.v. induces. It is also called the *mean of the distribution*.

The second moment (about the origin) of a probability distribution P_X is the expected value of X^2. If X is a continuous r.v., it is given by

$$\mathcal{E}[X^2] = \int_{-\infty}^{\infty} x^2 f_x(x)\, dx \tag{A.3}$$

Higher-order moments are defined accordingly. Also of interest are the *central moments* of a probability distribution P_X. These are moments about the mean. The kth central moment is given by

$$\mu_k = \int_{-\infty}^{\infty} (x - m_1)^k f_x(x)\, dx \tag{A.4}$$

The second order central moment is also called the *variance* of the distribution and its square root is called the *standard deviation*. The standard deviation expresses the approximate spread of the distribution relative to the mean. In many practical applications these first two moments can provide enough information to allow a problem to be solved. Higher-order moments are also sometimes needed. For example, the third moment for a Gaussian distribution is zero. Hence, third-order moments can be useful if one assumes that the underlying noise process in a system is Gaussian. This assumption implies that if a signal is contaminated by additive Gaussian noise, the third moment of this composite signal is equal to the third moment of the signal only. Therefore, using third moments one can develop various nice results. But in practice no noise is strictly Gaussian in nature since a Gaussian distribution has infinite support. Hence, when one makes a Gaussian assumption for real data, it may be difficult to assess the degree of suitability of that assumption.

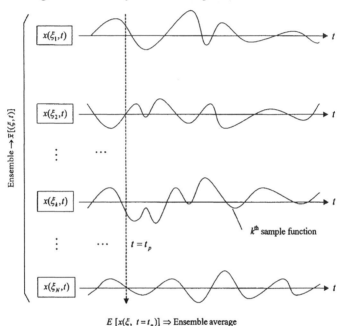

$$E\,[x(\xi,\ t=t_p)] \Rightarrow \text{Ensemble average}$$

Figure A.4.   Example of an ensemble.

A.4   RANDOM PROCESS

Consider again a basic experiment whose various possible outcomes are represented by sample points ξ_k in a sample space S. Therefore, ξ_k denotes the kth possible outcome of the basic experiment. A random variable would assign a numerical value to each sample point. Instead, suppose that the experiment involves observing some phenomenon over time. Then we want to assign to each sample point ξ_k a time function $x(\xi_k, t)$. In this way an ensemble of time functions is created. The ensemble, along with the mapping that created it, is referred to as a stochastic or random process. Papoulis [5–7] denotes the random process by $\bar{x}(\xi,t)$. A possible ensemble is shown in Figure A.4 [8]. In this ensemble, the sample functions drawn in a random manner are such that future values cannot be predicted exactly from the observed past values. Such a random process is said to be *regular* or *nondeterministic*. Regular stochastic processes must be interpreted with care. Even though each sample function is drawn in a noiselike manner, the waveforms are fixed. For example, the identical waveform $x(\xi_2,t)$ occurs whenever the outcome of the basic experiment results in the sample point ξ_2.

A random process can have sample functions for which future values can be predicted exactly from a knowledge of the past values. Such a random process is said to be *predictable* or *deterministic*. An example of a predictable random process is $\bar{x}(\xi,t) = A\cos[\omega t + \bar{\theta}(\xi)]$, where A and ω are constants and θ is a random variable uniformly distributed between 0 and 2π. It provides an ensemble that is highly structured. The only randomness occurs in the phase of the sample functions. A different example of a random process is one whose ensemble consists of various noise-like waveforms, where the amplitude at any given time t has a Gaussian distribution. However, in the ensemble, all these noiselike waveforms are considered fixed and deterministic. So, for a random process, we should note that it is not the waveform that is random. Rather, it is the selection of a particular waveform or *sample function* from the ensemble that is random. Performing the basic experiment causes the selection that determines which possible outcome actually occurs.

The value which a random process $\bar{x}(\xi,t)$ takes on or assigns at the time instant $t = t_k$ can be denoted $\bar{x}(\xi,t_k) = \bar{x}(t_k) = \bar{x}_{t_k} = \bar{x}_k$ [7, 8]. One may try to characterize this random process by the first-order density function $f_{\bar{x}_k}(x_k)$ for all possible values of t_k. However, this is not enough. It is also necessary to specify the joint density function $f_{\bar{x}_1\bar{x}_2}(x_1, x_2)$ for all possible pairs of values of t_1 and t_2. This is not all! The random process $\bar{x}(t)$ is completely specified only if one knows the nth-order density function $f_{\bar{x}_1\bar{x}_2...\bar{x}_n}(x_1, x_2,..., x_n)$ for all possible values of n and $t_1, t_2, ..., t_n$ [8].

In general, this is a formidable task, which is difficult to accomplish. However, for some processes, such as the white noise process, the Gaussian process, and the Markoff process, knowledge of the first- and second-order density functions $f_{\bar{x}_k}(x_k)$ and $f_{\bar{x}_j\bar{x}_k}(x_j,x_k)$ are sufficient to generate all the higher-order density functions [8]. The solutions to many problems of interest require averages based upon only first- and second-order statistics, that is, parameters computed from the first- and second-order density functions. One such parameter is the first-order moment or mean, which is given by

$$m_1(t) = \mathcal{E}[\bar{x}_t] = \int_{-\infty}^{\infty} x_t\, f_{\bar{x}_t}(x_t)\, dx_t \qquad (A.5)$$

For each value of t, $m_1(t)$ is the mean of the random variable $\bar{x}_t$. In general, this can vary from one time instant to the next. It should be noted that $m_1(t)$ is computed over the ensemble or over all the possible outcomes. Hence, in Figure A.4, the average is carried out along a vertical line for a fixed t_k. To find this ensemble average for all t requires that $f_{\bar{x}_t}(x_t)$ is known for all time t.

A random process $\bar{x}(t)$ is called *stationary in the strict sense* if all its statistics remain unchanged by a shift in the time origin. More specifically, if c is a shift in the time origin, $\bar{x}(t)$ is stationary in the strict sense provided that irrespective of the value of c, the two processes $\bar{x}(t)$ and $\bar{x}(t+c)$ have identical density functions of any order. In other words, the requirement is that [8]

$$f(x_{t_1}, x_{t_2}, ..., x_{t_n}) = f(x_{t_1}+c, x_{t_2}+c, ..., x_{t_n}+c) \qquad (A.6)$$

For $n = 1$, stationarity in the strict sense requires that for all c,

$$f(x_t) = f(x_{t+c}) \qquad (A.7)$$

That is, $f(x_t)$ does not depend on t. For $n = 2$, stationarity in the strict sense requires for all c, and

$$f(x_{t_1}, x_{t_2}) = f(x_{t_1}+c, x_{t_2}+c) \qquad (A.8)$$

for all values of t_1 and t_2. In other words, the second-order density functions are independent of the choice of the time origin: they depend only on the separation between the two observation instants involved. Therefore, if $t_1 = t + \tau$ and $t_2 = t$, so that τ is the separation between the two observation instants, the function $f(x_{t_1}, x_{t_2})$ depends only on the variable $\tau = t_1 - t_2$. Strict-sense stationarity is a convenient property of a random process because it simplifies the representation and analysis of the random process.

A simpler form of stationarity, called *wide-sense stationarity*, is more widely applicable and also provides analytic conveniences. For $\bar{x}(t)$ to be wide-sense stationary, its first- and second-order moments must be independent of the time origin, that is,

$$\mathcal{E}\,[\bar{x}_t] = \text{constant} \qquad (A.9)$$

$$\mathcal{E}[\bar{x}_1, \bar{x}_2] \text{ is only a function of } \tau = t_1 - t_2 \qquad (A.10)$$

$\mathcal{E}[\bar{x}_1, \bar{x}_2]$ is also called the *autocorrelation function* of the random process $\bar{x}(t)$ and is denoted

$$R_{xx}(t_1, t_2) = \mathcal{E}[\bar{x}_1, \bar{x}_2] = \int\!\!\!\int_{-\infty}^{\infty} x_1\, x_2\, f_{\bar{x}_1 \bar{x}_2}(x_1, x_2)\, dx_1\, dx_2 \qquad (A.11)$$

Wide-sense stationarity does not imply strict-sense stationarity, but strict-sense stationarity does imply wide-sense stationarity. If $\bar{x}(t)$ is at least wide-sense stationary, its autocorrelation function is denoted by $R_{xx}(\tau)$.

The development of stochastic modeling that has now been presented, beginning with a probabilistic experiment and culminating with the random process, should make it clear that this kind of model requires detailed and specific information about the phenomenon to be modeled. The purpose of probabilistic analysis is to take into account the uncertainty that exists prior to performing an experiment, that is, prior to making an observation. It does not compensate for a lack of knowledge about the phenomenon to be observed. Uncertainty is not the same as lack of knowledge about the nature of the system. Rather, it reflects the existence of a source of randomness that influences the phenomenon that is being modeled.

A.5 ERGODICITY

Equations (A.9) to (A.11) involve averaging over the ensemble at one or two specified time instants. However, not enough may be known about the ensemble of a given random process to be able to carry out these computations. This difficulty can be circumvented if the property of ergodicity applies. A random process $\bar{x}(t)$ is called *ergodic* if all its ensemble averages equal the corresponding time averages with probability 1. Thus, if $\bar{x}(t)$ is ergodic (A.5) becomes

$$\mathcal{E}[\bar{x}(t)] = \int_{-\infty}^{\infty} x_t f_{\bar{x}_t}(x_t) \, dx_t = \lim_{T \to \infty} \frac{1}{2T} \int_{-T}^{T} x(t) \, dt \qquad (A.12)$$

which is a constant. Also, (A.11) combined with (A.10) becomes

$$\begin{aligned} \mathcal{E}[\bar{x}_{t+\tau}, \bar{x}_t] &= \int\int_{-\infty}^{\infty} x_{t+\tau} x_t f_{\bar{x}_{t+\tau} \bar{x}_t}(x_{t+\tau}, x_t) \, dx_{t+\tau} \, dx_t \\ &= \lim_{t \to \infty} \frac{1}{2T} \int_{-T}^{T} x(t+\tau) \, x(t) \, dt \end{aligned} \qquad (A.13)$$

As can be seen, conditions (A.9) and (A.10) must be satisfied. In fact, $\bar{x}(t)$ must be a stationary process in the strict sense if ergodicity is to apply. On the other hand, stationarity does not guarantee ergodicity.

A simpler kind of ergodicity is called *ergodicity of the mean* and requires only (A.12) to hold, in which case $\bar{x}(t)$ is a mean ergodic process. Thus, for a mean ergodic process, the time average of any one sample function equals the ensemble average with probability 1. If ergodicity of the mean does not apply, neither the mean nor any other ensemble average can be determined by performing a time average on a single sample function.

A random process $\overline{x}(t)$ is ergodic in the mean provided that the variance σ^2 of the random variable $\overline{\eta}_T$ defined by $\quad \overline{\eta}_T = \dfrac{1}{2T} \displaystyle\int_{-T}^{T} \overline{x}(t) \, dt \quad$ satisfies

$$
\begin{aligned}
\lim_{T \to \infty} \sigma_{\eta_T}^2 &= \lim_{T \to \infty} \{ \mathcal{E}[\overline{\eta}_T^2] - \mathcal{E}^2[\overline{\eta}_T] \} \\
&= \lim_{T \to \infty} \frac{1}{2T} \int_{-T}^{T} [R_{xx}(\tau) - m_1^2]\left(1 - \frac{\tau}{2T}\right) d\tau = 0
\end{aligned} \tag{A.14}
$$

Ergodicity implies that all the various moments of the random process can be approximated from measurements performed on a single sample function [8]. If ergodicity holds, one can therefore bypass all the stochastic modeling described so far and carry out all analysis using a purely deterministic model based on a single observed waveform. Typically, of course, such a waveform would be difficult if not impossible to describe analytically for all t. However, an approximate representation over a finite interval can be obtained by means of a model-based parameter estimation technique [9–16]. The resulting estimates can then be used in equations (A.12) and (A.13).

So we see that when we come to ergodic random processes, we arrive at an intersection between probabilistic analysis and deterministic analysis. Either type of model and mathematics can be used in that case. This is addressed at some length in comments by various authors that are introduced in [13]. Which is to be preferred? That depends on what it is that one wants to accomplish. Unfortunately, no theory tells us when to use it and when not. Broadly speaking, however, a probabilistic approach tends to be more useful for analysis, whereas a deterministic approach tends to be more useful for system design—that is, designing a new type of system rather than merely optimizing some parameters in a given system configuration.

To illustrate this point, we return to the simple example of throwing a die. Suppose that you find yourself in a game of dice with a professional gambler, except we will assume for simplicity that only a single die is thrown, with the aim of getting as many "sixes" as possible. You check the die provided by the gambler, and it appears to be well balanced, and all six sides seem to come up with more or less equal likelihood when you throw the die repeatedly. Yet, as you play against the gambler, your opponent seems to get many more sixes than you.

You have in your mind the probabilistic experiment described earlier as the conceptual model for the die throws. This allows you to carry out the analysis, which tells you that it is extremely unlikely that your opponent's frequent sixes is simply due to chance. It even allows you to compute just how unlikely it is. But it gives you absolutely no clue what sort of foul play could possibly be involved—or how you might device a scheme that could achieve similar results.

The gambler has, however, been thinking about the die throws "deterministically," that is, by considering each throw of the die as a distinct phenomenon. This led him to design the following system: A hidden high-speed video camera feeds information to a picture processor that extracts parameters of motion of the die in real time. This data is fed to a learning automaton that sends signals to an array of small electromagnets attached to the underside of the table where the die, which contains a magnetic material, is thrown. The learning automaton has, furthermore, been trained over a long period of time to influence the motion of the die so as to favor a "six." Also, the system recognizes when it is the gambler's throw by the ring he wears on one of the fingers.

The *behavior* of the gambler's adaptive control system can still be expressed probabilistically, but it is too complicated to do this by analysis. Instead, considerable data would have to be collected to arrive at a reasonable precise estimate of the probabilities with which the various faces of the die now appear. However, in order to *design* such a system, thinking in terms of a probabilistic experiment is not very useful. The problem is one of characterizing, extrapolating, and influencing the irregular motion of the die—not the average motion but the motion that is actually occurring in each specific throw of the die.

Thus, in the case of a channel in which the signal can be represented as an ergodic process, and one is interested in performance analysis or optimization based on a probabilistic criterion, either a stochastic or a deterministic model can be used to obtain the same results—precisely because the process is ergodic. Unless there are other factors involved, it is then merely a matter of which type of calculation one feels more comfortable with. If one is interested in designing a new type of processor, especially if it is to be a time-varying system, a deterministic model provides greater insight and flexibility, and it does not really matter whether or not one is dealing with an ergodic process. If ergodicity does not hold, a deterministic model cannot be used to carry out the stochastic analysis or optimization. However, deterministic modeling is still the approach that is likely to be most effective for designing new kinds of processors.

For example it is shown in [13, 14] that for direction finding using a 2D array, a deterministic model has led to a better processor for a given finite data set, yielding a lower Cramer–Rao bound than that for a processor based on a random model. The Cramer–Rao bound expresses the smallest possible value for the variance of an unbiased estimator for a given functional form of the log-likelihood function. If an unbiased estimator can be found whose variance equals the Cramer–Rao lower bound, there is no other unbiased estimator with a smaller variance. Now, for a situation where an array consisting of identical omnidirectional sensors receives a single plane wave in the presence of additive white Gausssian noise, it has been shown that a processor design based on a deterministic model yields a lower Cramer–Rao bound (which is achievable for high signal-to-noise ratio) than the Cramer–Rao bound obtained analytically for the random waveform model (which is reachable for high signal-to-noise ratio as well as for large values of the number of data points). The reason why the Cramer–Rao bound of the deterministic model is not achievable by increasing the number of data points is because in this case the number of unknowns also increases with N. This is consistent with the observations made in [14] and does not detract from the usefulness of the processor design [15, 16].

A.6 APPLICATION OF RANDOM PROCESSES IN FILTER THEORY

We illustrate now how a stochastic methodology can lead to results that are not obtainable by a deterministic approach. The example that will be considered is filtering of a signal in the presence of noise [8]. Consider a noisy signal represented by the random process $\overline{x}(t)$, so that

$$\overline{x}(t) = \overline{s}(t) + \overline{n}(t); \quad a \leq t \leq b \qquad (A.15)$$

where $\overline{s}(t)$ is the signal of interest and $\overline{n}(t)$ is noise, and both are wide-sense stationary. Note that if the signal and the noise spectrum are nonoverlapping, one can use an ordinary filter to extract the signal in the presence of noise. However, if the spectra of the signal $S(\omega)$ and noise $N(\omega)$ are overlapping, a filtering operation may not be able to provide the best estimate for the signal. A good example of this is a signal embedded in white noise. Here, white noise is meant to imply that the noise spectrum is flat over the bandwidth of the signal—the signal has a finite bandwidth and the noise spectrum is flat over a very wide bandwidth, including that of the signal [8]. Next, we would like to obtain the optimum linear time-invariant filter that yields a minimum mean-squared error estimate of $\overline{s}(t)$ in the presence of white noise. It should be noted that this is not a system design problem but a system optimization problem. The nature of the system, a linear time-invariant filter, is specified in advance.

Since the system is required to be time invariant, the estimator assumes the form [8]

$$\overline{\hat{g}}(t) = \int_a^b h(t-v)x(v)\, dv \qquad (A.16)$$

where $h(t)$ is the impulse response of the filter. Note that t can be of any value. We do not require $h(t-v)$ to be zero for $t < v$. So the optimum filter may not be causal and hence could be unrealizable in practice. Since $\overline{\hat{g}}(t) \approx \overline{s}(t)$ (an estimate of the signal), the orthogonality principle requires that $h(t)$ satisfy [8]

$$\mathcal{E}\left\{\left[\overline{s}_t - \int_{-b+t}^{-a+t} \overline{x}(t-\alpha)\, h(\alpha)\, d\alpha\right]\overline{x}_v\right\} = 0 \quad a \leq v \leq b \qquad (A.17)$$

Therefore,

$$\mathcal{E}\left[\overline{s}_t, x_v\right] = R_{sx}(t-v) = \int_{-b+t}^{-a+t} R_{xx}(t-v-a)h(a)\, da \qquad (A.18)$$

Hence,

$$R_{sx}(\tau) = \int_{-b+t}^{-a+t} R_{xx}(\tau - \alpha)h(\alpha)\, d\alpha \tag{A.19}$$

If the observed process is available over the infinite time interval $(-\infty, \infty)$, then $a = -\infty$, $b = +\infty$. Then the integral equation defining the optimum filter becomes

$$
\begin{aligned}
R_{sx}(\tau) &= \int_{-\infty}^{\infty} R_{xx}(\tau - a)h(a)\, da \\
&= R_{xx}(\tau) \; ❂ \; h(\tau) \qquad -\infty \leq \tau \leq \infty
\end{aligned}
\tag{A.20}
$$

where ❂ denotes a convolution. The autocorrelation function R_{xx} is related to the power spectral density S_{xx} through the Fourier transform, that is,

$$S_{xx}(j\omega) = \int_{-\infty}^{\infty} R_{xx}(t)e^{-j\omega t}\, dt \tag{A.21}$$

The transfer function of the filter required to extract the signal of interest in the presence of white noise is given [from (A.20), by taking the Fourier transform of both sides] by

$$H(j\omega) = \frac{S_{sx}(j\omega)}{S_{xx}(j\omega)} \tag{A.22}$$

Now if the signal and the noise are statistically independent, then

$$
\begin{aligned}
R_{sx}(t) = \mathcal{E}[s(t), x(t)] &= \mathcal{E}[\bar{s}(t), \{\bar{s}(t)+\bar{n}(t)\}] \\
&= \mathcal{E}[\bar{s}(t), \bar{s}(t)] + \mathcal{E}[\bar{s}(t), \bar{n}(t)] \\
&= R_{ss}(t) + 0
\end{aligned}
\tag{A.23a}
$$

$$
\begin{aligned}
R_{xx}(t) = \mathcal{E}[x(t), x(t)] &= \mathcal{E}[\{\bar{s}(t) + \bar{n}(t)\}\,\{\bar{s}(t) + \bar{n}(t)\}] \\
&= \mathcal{E}[\bar{s}(t), \bar{s}(t)] + \mathcal{E}[\bar{n}(t), \bar{n}(t)] = R_{ss}(t) + R_{nn}(t)
\end{aligned}
\tag{A.23b}
$$

Hence from (A.22) and (A.23), one obtains

$$H(j\omega) = \frac{S_{ss}(j\omega)}{S_{ss}(j\omega) + S_{nn}(j\omega)} \tag{A.24}$$

which implies that the transfer function of the filter which provides the best estimate of the signal in the presence of white noise has the form of the signal spectrum divided by the signal plus the noise spectrum. Here the signal and the noise spectrum overlap and hence the conventional deterministic filtering cannot separate them. It is important to point out that this filter defined by (A.24) is optimum in a probabilistic sense. Namely, if one attempts to estimate the signal in the presence of noise, this filter will provide the optimum output on the average for all the waveforms catalogued in the ensemble. Hence for these classes of problems where a deterministic methodology cannot separate the desired from the unwanted because their characteristics overlap, the introduction of the probability concepts will provide a solution, which cannot be achieved by any other technique.

When attempting to use a stochastic analysis, one has to be mindful of the following:

- The nature of the signals and the nature of any processing that is applied to these signals must be specified. This is necessary in order to formulate the probabilistic experiment which underlies the mathematical analysis. If various different processing schemes are to be considered and compared, each requires its own probabilistic experiment to be formulated.
- An adequate probabilistic description of the signals (and noise) must be available so that the necessary calculations can be carried out.
- The time duration over which the probabilistic description of the signals is valid must exceed the range over which the results are to be applicable. Otherwise, transient effects have to be taken into account.
- If signals pass through a time-varying and/or nonlinear systems, as is typically the case with adaptive processors, a stochastic analysis is very difficult to carry out. Such systems are usually designed using a deterministic model, and analysis of their performance has to be based on this deterministic model.

In a real-life signal processing environment, the density functions or the actual ensemble may not be known, and therefore it becomes difficult to apply a stochastic analysis so as to yield meaningful results. This is true for most radar detection problems [15, 16] and adaptive techniques [13] as well. In such cases, one can arrive at a processor design by applying a direct deterministic least squares method, and use a deterministic performance analysis.

A.7 CONCLUSION

In conclusion, it appears that the areas of primary applicability of stochastic and deterministic modeling are somewhat different. Stochastic modeling is more relevant for the analysis and optimization of a specified system. This is because the probability paradigm requires us to have a description of a particular system in mind in order to be able to formulate the probabilistic experiment and to define random variables and random processes. Deterministic modeling is more flexible because it is simpler and because it leads us more readily to think about possible new configurations for carrying out a specified task. Furthermore, the deterministic modeling is not significantly constrained by processing complexity, while probabilistic modeling easily becomes computationally intractable in the case of time-invariant, nonlinear processing.

REFERENCES

[1] C. E. Pearson, *Handbook of Applied Mathematics*, Van Nostrand Reinhold, New York, 1974.

[2] G. James and R. James, *Mathematics Dictionary*, Van Nostrand Reinhold, New York, 1968.

[3] I. Todhunter, *A History of Mathematical Theory*, Chelsea Publishing, New York, 1865.

[4] J. R. Newman, *The World of Mathematics*, Vols. 2 and 3, Simon and Schuster, New York, 1956.

[5] H. Schwarzlander, *EE 606: Notes on Probability*, Department of Electrical and Computer Engineering, Syracuse University, Syracuse, NY, 1972.

[6] W. Feller, *An Introduction to Probability Theory and Its Applications*, Vol. 1, Wiley, New York, 1968.

[7] A. Papoulis, *Probability, Random Variables and Stochastic Processes*, (2nd ed.), McGraw-Hill, New York, 1984.

[8] D. D. Weiner, *EE 756: Notes on Random Processes*, 2nd ed., Department of Electrical and Computer Engineering, Syracuse University, Syracuse, NY, 1973.

[9] S. Haykin, *Adaptive Filter Theory*, Prentice Hall, Upper Saddle River, NJ, 1996.

[10] E. M. Hofstetter, "Random Processes," in H. Margenau and G. M. Murphy (eds.), *The Mathematics of Physics and Chemistry*, Vol. II, Chap. 3, Van Nostrand, Princeton, NJ, 1964.

[11] W. A. Gardner, *Statistical Spectral Analysis: A Nonprobabilistic Theory*, Prentice Hall, Englewood Cliffs, NJ, 1987.

[12] W. A. Gardner, *Cyclostationarity in Communications and Signal Processing*, IEEE Press, Piscataway, NJ, 1994.

[13] T. K. Sarkar, J. Koh, R. Adve, R. A. Schneible, M. C. Wicks, S. Choi, and M. Salazar-Palma, "A Pragmatic Approach to Adaptive Antennas," *IEEE Antennas and Propagation Magazine*, Vol. 42, No. 2, pp. 39–55, Apr. 2000.

[14] Y. Hua and T. K. Sarkar, "A Note on the Cramer–Rao Bound for 2-D Direction Finding Based on a 2-D Array," *IEEE Transactions on Signal Processing*, Vol. 39, No. 5, pp. 1215–1218, May 1991.

[15] R. S. Adve and T. K. Sarkar, "Compensation for the Effects of Mutual Coupling on Direct Data Domain Adaptive Algorithms," *IEEE Transactions on Antennas and Propagation*, Vol. 48, No. 1, pp. 86–94, Jan. 2000.

[16]   T. K. Sarkar, H. Wang, S. Park, R. Adve, J. Koh, Y. Zhang, M. C. Wicks, and R. D. Brown, "A Deterministic Approach to Space Time Adaptive Processing," *IEEE Transactions on Antennas and Propagation*, Vol. 49, No. 1, Jan. 2001.

APPENDIX B

A BRIEF SURVEY OF THE CONJUGATE GRADIENT METHOD

SUMMARY

In this appendix we provide an overview of the conjugate gradient method and provide a list of Fortran computer programs to numerically implement this method on a digital computer.

B.1 INTRODUCTION

In this appendix we provide many different variants of the conjugate gradient method, which is quite suitable for real-time implementations. However, out of these many versions, there are two classes which are quite amenable to real-time adaptive processing. The first class produces a monotonic decrease in the magnitude of the residuals as we iterate. However, the difference between the actual solution and its estimate may oscillate from iteration to iteration. For the second class, the error between the true solution and its estimates decreases monotonically with the number of iteration, whereas the magnitude of the residuals may oscillate. This is quite significant as the error in the solution is guaranteed to reduce monotonically with the number of iterations even though we do not know the exact solution. After describing the theory, Fortran computer programs for three different versions belonging to the two classes are described for which the actual codes are provided.

B.2 DEVELOPMENT OF THE CONJUGATE GRADIENT METHOD

We now develop a general approach for the solution of a matrix equation [1–4]

$$[A] [X] = [Y] \tag{B.1}$$

423

where $[A]$ is a square matrix of dimension $N \times N$ with known computed coefficients, $[X]$ is a column matrix of dimension $N \times 1$ unknowns, and $[Y]$ is a column matrix of dimension $N \times 1$ known excitations. In an iterative method, instead of solving the equation $[A] [X] = [Y]$ directly, we try to minimize a functional $F([X])$ defined by

$$F([X]) = <[S] [R]; [R]> = [R]^H [S] [R] \qquad (B.2)$$

where ^{H} denotes the conjugate transpose of a matrix, $[S]$ is a Hermitian positive definite operator, which is assumed to be known, and the residual $[R]$ is given by

$$[R] = [Y] - [A] [X] \qquad (B.3)$$

The inner product and the norms are defined as

$$<[R]; [R]> = [R]^H [R] = \| [R] \|^2 \qquad (B.4)$$

In an iterative method the objective is to start from an initial guess $[X]_i$ and proceed along the direction $[P]_i$ a distance t_i to arrive at the updated point $[X]_{i+1}$. The distance t_i is selected in such a way that $F([X]_{i+1})$ is minimized. Hence,

$$[X]_{i+1} = [X]_i + t_i [P]_i \qquad (B.5)$$

Since in our case $F([X])$ is a quadratic function, it has only one stationary point and that point is the unique global minimum that is sought. There are no other local minima at all. So we select t_i such that

$$F([X]_{i+1}) = <[S] [R]_{i+1} ; [R]_{i+1}> = <[S] \{[R]_i - t_i [A] [P]_i\}; [R]_i - t_i [A] [P]_i> \qquad (B.6)$$

is minimized. In other words, we require that

$$\frac{\partial F([X]_{i+1})}{\partial a_i} = 0, \qquad \frac{\partial F([X]_{i+1})}{\partial b_i} = 0 \qquad (B.7)$$

where a_i and b_i denotes the real and imaginary parts of the length of the search directions t_i, respectively. Note that

$$\begin{aligned} F([X]_{i+1}) &= <[R]_{i+1} ; [S] [R]_{i+1}> \\ &= <[R]_i ; [S] [R]_i> - t_i <[A] [P]_i ; [S] [R]_i> - t_i^* <[R]_i ; [S] [A] [P]_i> \\ &\quad + |t_i|^2 <[A] [P]_i ; [S] [A] [P]_i> \qquad (B.8) \end{aligned}$$

where $*$ denotes the conjugate of a complex number and $| \bullet |$ denotes the magnitude of a complex number. Application of (B.7) yields

$$t_i = a_i + jb_i = \frac{< [R]_i \,;\, [S][A][P]_i >}{< [A][P]_i \,;\, [S][A][P]_i >} \tag{B.9}$$

Since it is clear from (B.3) and (B.5) that the residuals can be generated recursively as follows:

$$[R]_{i+1} = [Y] - [A][X]_{i+1} = [R]_i - t_i[A][P]_i \tag{B.10}$$

it is seen from (B.9) and (B.10) that

$$< [R]_{i+1} \,;\, [S][A][P]_i > \, = 0 \tag{B.11}$$

The search directions $[P]_i$ are also generated recursively from the gradient of $F([X])$ as follows:

$$[P]_{i+1} = s_{i+1} ([K][A]^H [S][R]_{i+1} + q_i[P]_i) \tag{B.12}$$

where $[A]^H$ is the adjoint operator (conjugate transpose for a matrix) and s_{i+1} is a scale factor chosen to make the vector $[P]_{i+1}$ a unit vector. Scaling can be quite effective for a certain class of problems to be described later. $[K]$ denotes another Hermitian positive definite operator to be specified. The constant q_i is chosen for the conjugate gradient method such that $[A][P]_i$ are orthogonal with respect to the following inner product:

$$< [A][P]_i \,;\, [S][A][P]_j > \, = 0 \quad \text{for } i \neq j \tag{B.13}$$

Condition (B.13) guarantees that the method will converge in a finite number of steps barring round-off error accumulated though the computation of $[A][P]$ or $[A]^H[R]$ and the numerical round-off error propagated through the recursive computations of $[R]_i$ and $[P]_i$. However, it has been our experience that numerical errors accumulated through recursive computations of $[R]_{i+1}$ and $[P]_{i+1}$ are negligible. There are, however, some numerical errors in the evaluation of $[A][P]$ and $[A]^H[R]$.

By utilizing (B.12) and (B.13), we obtain q_i as

$$q_i = \frac{< [A][K][A]^H [S][R]_{i+1} \,;\, [S][A][P]_i >}{< [P]_i \,;\, [S][A][P]_i >} \tag{B.14}$$

By using (B.12) and (B.11),

$$t_i = \frac{< [R]_i \,;\, [S][A][K][A]^H [S][R]_i >}{< [A][P]_i \,;\, [S][A][P]_i >} \tag{B.15}$$

and

$$< [R]_{i+1} \; ; \; [S]\,[A]\,[K]\,[A]^H [S]\,[R]_i > \; = \; 0 \tag{B.16}$$

The equations above thus describe a class of conjugate gradient methods. The class is generated by how one selects the possible Hermitian definite operators $[S]$ and $[K]$ and the scale factor s_{i+1}. It is easy to see that the functional $F([X])$ (the error function) decreases at each iteration. This is because

$$F([X]_i) \; - \; F([X]_{i+1}) \; = \; \frac{\big| < [R]_i \, ; \, [S][A]\,[P]_i > \big|^2}{< [A][P]_i \, ; \, [S]\,[A][P]_i >} \tag{B.17}$$

$$= \; |t_i|^2 < [A][P]_i \, ; \, [S][A][P]_i >$$

is always positive since $[S]$ is a positive definite operator.

We now consider five special cases of the conjugate gradient method.

Case A. We choose

$$[S] \; = \; [K] \; = \; [I] \tag{B.18}$$

where $[I]$ is the identity matrix, and so

$$F([X]_i) \; = \; \big\| \, [R]_i \big\|^2 \tag{B.19}$$

$$[X]_{i+1} \; = \; [X]_i \; + \; t_i\,[P]_i \tag{B.20}$$

$$[R]_{i+1} \; = \; [R]_i \; - \; t_i\,[A]\,[P]_i \tag{B.21}$$

$$t_i \; = \; \big\| \, [A]\,[P]_i \big\|^{-2} \tag{B.22}$$

We select

$$s_{i+1} \; = \; \big\| \, [A]^H [R]_{i+1} \big\|^{-2} \tag{B.23}$$

which yields

$$q_i \; = \; \big\| \, [A]^H [R]_{i+1} \big\|^2 \tag{B.24}$$

Therefore,

$$[P]_{k+1} \; = \; \frac{[A]^H [R]_{i+1}}{\big\| [A]^H [R]_{i+1} \big\|^2} \; + \; [P]_k \tag{B.25}$$

with

$$[P]_0 \; = \; \frac{[A]^H [R]_0}{\big\| [A]^H [R]_0 \big\|^2} \; + \; \frac{[A]^H ([Y] - [A][X]_0)}{\big\| [A]^H [R]_0 \big\|^2} \tag{B.26}$$

This is the description of the first algorithm.

Case B. For this case, we choose

$$[S] = [K] = [I] \tag{B.27}$$

and

$$s_{i+1} = 1 \quad \text{for all } i \tag{B.28}$$

This results in

$$F([X]_i) = \left\| [R]_i \right\|^2 \tag{B.29}$$

$$[X]_{i+1} = [X]_i + t_i [P]_i \tag{B.30}$$

$$[R]_{i+1} = [R]_i - t_i [A] [P]_i \tag{B.31}$$

$$t_i = \frac{\left\| [A]^H [R]_i \right\|^2}{\left\| [A][P]_i \right\|^2} \tag{B.32}$$

$$[P]_{i+1} = [A]^H [R]_{i+1} + q_i [P]_i \tag{B.33}$$

with

$$q_i = \frac{\left\| [A]^H [R]_{i+1} \right\|^2}{\left\| [A]^H [R]_i \right\|^2} \tag{B.34}$$

$$[P]_0 = [A]^H [R]_0 = [A]^H ([Y] - [A] [X]_0) \tag{B.35}$$

Case C. For this case

$$[S] = ([A]^H)^{-1} ([A])^{-1} \tag{B.36}$$

$$[K] = [A]^H [A] \tag{B.37}$$

$$s_{i+1} = 1 \tag{B.38}$$

and so

$$F([X]_i) = \left\| [X]_{\text{exact}} - [X]_i \right\|^2 \tag{B.39}$$

$$[X]_{i+1} = [X]_i + t_i [P]_i \tag{B.40}$$

$$[R]_{i+1} = [R]_i - t_i [A] [P]_i \tag{B.41}$$

$$t_i = \frac{\| [R]_i \|^2}{\| [P]_i \|^2} \tag{B.42}$$

$$[P]_{i+1} = [A]^H [R]_{i+1} + q_i [P]_i \tag{B.43}$$

with

$$q_i = \frac{\| [R]_{i+1} \|^2}{\| [R]_i \|^2} \tag{B.44}$$

Case D. For this case

$$[S] = [I] \tag{B.45}$$

$$[K] = \text{a Hermitian matrix} \tag{B.46}$$

(a good choice for $[K]$ is the square of the inverse diagonal elements of $[A]^H$) and

$$s_{i+1} = 1 \quad \text{for all } i \tag{B.47}$$

This results in

$$F([X]_i) = \| [R]_i \|^2 \tag{B.48}$$

$$[X]_{i+1} = [X]_{i+1} + t_i [P]_i \tag{B.49}$$

$$[R]_{i+1} = [R]_i - t_i [A] [P]_i \tag{B.50}$$

$$t_i = \frac{<[A]^H [R]_i ; [K][A]^H [R]_i >}{\| [A][P]_i \|^2} \tag{B.51}$$

$$[P]_{i+1} = [K] [A]^H [R]_{i+1} + q_i [P]_i \quad \text{with } q_0 = 0 \tag{B.52}$$

Here $[K]$ is a preconditioning operator, with

$$q_i = \frac{<[A]^H [R]_{i+1} ; [K][A]^H [R]_{i+1} >}{<[A]^H [R]_i ; [K][A]^H [R]_i >} \tag{B.53}$$

Case E. For this case

$$[K] = \text{a Hermitian matrix} \tag{B.54}$$

$$[S] = [I] \tag{B.55}$$

$$s_i = <[A]^H[R]_i \,;\, [K]\,[A]^H[R]_i >^{-1} \tag{B.56}$$

Therefore,

$$F([X]_i) = \left\| [R]_i \right\|^2 \tag{B.57}$$

$$[X]_{i+1} = [X]_i + t_i\,[P]_i \tag{B.58}$$

$$[R]_{i+1} = [R]_i - t_i\,[A]\,[P]_i \tag{B.59}$$

$$t_i = \left\| [A]\,[P]_i \right\|^{-2} \tag{B.60}$$

$$[P]_{i+1} = \frac{[K][A]^H\,[R]_{i+1}}{<[A]^H\,[R]_{i+1}\,;\,[K][A]^H\,[R]_{i+1}>} + [P]_i \tag{B.61}$$

B.3   QUALITATIVE ASSESSMENT

In this section we qualitatively discuss the strong and weak points of the various algorithms presented. The algorithms discussed so far can be characterized as follows:

Case A: SRCG (the search directions are scaled at each iteration and the residuals are minimized).

Case B: RCG (the residuals are minimized at each iteration; no scaling has been introduced).

Case C: XCG (the error between the true solution and the approximate solution is minimized at each iteration).

Case D: RPCG (the residuals are minimized at each iteration, although, a preconditioning operator K is introduced; its significance will be described later).

Case E: XPCG (the error between the true solution and the approximate solution is minimized at each iteration, although, a preconditioning operator K is introduced).

All five algorithms (**SRCG**, **RCG**, **XCG**, **RPCG**, and **XPCG**) minimize a meaningful error at each iteration. All the methods are theoretically guaranteed to converge in at most N steps. Numerically, they generally converge in much less than N steps. The rate of convergence depends on how the eigenvalues of the matrix $[A]$ are distributed. If they are bunched together, the convergence is really fast. It has been observed that the methods provide a faster convergence for various signal-processing applications, as these eigenvalues are related to the power of the various signals and hence their effective number is quite small. However, a disadvantage associated with all these techniques is that in solving the original equation, we have actually squared the condition number, as in practice we are solving the normal equations $[A]^H[A][X] = [A]^H[Y]$. For some problems we have observed that the rate of convergence becomes slow and that the residuals $\|[R]\|$ do not decrease at a very fast rate even though $\|[A]^H[R]\|$ is quite small. The reason for this is that since $\|[A]^H[R]\|$ is small, the search directions are not very different from iteration to iteration. When this happens there are two alternatives:

1.) Accept the solution as a least squares solution because $\|[A]^H[R]\|$ is small.
2.) Restart the iteration with a new initial guess, $[X]_1 = [X]_{old} + [Z]$ (a new vector), and continue.

The five algorithms attempt to address in different ways the problem of $\|[A]^H[R]\|$ being small and $\|[R]\|$ being relatively large. In **SRCG**, each search direction is scaled and we have found that for a deconvolution problem [4, 5], this algorithm is quite suitable. **RCG** and **XCG** works equally well, and both of them yield slow convergence when $\|[A]^H[R]\|$ is small even though $\|[R]\|$ is large. A different way to approach the problem is to use a preconditioning matrix which derives the search directions from vectors $[K][A]^H[R]_i$ rather than from $[A]^H[R]_i$. Hence, if $[K]$ can be chosen judiciously, the pitfalls of **XCG** and **RCG** can be rectified. However, the fundamental problem is that it is not possible *a priori* to select a good preconditioner for any problem! For all the examples tested, the fundamental problem associated with **XCG** and **RCG** which we have encountered is a slow rate of convergence due to $\|[A]^H[R]\|$ being small. When that occurs, one possibility is to use **SRCG**. If that fails, use **XPCG** or **RPCG**, provided that a judicious choice of K can be made. If a good choice for K is not available, an alternative is to use the biconjugate or the augmented conjugate gradient algorithms [1, 4].

B.4 IMPLEMENTATION OF FFT IN A CONJUGATE GRADIENT ALGORITHM

It is seen that in all five algorithms, the computational bottleneck lies in carrying out a matrix-vector product such as $[A][P]$ and $[A]^H[R]$ associated with equations (B.22), (B.23), (B.32), (B.41), (B.42), (B.50), (B.51), (B.59), and (B.61). This computational bottleneck can be eliminated by using the FFT to carry out the matrix vector product as outlined in Chapter 4 [6, 7]. This significantly enhances the performance of all five algorithms.

B.5 CONCLUSION

This appendix has reviewed several conjugate gradient algorithms that have found use in solving complex matrix equations arising in electromagnetic wave interactions and signal analysis. Both the strong and weak points of each version of the conjugate gradient method have been presented.

REFERENCES

[1]   M. Hestenes, *Conjugate Direction Methods in Optimization*, Springer-Verlag, New York, 1980.

[2]   J. M. Daniel, "The Conjugate Gradient Method for Linear and Nonlinear Operator Equations," Ph.D. dissertation, Stanford University, Stanford, CA, 1965.

[3]   T. K. Sarkar, (ed.), "Application of Conjugate Gradient Method to Electromagnetics and Signal Analysis," Vol. 5 in *Progress in Electromagnetics Research*, J. A. Kong, (ed.), Elsevier, New York, 1991.

[4]   T. K. Sarkar, X. Yang, and E. Arvas, "A Limited Survey of Various Conjugate Gradient Methods for Solving Complex Matrix Equations Arising in Electromagnetic Wave Interactions," in *Wave Motion*, Vol. 10, pp. 527–547, North Holland, New York, 2000.

[5]   T. K. Sarkar, F. I. Tseng, S. A. Dianat, and B. Z. Hollman, "L_2 Approximation of Impulse Response from Time-Limited Input and Output: Theory and Experiment," *IEEE Transactions on Instrumentation and Measurement*, Vol. 34, pp. 546–561, 1985.

[6]   F. I. Tseng and T. K. Sarkar, "Deconvolution of the Impulse Response of a Conducting Sphere by the Conjugate Gradient Method," *IEEE Transactions on Antennas and Propagation*, Vol. 35, pp. 105–110, 1987.

[7]   T. K. Sarkar, E. Arvas, and S. M. Rao, "Application of FFT and the Conjugate Gradient Method for the Solution of Electromagnetic Radiation from Electrically Large and Small Conducting Bodies," *IEEE Transactions on Antennas and Propagation,* Vol. 34, pp. 635–640, 1986.

COMPUTER PROGRAMS

```
      SUBROUTINE SRCG(CA,CY,CX,N)
C     Case A: direction scaled, residual minimized
      IMPLICIT REAL*8 (A-B,D-H,O-Z)
      IMPLICIT COMPLEX*16 (C)
```

```
      DIMENSION CA(1700,1700),CX(1700),CY(1700),CR(1700)
      DIMENSION CP(1700),CPROD(1700)
      K=0
      AY=0.
      DO 10 I=1,N
      CPROD(I)=(0.,0.)
        AY=AY+CY(I)*CONJG(CY(I))
10    CONTINUE
      DO 20 J=1,N
        CXJ=CX(J)
        DO 20 I=1,N
          CPROD(I)=CPROD(I)+CA(I,J)*CXJ
20    CONTINUE
      DO 30 I=1,N
        CR(I)=CY(I)-CPROD(I)                    !R(0)
30    CONTINUE
      SK=0.
      DO 40 I=1,N
        CPROD(I)=(0.,0.)
        DO 38 J=1,N
38        CPROD(I)=CPROD(I)+CONJG(CA(J,I))*CR(J)   !A(*)R
      SK=SK+CPROD(I)*CONJG(CPROD(I))
40    CONTINUE
      DO 45 I=1,N
        CP(I)=CPROD(I)/SK                      !Eq.(B.26)
45      CPROD(I)=(0.,0.)
700   CONTINUE
      DO 50 J=1,N
        CPJ=CP(J)
        DO 48 I=1,N
48        CPROD(I)=CPROD(I)+CA(I,J)*CPJ         !AP
50    CONTINUE
      AK=0.
      DO 52 I=1,N
52    AK=AK+CPROD(I)*CONJG(CPROD(I))
      AK=1./AK                                 !Eq.(B.22)
      EK=0.
      DO 60 I=1,N
        CX(I)=CX(I)+AK*CP(I)                    !Eq.(B.20)
        CR(I)=CR(I)-AK*CPROD(I)                 !Eq.(B.21)
        EK=EK+CR(I)*CONJG(CR(I))
60    CONTINUE
      PRINT*,'SRCG',K,EK/AY
      WRITE(22,*) 'SRCG',K,EK/AY
      IF((EK/AY) .LT. 1.0D-8) GO TO 100
      SK=0.
```

```
      DO 70 I=1,N
        CPROD(I)=(0.,0.)
        DO 68 J=1,N
68        CPROD(I)=CPROD(I)+CONJG(CA(J,I))*CR(J)   !A(*)R
      SK=SK+CPROD(I)*CONJG(CPROD(I))
70    CONTINUE
      DO 80 I=1,N
        CP(I)=CPROD(I)/SK+CP(I)                          !Eq.(B.25)
        CPROD(I)=(0.,0)
80    CONTINUE
      K=K+1
      GO TO 700
100   CONTINUE
      RETURN
      END




      SUBROUTINE RCG(CA,CY,CX,N)
C     Case B: residual minimized
      IMPLICIT REAL*8 (A-B,D-H,O-Z)
      IMPLICIT COMPLEX*16 (C)
      DIMENSION CA(1700,1700),CX(1700),CY(1700),CR(1700)
      DIMENSION CP(1700),CPROD(1700)
      K=0
      AY=0.
      DO 10 I=1,N
        CPROD(I)=0.
        AY=AY+CY(I)*CONJG(CY(I))
10    CONTINUE
      DO 20 J=1,N
        CXJ=CX(J)
        DO 20 I=1,N
          CPROD(I)=CPROD(I)+CA(I,J)*CXJ
20    CONTINUE
      DO 30 I=1,N
      CR(I)=CY(I)-CPROD(I)                               !R(0)
30    CONTINUE
      SK=0.
      DO 40 I=1,N
        CP(I)=(0.,0.)
        DO 38 J=1,N
38        CP(I)=CP(I)+CONJG(CA(J,I))*CR(J) !EQ.(B.35)A(*)R
      SK=SK+CP(I)*CONJG(CP(I))
40      CPROD(I)=(0.,0.)
700   CONTINUE
```

```
      DO 50 J=1,N
        CPJ=CP(J)
        DO 48 I=1,N
48        CPROD(I)=CPROD(I)+CA(I,J)*CPJ          !AP
50    CONTINUE
      AK=0.
      DO 52 I=1,N
52      AK=AK+CPROD(I)*CONJG(CPROD(I))
      AK=SK/AK                                   !Eq.(B.32)
      EK=0.
      DO 60 I=1,N
        CX(I)=CX(I)+AK*CP(I)                     !Eq.(B.30)
        CR(I)=CR(I)-AK*CPROD(I)                  !Eq.(B.31)
        EK=EK+CR(I)*CONJG(CR(I))
60    CONTINUE
      PRINT*,'RCG',K,EK/AY
      WRITE(22,*)'RCG',K,EK/AY
      IF((EK/AY) .LT. 1.0D-8) GO TO 100
      SK2=0.
      DO 70 I=1,N
        CPROD(I)=(0.,0.)
        DO 68 J=1,N
68        CPROD(I)=CPROD(I)+CONJG(CA(J,I))*CR(J)  !A(*)R
        SK2=SK2+CPROD(I)*CONJG(CPROD(I))
70    CONTINUE
      QK=SK2/SK                                  !Eq.(B.34)
      DO 80 I=1,N
       CP(I)=CPROD(I)+QK*CP(I)                   !Eq.(B.33)
       CPROD(I)=(0.,0)
80    CONTINUE
      K=K+1
      SK=SK2
      GO TO 700
100   CONTINUE
      RETURN
      END



      SUBROUTINE XCG(CA,CY,CX,N)
C     Case C: error of solution X minimized
      IMPLICIT REAL*8 (A-B,D-H,O-Z)
      IMPLICIT COMPLEX*16 (C)
      DIMENSION CA(1700,1700),CX(1700),CY(1700),CR(1700)
      DIMENSION CP(1700),CPROD(1700)
      K=0
```

```
      AY=0
      DO 10 I=1,N
        CPROD(I)=(0.,0.)
        AY=AY+CY(I)*CONJG(CY(I))
10    CONTINUE
      DO 20 J=1,N
        CXJ=CX(J)
        DO 20 I=1,N
          CPROD(I)=CPROD(I)+CA(I,J)*CXJ          !AX
20    CONTINUE
      SK=0.
      DO 30 I=1,N
        CR(I)=CY(I)-CPROD(I)                     !R(0)
        SK=SK+CR(I)*CONJG(CR(I))
30    CONTINUE
      DO 40 I=1,N
        CP(I)=(0.,0.)
        DO 38 J=1,N
38        CP(I)=CP(I)+CONJG(CA(J,I))*CR(J)       !A(*)R
40    CPROD(I)=(0.,0.)
700   CONTINUE
      AK=0.
      DO 50 J=1,N
        CPJ=CP(J)
        DO 48 I=1,N
48        CPROD(I)=CPROD(I)+CA(I,J)*CPJ          !AP
        AK=AK+CPJ*CONJG(CPJ)
50    CONTINUE
      AK=SK/AK                                   !Eq.(B.42)
      SK2=0
      DO 60 I=1,N
        CX(I)=CX(I)+AK*CP(I)                     !Eq.(B.40)
        CR(I)=CR(I)-AK*CPROD(I)                  !Eq.(B.41)
        SK2=SK2+CR(I)*CONJG(CR(I))
60    CONTINUE
      PRINT*,'XCG',K,SK2/AY
      WRITE(22,*)'XCG',K,SK2/AY
      IF ((SK2/AY) .LT. 1.0D-8) GO TO 100
      DO 70 I=1,N
      CPROD(I)=(0.,0.)
      DO 68 J=1,N
68      CPROD(I)=CPROD(I)+CONJG(CA(J,I))*CR(J)   !A(*)R
70    CONTINUE
      QK=SK2/SK                                  !Eq.(B.44)
      DO 80 I=1,N
        CP(I)=CPROD(I)+QK*CP(I)                  !Eq.(B.43)
```

```
      CPROD(I)=(0.,0.)
80    CONTINUE
      K=K+1
      SK=SK2
      GO TO 700
100   CONTINUE
      RETURN
      END
```

APPENDIX C

ESTIMATION OF THE DIRECTION OF ARRIVAL IN ONE AND TWO DIMENSIONS USING THE MATRIX PENCIL METHOD

SUMMARY

In this appendix we present the Matrix Pencil method and illustrate how to use it for 1D and 2D direction of arrival estimation.

C.1 ESTIMATION OF THE DIRECTION OF ARRIVAL IN ONE DIMENSION USING THE MATRIX PENCIL METHOD

The objective of this appendix is to illustrate how to estimate the directions of arrival (DOA) of several signals using the Matrix Pencil (MP) method in one dimension [1, 2]. If we have a uniformly spaced array of omnidirectional isotropic point sensors located along the z-axis and the distance between any two of them is Δ, one can write the voltage $x(n)$ induced in each of the n antenna elements, for $n = 0, 1, 2, ..., N$, as

$$x(n) = \sum_{p=0}^{P} A_p \exp(j\gamma_p + j2\pi\Delta n \sin\varphi_p) \qquad 0 \leq n \leq N \qquad (C.1)$$

where A_p, γ_p, and φ_p are the amplitude, phase and direction of arrival of each of the $P + 1$ plane waves incident on the array, and Δ is the sampling interval. In (C.1), $V(n)$ represents the voltages measured at each of the $N + 1$ elements of the linear array and is assumed to be known. Our goal is to find P and the characteristics of each of the P signals (i.e., to obtain the values for A_p, γ_p, and φ_p). Equivalently, we can write (C.1) as

$$x(n) = \sum_{p=0}^{P} c_p y_p^n$$

$$c_p = A_p \exp(j\gamma_p) \tag{C.2}$$

$$y_p = \exp(j2\pi \Delta \sin\theta_p)$$

and use the following matrix notation:

$$[G] = [Y][C] \tag{C.3}$$

$$[G] = \begin{bmatrix} x(0) \\ x(1) \\ \vdots \\ x(N) \end{bmatrix} \tag{C.4}$$

$$[Y] = \begin{bmatrix} 1 & 1 & \cdots & 1 \\ y_0 & y_1 & \cdots & y_P \\ \vdots & \vdots & \ddots & \vdots \\ y_0^N & y_1^N & \cdots & y_P^N \end{bmatrix} \tag{C.5}$$

$$[C] = \begin{bmatrix} c_0 \\ c_1 \\ \vdots \\ c_P \end{bmatrix} \tag{C.6}$$

As the first step in the Matrix Pencil method, $x(n)$ is partitioned to exploit the structure inherent to the setting up of the problem of the original data that is illustrated in the following equation:

$$[X] = \begin{bmatrix} x(0) & x(1) & \cdots & x(N-L+1) \\ x(1) & x(2) & \cdots & x(N-L+2) \\ \vdots & \vdots & \ddots & \vdots \\ x(L-1) & x(L) & \cdots & x(N) \end{bmatrix} \tag{C.7}$$

$[X]$ is an $L \times (N-L+2)$ Hankel matrix, and each column of $[X]$ is a windowed segment of the original data $\{x(0) \; x(1) \; \cdots \; x(N)\}$ with window length L. The parameter L, often called the *pencil parameter*, must satisfy the following bounds:

$$N + 1 - P \geq L \geq P + 2 \tag{C.8}$$

Next, a singular value decomposition (SVD) of the matrix $[X]$ is performed:

$$[X] = [U][\Sigma][V]^H \tag{C.9}$$

where $[U]$ is the $L \times L$ unitary matrix whose columns are the eigenvectors of $[X][X]^H$, $[\Sigma]$ is the $L \times (N - L + 2)$ diagonal matrix with singular values of $[X]$ located along its main diagonal in descending order $\sigma_1 \geq \sigma_2 \geq \cdots \geq \sigma_{min}$ and $[V]$ is the $(N - L + 2) \times (N - L + 2)$ unitary matrix whose columns are the eigenvectors of $[X]^H[X]$. The superscript H denotes the conjugate transpose of a matrix.

If we consider that the data $x(n)$ are not contaminated by any noise, only the first $P + 1$ singular values are nonzero. Hence, $\sigma_i > 0$ for $i = 0, \ldots, P$ and $\sigma_i = 0$ for $i = P + 1, \ldots, L$.

If the data are corrupted by additive noise, the parameter P is estimated by observing the ratio of the various singular values to the largest one as defined by

$$\frac{\sigma_c}{\sigma_{max}} \approx 10^{-w} \tag{C.10}$$

where w is the number of accurate significant decimal digits of the data $x(n)$ [1].

Next, we define the following three submatrices based on the first $P + 1$ dominant singular values:

- $[U]_s$: the first $P + 1$ columns of $[U]$
- $[\Sigma]_s$: the first $P + 1$ columns of $[\Sigma]$
- $[V]_s$: the first $P + 1$ columns of $[V]$

Finally, we form the matrix pencil to estimate the parameters of the problem. We now define the following matrices:

- $[U_1] = [U]_s$ with the last row deleted
- $[U_2] = [U]_s$ with the first row deleted

and form

$$[U_2] - \lambda [U_1] = 0 \tag{C.11}$$

$$([U_2] - \lambda [U_1])[X] = 0$$

$$[U_1]^H [U_2][X] = \lambda [U_1]^H [U_1][X]$$

$$([U_1]^H [U_1])^{-1} [U_1][U_2][X] = \lambda [X]$$ (C.12)

$$[U_1]^\dagger [U_2][X] = \lambda [X]$$

The eigenvalues of $[U_1]^\dagger [U_2]$ provide values for the exponents $\{y_p: p = 0, \ldots, P\}$, where $[U_1]^\dagger$ is the Moore–Penrose pseudo-inverse of $[U_1]$ and is defined by

$$[U_1]^\dagger = ([U_1]^H [U_1])^{-1} [U_1]$$ (C.13)

The directions of arrival φ_p are obtained from the values of the exponents as

$$\varphi_p = \arcsin \left[\frac{\ln(y_p)}{j2\pi \Delta} \right]$$ (C.14)

The amplitudes and phases of the $P + 1$ signals can be obtained by solving (D.3) using the principle of least squares as

$$[C] = ([Y]^H [Y])^{-1} [Y]^H [G]$$ (C.15)

C.2 ESTIMATION OF THE DIRECTION OF ARRIVAL IN TWO DIMENSIONS USING THE MATRIX PENCIL METHOD

Estimation of the direction of arrival in two-dimensions of several signals that simultaneously impinge on a two-dimensional planar array can also be estimated using the Matrix Pencil method. As an example, consider an array consisting of omnidirectional isotropic point radiators that are uniformly spaced in a two-dimensional grid along the x and y-axes. The total number of elements is of dimensions $(M + 1) \times (N + 1)$. Let the spacing between two antenna elements that lie parallel to the x-axis be Δ_x. Similarly, let the spacing between the antenna elements that lie parallel to the y-axis be Δ_y. The total number of signals impinging on the array is $P + 1$. Each of the $P + 1$ signals has associated with them an azimuth and an elevation angle of incidence. They are ϕ_p and θ_p, representing the azimuth and elevation angle, respectively of the pth signal. Hence, the voltage $x(m; n)$ induced at the feed point of the antenna elements will be given by

$$x(m; n) = \sum_{p=0}^{P} A_p \exp(j\gamma_p + j2\pi m \Delta_y \sin\theta_p \sin\phi_p + j2\pi n \Delta_x \sin\theta_p \cos\phi_p)$$

$$\text{for} \quad 0 \le m \le M \quad \text{and} \quad 0 \le n \le N$$

$$(C.16)$$

Here $x(m; n)$, representing the voltages induced in each element of the two-dimensional array, is assumed to be known. The goal is to estimate the parameters that define the azimuth and elevation direction of arrival, ϕ_p and θ_p, respectively, along with their amplitudes and phases, A_p and γ_p, respectively. The number of signals $P + 1$ is also to be estimated.

In matrix form (D.16) can be written as

$$x(m;n) = \sum_{p=0}^{P} c_p y_p^m z_p^n$$

$$c_p = A_p \exp(j\gamma_p)$$

$$y_p = \exp(j2\pi \Delta_y \sin\theta_p \sin\phi_p)$$

$$(C.17)$$

$$z_p = \exp(j2\pi \Delta_x \sin\theta_p \cos\phi_p)$$

or equivalently, as

$$[X] = [Y][C][Z] \qquad (C.18)$$

These matrices are in bold letters so as to differentiate them from the one-dimensional case presented in Section C.1.

$$[X] = \begin{bmatrix} x(0;0) & x(0;1) & \cdots & x(0;N) \\ x(1;0) & x(1;1) & \cdots & x(1;N) \\ \vdots & \vdots & \ddots & \vdots \\ x(M;0) & x(M;1) & \cdots & x(M;N) \end{bmatrix} \qquad (C.19)$$

$$[Y] = \begin{bmatrix} 1 & 1 & \cdots & 1 \\ y_0 & y_1 & \cdots & y_P \\ \vdots & \vdots & \ddots & \vdots \\ y_0^M & y_1^M & \cdots & y_P^M \end{bmatrix} \qquad (C.20)$$

$$[C] = \text{diag}[c_0, c_1, \cdots, c_P] \qquad (C.21)$$

$$[Z] = \begin{bmatrix} 1 & z_0 & \cdots & z_0^N \\ 1 & z_1 & \cdots & z_1^N \\ \vdots & \vdots & \ddots & \vdots \\ 1 & z_P & \cdots & z_P^N \end{bmatrix} \tag{C.22}$$

We now form the various matrix pencils in order to extract the poles associated with the two different dimensions. Let us consider the first matrix pencil, defined by

$$[A_1] - \lambda_1[B_1] = 0 \tag{C.23}$$

where $[A_1]$ and $[B_1]$ are defined as follows:

$$[A_I] = \begin{bmatrix} x(1;0) & x(1;1) & \cdots & x(1;N) \\ x(2;0) & x(2;1) & \cdots & x(2;N) \\ \vdots & \vdots & \ddots & \vdots \\ x(M;0) & x(M;1) & \cdots & x(M;N) \end{bmatrix} \tag{C.24}$$

$$[B_1] = \begin{bmatrix} x(0;0) & x(0;1) & \cdots & x(0;N) \\ x(1;0) & x(1;1) & \cdots & x(1;N) \\ \vdots & \vdots & \ddots & \vdots \\ x(M-1;0) & x(M-1;1) & \cdots & x(M-1;N) \end{bmatrix} \tag{C.25}$$

The eigenvalues of $[B_1]^\dagger[A_1]$ are the solution for the first set of exponents associated with one dimension. Let us define them to be $\{y_i: i = 1, \ldots, I\}$. Here $[B_1]^\dagger$ is the Moore–Penrose pseudo-inverse of $[B_1]$ and is defined by

$$[B_1]^\dagger = \left([B_1]^H [B_1]\right)^{-1}[B_1]^H \tag{C.26}$$

Consider the second matrix pencil to be of the form

$$[A_2] - \lambda_2[B_2] = 0 \tag{C.27}$$

where $[A_2]$ and $[B_2]$ are defined as follows:

$$
[A_2] = \begin{bmatrix} x(0;1) & x(0;2) & \cdots & x(0;N) \\ x(1;1) & x(1;2) & \cdots & x(1;N) \\ \vdots & \vdots & \ddots & \vdots \\ x(M;1) & x(M;2) & \cdots & x(M;N) \end{bmatrix} \tag{C.28}
$$

$$
[B_2] = \begin{bmatrix} x(0;0) & x(0;1) & \cdots & x(0;N-1) \\ x(1;0) & x(1;1) & \cdots & x(1;N-1) \\ \vdots & \vdots & \ddots & \vdots \\ x(M;0) & x(M;1) & \cdots & x(M;N-1) \end{bmatrix} \tag{C.29}
$$

The eigenvalues of $[B_2]^{\dagger}[A_2]$ are the solution for the second set of exponents along the other dimension, which are defined to be $\{z_j: j = 1, \ldots, J\}$, where $[B_2]^{\dagger}$ is the Moore–Penrose pseudo-inverse of $[B_2]$ defined in (C.26).

We now assume that the complete solution for all the exponents is given by the tensor product of these two single exponents found before (i.e., the exponents are a possible product combination of the different pairs $\{(y_i, z_j), i = 1, \ldots, I; j = 1, \ldots, J\}$ that define the direction of arrival of each signal). Of course, some of them will not be related to the actual signal. Those components are eliminated when we look at the final residue.

Based on the tensorial product of the two sets of the one-dimensional solution, we form all possible pairs or combinations. If the number of one-dimensional solutions for the first matrix pencil are I and for the second are J, the total number of combinations or possible pairs will be $I \times J = R$. Here we estimate the amplitude for all possible combination pairs by solving the following matrix equation for the residues or complex amplitudes $[C]$, which is defined by the column vector $[c_1, c_2, \ldots, c_R]$:

$$
\begin{bmatrix}
1 & 1 & \cdots & 1 \\
z_1 & z_2 & \cdots & z_R \\
\vdots & \vdots & \ddots & \vdots \\
z_1^N & z_2^N & \cdots & z_R^N \\
y_1 & y_2 & \cdots & y_R \\
y_1 z_1 & y_2 z_2 & \cdots & y_R z_R \\
\vdots & \vdots & \ddots & \vdots \\
y_1 z_1^N & y_2 z_2^N & \cdots & y_R z_R^N \\
\vdots & \vdots & \vdots & \vdots \\
y_1^M & y_2^M & \cdots & y_R^M \\
y_1^M z_1 & y_2^M z_2 & \cdots & y_R^M z_R \\
\vdots & \vdots & \ddots & \vdots \\
y_1^M z_1^N & y_2^M z_2^N & \cdots & y_R^M z_R^N
\end{bmatrix}
\begin{bmatrix}
c_1 \\
c_2 \\
\vdots \\
c_R
\end{bmatrix}
=
\begin{bmatrix}
x(0;0) \\
x(0;1) \\
\vdots \\
x(0;N) \\
x(1;0) \\
x(1;1) \\
\vdots \\
x(1;N) \\
\vdots \\
x(M;0) \\
x(M;1) \\
\vdots \\
x(M;N)
\end{bmatrix}
\qquad (C.30)
$$

This matrix is solved in a least squares fashion for all complex amplitudes. Once we have the amplitudes for all possible pairs we fix a threshold to eliminate the undesired pairs and take only those signals as possible solutions whose amplitudes are greater than this threshold. The number of signals over this threshold must be $P+1$, and that is equal to the total number of signals that impinge on the array.

The azimuth and elevation angle (ϕ and θ) can be obtained using the following equations obtained from the ordered set of R pairs $\{(y_i, z_j), i = 1, \ldots, I; j = 1, \ldots, J\}$:

$$
\phi_k = \arctan\left[\left(\frac{\ln y_i}{\ln z_j}\right)\frac{\Delta_x}{\Delta_y}\right] \qquad (C.31)
$$

$$
\theta_k = \arcsin\left[\left(\frac{1}{j2\pi\,\Delta_x}\right)\sqrt{\left(\frac{\Delta_x}{\Delta_y}\ln y_i\right)^2 + \left(\ln z_j\right)^2}\right] \qquad (C.32)
$$

The complex amplitudes associated with them are given by c_k, $k = 1, 2, \ldots, R$. This completes the solution process.

However, the method outlined in [3] is more accurate when the signal-to-noise ratio of the data is low. The method in [3] takes more time than the method outlined here.

C.3 CONCLUSION

This appendix described the Matrix Pencil method for estimation of the direction of arrival of signals impinging on a one-dimensional and then a two-dimensional array. This is a direct data domain method and hence does not require formation of a covariance matrix. This implies that we can deal with coherent signals using a single snapshot (i.e., the data voltages received at the elements of the antenna array).

REFERENCES

[1] T. K. Sarkar and O. Pereira, "Using the Matrix Pencil Method to Estimate the Parameters of a Sum of Complex Exponentials," *IEEE Antennas and Propagation Magazine*, Vol. 37, No. 1, Feb. 1995.

[2] Y. Hua and T. K. Sarkar, "Matrix Pencil Method for Estimating Parameters of Exponentially Damped/Undamped Sinusoids in Noise," *IEEE Transactions on Acoustics, Speech, and Signal Processing*, Vol. 35, No. 5. May 1990.

[3] Y. Hua, "Estimating Two-Dimensional Frequencies by Matrix Enhancement and Matrix Pencil," *IEEE Transactions on Signal Processing*, Vol. 40, No. 9, Sept. 1992.

INDEX